Algebra and Geometry with Python

Sergei Kurgalin • Sergei Borzunov

Algebra and Geometry with Python

 Springer

Sergei Kurgalin
Digital Technologies Department
Voronezh State University
Voronezh, Russia

Sergei Borzunov
Digital Technologies Department
Voronezh State University
Voronezh, Russia

ISBN 978-3-030-61543-7   ISBN 978-3-030-61541-3   (eBook)
https://doi.org/10.1007/978-3-030-61541-3

This Springer imprint is published by the registered company Springer Nature Switzerland AG.
The registered company address is: Gewerbestrasse 11, 6330 Cham, Switzerland

Preface

The rapid development of computing places special demands on the training of young specialists in this field. The complexity of the problems that must be solved to satisfy the needs of science, technology, industry and the economy also grows, simultaneously with the rapid growth in the power of modern computer systems. In this connection, we believe that it is essential that computing specialists have fundamental knowledge about the development of the related mathematical frameworks and about how to create methods for solving the above problems.

Algebra and geometry are deemed important areas whose ideas and results are actively used in the development of information systems, as well as in software developed for business projects. The basic notions of algebra are numerical matrices and the methods for working with matrix algorithms. They may be used extensively in scientific and technical problems and in the game industry. The rapid development of game technologies, as well as augmented and alternative reality technologies, means that we must pay special attention to university courses in analytical geometry and linear algebra, pattern properties in 3D space and fast algorithms for working with two- and three-dimensional objects.

Another promising area of application for linear algebra algorithms that has seen rapid development in recent years is Big Data. Analysis of extremely large arrays requires not only knowledge and use of the known methods, but it also issues the challenge to develop new approaches and high-performance algorithms.

This textbook is an introduction to linear algebra and analytical geometry for higher-education students in the natural sciences. It is based on the courses Algebra and Geometry, Analytical Geometry and Fundamental and Computer Algebra, which are taught to first-year students of the Faculty of Computer Sciences at the Voronezh State University. The teaching is meant for theoretical training, as a supplement to the existing textbooks, for practical and laboratory classes, and also for self-study. Going forward, the terms "Algebra" and "Linear Algebra" will be considered equivalent, as well as "Geometry" and "Analytical Geometry".

The authors have attempted to lay the material down in the most comprehensible form while not sacrificing strictness in definitions and theorems. The statements (theorems, properties) are accompanied with proofs, or references to specialist literature for advanced study of the materials.

The fundamentals of algebra and geometry are presented in the form most suitable for future specialists in computing. We have considered the basic algorithms for working with matrices, vectors and systems of linear equations. The theoretical material contains solutions of most types of problems and is supplemented with plenty of analysed examples. The end of an example is designated by the symbol □. Each chapter ends with problems for self-study. Many of them are provided not only with full answers but also with detailed solutions. The asterisk sign ($*$) marks the advanced (enhanced complexity) problems.

Apart from the sections traditionally included in algebra and geometry courses, one of the chapters is devoted to the mathematical fundamentals of the modern section of cryptography, namely elliptic curve cryptography. The availability of this chapter will be a connecting link between the mathematical courses and methods applied in practice by the application software developers.

The section about quantum computing is devoted to one of the examples of the application of algebra. It demonstrates that the notions of linear algebra are used for constructing new algorithms, whose computation capacity exceeds the existing ones considerably.

Let us briefly summarize the content of this textbook. The first four chapters are devoted to classical divisions of linear algebra; they consider matrices and determinants, and systems of linear equations; definitions are given for the notion of vector space and the fundamental solution of a homogeneous system. The next few chapters introduce the fundamentals of vector algebra and the coordinate method on a plane and in a 3D space. The following subjects are considered: vectors in three-dimensional space, the equation of a line on a plane, the equation of a plane in space and the equation of a line in space. Second-order curves are analysed. Material on elliptic curves is usually not included in a "traditional" algebra and geometry course. However, its presence in this book, in our opinion, contributes to a deeper understanding of the methods of linear algebra and analytical geometry and provides an example of the implementation of such methods for solving problems in theoretical and practical cryptography.

We use the Python programming language for illustration of the considered algorithms. This allows us to familiarize readers with implementation at the initial stage of study. Python was selected because it is a universal and widely used general-purpose programming language, suitable for the successful realization of numerical algorithms; Python is a continuously evolving language; and many of its realizations are open source. Python has the necessary tools to automatically check for the errors that might appear in the program code in the process of its creation. The availability of a great number of additional libraries (such as *NumPy, SciPy, pandas*) substantially expands the programmer's capabilities. Thus, this language is quite suitable for teaching linear algebra and analytical geometry algorithms.

As the time of writing, the version of Python known as "Python 2" is still being used in many significant projects and in the literature. However, official support for Python 2 is diminishing and is scheduled to end. So, we use the latest major version, "Python 3", in this book. Note that there are significant differences between Python 2 and Python 3; however, extended support documentation and tools are available for conversion between the two major versions. Refer to the official Python webpage (https://www.python.org/) for more details.

The book offers a list of training literature on linear algebra and analytical geometry, which may be used for a more detailed study on the issues touched upon in this textbook.

The appendices contain reference information, including basic operators in Python and C, trigonometric formulae and the Greek alphabet. These reduce the necessity to address reference literature.

Below you can see the chart of the chapter information dependence in the form of an oriented graph reflecting the preferable order of covering the academic material. For instance, after having studied Chaps. 1, 2 and 3, you can move to one of the two chapters, Chap. 4 or Chap. 6, the contents of which are relatively independent. After Chap. 9, we think Chaps. 11 and 12 can be mastered in any order.

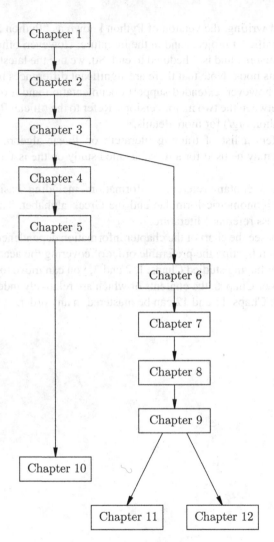

The chapter dependency chart

Acknowledgements

The authors express sincere gratitude to their colleagues for useful discussions and critical remarks. Very useful to the improvement of the text were the valuable remarks of Artem Atanov, Alexey Bukhovets, Tatiana Churakova, Andrei Fediakov, Evgenii Kiselev, Alexander Loboda, Pavel Lukin, Peter Meleshenko, Leonid Minin, Mikhail Semenov and Sergei Zapryagaev.

Enormous help at all stages of working on this book was provided by the experts at Springer. We are especially grateful to editors Aliaksandr Birukou, Ronan Nugent and Wayne Wheeler in collaboration with whom we have been fruitfully working on this and our previous books. Checking of the program code located in the book was performed with the assistance of students of the Faculty of Computer Sciences, Voronezh State University—Anna Danilova, Artem Konovskoy, Nikolai Paukov, Margarita Tepliakova, Vladimir Ushakov and Sergei Zaitsev.

The manuscript of this book was used as the basic teaching for the students at the Voronezh State University, for several semesters. Their interest and enthusiasm had a positive influence on the content of the training course. We thank all the students who took the course in algebra and geometry, and for helping to eradicate some of the misprints and mistakes found in the first versions of the manuscript. Great help in checking the solutions and answers to the problems and exercises was rendered by students Pavel Burdyug, Dmitry Demianov, Kirill Ganigin, Elizaveta Koroteyeva, Zakhar Korsakov, Denis Makushin, Pavel Nekrasov, Natalya Salova, Anna Tsybulskaia, Dmitry Tynyanyy and Anna Yankevich.

Of course, all other mistakes remaining in the text are exclusively on the authors' conscience.

Sergei Kurgalin is especially grateful to Olga Kurgalina and Alexander Shirokov for their constant support during the work on the book.

Voronezh, Russia   Sergei Kurgalin
Voronezh, Russia   Sergei Borzunov
July 2019

Contents

Notation

$\mathbb{N}$	The set of natural numbers
$\mathbb{Z}$	The set of integers
$\mathbb{Q}$	The set of rational numbers
$\mathbb{R}$	The set of real numbers
$\mathbb{C}$	The set of complex numbers
$\mathbb{R}^n$	n-dimensional vector space
$\varnothing$	The empty set
$A \Rightarrow B$	Logical consequence, or implication
$A \Leftrightarrow B$	Logical equivalence
$\forall x (P(x))$	For all x, the statement $P(x)$ is true
$\exists x (P(x))$	There exists such x that the statement $P(x)$ is true
A **and** B	Conjunction of logical expressions A and B
A **or** B	Disjunction of logical expressions A and B
$A \equiv B$	Equivalency
$\{a_1, a_2, \ldots, a_n\}$	The set consisting of the elements $a_1, a_2, \ldots, a_n$
$\displaystyle\sum_{i=1}^{n} a_i$	The sum $a_1 + a_2 + \cdots + a_n$
$\displaystyle\prod_{i=1}^{n} a_i$	The product $a_1 a_2 \ldots a_n$
$A = (a_{ij})$	Matrix formed by the elements a_{ij}
A^T	Matrix transposed relative to A
I	Identity matrix
O	Zero matrix
δ_{ij}	Kronecker delta
$[A, B]$	Commutator of the matrices A and B
$\operatorname{tr} A$	Trace of the matrix A
$O(g(n))$	Class of functions growing not faster than the function $g(n)$
$G(V, E)$	G is a graph with vertex set V and edge set E
$d(v)$	Degree of vertex v of a graph
$D(V, E)$	D is a directed graph with vertex set V and edge set E

$d^+(v)$	Out-degree of vertex v in a digraph
$d^-(v)$	In-degree of vertex v in a digraph
$\lfloor x \rfloor$	Floor function of x, i. e., the greatest integer less than or equal to the real number x (see definition on page 78)
M_{ij}	Additional minor of the matrix element placed at the intersection of the i-th row and the j-th column
$A_{ij} = (-1)^{i+j} M_{ij}$	Cofactor of the element a_{ij}
A^{-1}	Inverse of the matrix A
$M^{i_1,i_2,\ldots,i_k}_{j_1,j_2,\ldots,j_k}$	Minor of the k-th order (see page 59)
rk A	Rank of the matrix A
e^A or $\exp A$	Exponential of the matrix A
$\ln A$	Logarithm of the matrix A
$i = \sqrt{-1}$	Imaginary unit
z^*	Complex number conjugate of the complex number z
$|z|$	Modulus of the complex number z
$\arg z$	Argument of the complex number
Z^H	Hermitian conjugate matrix
$|\psi\rangle$	Quantum state
$|0\rangle, |1\rangle$	Basic quantum states of the qubit
$\sigma_1, \sigma_2, \sigma_3$	Pauli matrices
$x = [x_1, \ldots, x_n]^T$	Vector of the n-dimensional space $\mathbb{R}^n$
$\mathbf{0}$	Zero vector
$\|x\|$	Euclidean norm of the vector x
$X_{\text{gen.}}$	General solution of a homogeneous system of linear equations
$X_{\text{spec.}}$	Specific solution of a non-homogeneous system of linear equations
$\text{Pr}_L\, a$	Projection of the vector a onto the line L (see page 256)
i, j, k	Normalized vectors of the Cartesian coordinate system
$a \perp b$	Orthogonality of the vectors a and b
$(a \cdot b)$	Scalar or inner product of vectors
$a \times b$	Vectorial or outer product of vectors
(a, b, c)	Scalar triple product
$a \times (b \times c)$	Vector triple product
$\text{abs}(x)$	Absolute value of the real number x
$\text{sgn}(x)$	Sign of the real number x
μ	Normalizing factor (see pages 285 and 311)
δ	Deviation of a point from a line or a plane
$\mathcal{A}(x, y)$	Bilinear form
$\omega(x)$	Quadratic form
ε	Eccentricity of a curve of the second order
Γ	Elliptic curve with real points
Ξ	Elliptic curve with rational points
$\mathcal{O}$	Point at infinity of an elliptic curve
$A \oplus B$	The sum of two points A and B on an elliptic curve

Chapter 1
Matrices and Matrix Algorithms

1.1 Matrices and Operations with Them

Matrix of size $m \times n$ is a rectangular table of numbers with m rows and n columns. A matrix is written in the form

$$A = \begin{bmatrix} a_{11} & a_{12} & \ldots & a_{1n} \\ a_{21} & a_{22} & \ldots & a_{2n} \\ \ldots\ldots\ldots\ldots\ldots \\ a_{m1} & a_{m2} & \ldots & a_{mn} \end{bmatrix}. \tag{1.1}$$

Matrices are usually denoted by capital Latin letters, for example, $A, B, U, \ldots$

Numbers a_{ij}, included into the matrix, are its **elements**. An ordered set of elements $a_{i1}, a_{i2}, \ldots, a_{in}$ of the matrix A, having similar first index i, is referred to as the i-th **row** of the matrix, while an ordered set of elements $a_{1j}, a_{2j}, \ldots, a_{mj}$, having similar second index j, is referred to as the j-th **column**. Thus, the first index of an arbitrary element a_{ij} indicates the row number, while the second index indicates the column number, at the intersection of which this element is situated.

A brief matrix record is widely used:

$$A = (a_{ij}), \quad i = 1, 2, \ldots, m; \quad j = 1, 2, \ldots, n. \tag{1.2}$$

The column of n numbers is also called n-**vector**, or simply **vector**. So, the 1st vector represents a single number, or, in other words, **scalar**.

© Springer Nature Switzerland AG 2021
S. Kurgalin, S. Borzunov, *Algebra and Geometry with Python*,
https://doi.org/10.1007/978-3-030-61541-3_1

Note Matrices were initially introduced for a compact record of linear equations. Now, they are used in various divisions of mathematics and physics and their applications for simpler presentation of various mathematical operations on matrix elements.

Example 1.1 A point on a computer screen in RGB format is presented in the form of a 3-vector with components

$$P = \begin{bmatrix} p_R \\ p_G \\ p_B \end{bmatrix}, \tag{1.3}$$

where p_R, p_G, p_B are real numbers from interval $[0, 1]$, they characterize the intensity of red, green and blue colour components, respectively. Various combinations of the component values allow obtaining any colour. In particular, vectors

$$P_1 = \begin{bmatrix} 1 \\ 0 \\ 0 \end{bmatrix} \text{ and } P_2 = \begin{bmatrix} 0.2 \\ 0.2 \\ 0.6 \end{bmatrix} \tag{1.4}$$

determine red and dark-blue colours, respectively.   □

If the condition $m = n$ is met, then the matrix is called **square matrix of order** n. If the number of rows is not equal to the number of columns, and thus the inequality $m \neq n$ is met, then such a matrix is a **rectangular** one.

Note For presentation of matrices, the following notations are also used:

$$\begin{pmatrix} a_{11} & a_{12} & \dots & a_{1n} \\ a_{21} & a_{22} & \dots & a_{2n} \\ \dots & \dots & \dots & \dots \\ a_{m1} & a_{m2} & \dots & a_{mn} \end{pmatrix} \text{ or } \begin{Vmatrix} a_{11} & a_{12} & \dots & a_{1n} \\ a_{21} & a_{22} & \dots & a_{2n} \\ \dots & \dots & \dots & \dots \\ a_{m1} & a_{m2} & \dots & a_{mn} \end{Vmatrix}. \tag{1.5}$$

The elements of **real** matrices are real numbers from the set $\mathbb{R} = (-\infty, \infty)$, while the elements of **complex** matrices are complex numbers.

Note In a standard mathematical notation, the indices of the elements begin with one: $i, j = 1, 2, \dots$ In many programming languages, including Python and C, rows and columns are numbered from zero to $m - 1$ and $n - 1$, respectively. This difference should be paid attention to when realizing matrix algorithms in the mentioned languages.

For the matrix A we will build a new matrix B, where we transpose the rows and the columns:

$$B = \begin{bmatrix} a_{11} & a_{21} & \dots & a_{m1} \\ a_{12} & a_{22} & \dots & a_{m2} \\ \dots & \dots & \dots & \dots \\ a_{1n} & a_{2n} & \dots & a_{mn} \end{bmatrix}. \tag{1.6}$$

Such a matrix B is called **transposed** with respect to A and is denoted as A^T. As is easy to see, reapplication of the transposition operation returns to the initial matrix: $(A^T)^T = A$.

Example 1.2 Transposed with respect to the matrix $A = \begin{bmatrix} 5 & 0 & -4 \\ 2 & -1 & 3 \end{bmatrix}$ is the matrix

$$A^T = \begin{bmatrix} 5 & 2 \\ 0 & -1 \\ -4 & 3 \end{bmatrix}. \qquad \square$$

Let A be a square matrix. Its **main diagonal** is a set of elements $a_{11}, a_{22}, \dots, a_{nn}$, having the same indices, and **secondary diagonal**, or **cross-diagonal**, is the set of elements $a_{n1}, a_{(n-1)2}, \dots, a_{1n}$ of the matrix.

A square matrix is called **diagonal**, if all of her elements located outside the main diagonal are equal to zero:

$$\begin{bmatrix} d_1 & 0 & \dots & 0 \\ 0 & d_2 & \dots & 0 \\ \dots & \dots & \dots & \dots \\ 0 & 0 & \dots & d_n \end{bmatrix}. \tag{1.7}$$

If in a diagonal matrix of form (1.7) for all values $i = 1, 2, \dots, n$ the equalities $d_i = 1$ are true, then the matrix is called **identity matrix**, or **unit matrix**, and is denoted through I, while of all the elements $d_i = 0$, then it is called **zero matrix**, or **null matrix**, and is denoted by O:

$$I = \begin{bmatrix} 1 & 0 & \dots & 0 \\ 0 & 1 & \dots & 0 \\ \dots & \dots & \dots & \dots \\ 0 & 0 & \dots & 1 \end{bmatrix}, \quad O = \begin{bmatrix} 0 & 0 & \dots & 0 \\ 0 & 0 & \dots & 0 \\ \dots & \dots & \dots & \dots \\ 0 & 0 & \dots & 0 \end{bmatrix}. \tag{1.8}$$

For notation of the elements of an identity matrix, **Kronecker**[1] **symbol** is used, defined as follows:

$$\delta_{ij} = \begin{cases} 1, & \text{if } i = j, \\ 0, & \text{if } i \neq j. \end{cases} \tag{1.9}$$

Thus, in symbolic notations, we have $I = (\delta_{ij})$, where $i, j = 1, 2 \ldots, n$.

Note Often, in the notation of Kronecker symbol, the indices are divided by commas: $\delta_{i,j}$.

The matrix $A = (a_{ij})$ is called **upper triangular**, if $a_{ij} = 0$ at $i > j$, i.e. all the elements, positioned below the main diagonal, are equal to zero. Similarly the matrix $B = (b_{ij})$ is called **lower triangular**, if $b_{ij} = 0$ at $i < j$, i.e. all the elements above the main diagonal are equal to 0.

Upper and lower triangular matrices may schematically be denoted as shown in Fig. 1.1.

Square matrix $A = (a_{ij})$ is called **symmetric**, if for all values $i, j = 1, 2, \ldots, n$ elements $a_{ij} = a_{ji}$, in other words, all the elements symmetric with respect to the main diagonal are equal to each other.

Taking into account the notion of transposed matrix, the symmetry condition may be written in the form of the equality $A = A^T$.

For the **antisymmetric** matrix, the elements $a_{ij} = -a_{ji}$, where $i, j = 1, 2, \ldots, n$.

Let us turn to the notion of equality of matrices. Two matrices $A = (a_{ij})$ and $B = (b_{ij})$ of size $m \times n$ are **equal** to each other if and only if $a_{ij} = b_{ij}$ for all i and j. Thus, the property of equality can only be met for the matrices of the same size.

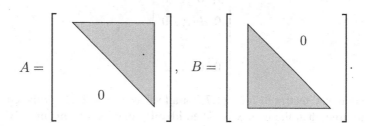

Fig. 1.1 Schematic notation for upper A and lower B triangular matrices. Highlighted is the position of the elements other than zero

---

[1]Leopold Kronecker (1823–1891), German mathematician.

Example 1.3 Consider two matrices C and D:

$$C = \begin{bmatrix} 1 & c^2 \\ -c^2 & c^4 \end{bmatrix}, \quad D = \begin{bmatrix} d^2 & -d \\ d & d^2 \end{bmatrix}, \tag{1.10}$$

where c and d are some real numbers.

Equality of matrices $C = D$ is equivalent to the system of equations reflecting equality of separate elements:

$$\begin{cases} d^2 = 1, \\ c^2 = -d, \\ -c^2 = d, \\ c^4 = d^2. \end{cases} \tag{1.11}$$

Then, the matrices C and D are equal if and only if the equalities $c = \pm 1$ and $d = -1$ are met. □

Example 1.4 **Binary matrix** or $(0, 1)$-**matrix** is called a matrix, whose elements take values 0 or 1. Let us calculate how many binary matrices of size $m \times n$ exist.

Each element of such a matrix may only take two values. Since the matrix consisting of m rows and n columns has a total of mn elements, then we obtain 2^{mn} ways to assign values to the elements. Hence, the number of binary matrices of size $m \times n$ is 2^{mn}. □

Consider the basic operations on matrices. Operations on matrices are introduced using the well-known arithmetic operations on their elements. Addition and multiplication of real numbers are naturally transferred to the matrices and form the basic operations of **matrix algebra**.

Sum of two matrices $A = (a_{ij})$ and $B = (b_{ij})$ of the same size $m \times n$ is the matrix $C = (c_{ij})$ of the same size, consisting of the elements $c_{ij} = a_{ij} + b_{ij}$. And, for the sum of matrices, it is written $C = A + B$.

Note, that one may only add square or rectangular matrices of the same size.

Example 1.5 Given two matrices A and B:

$$A = \begin{bmatrix} 2 & 0 & -1 \\ 1 & 3 & 4 \end{bmatrix}, \quad B = \begin{bmatrix} 0 & 5 & 3 \\ 2 & 1 & 4 \end{bmatrix}. \tag{1.12}$$

Find their sum $A + B$, having performed the operations of addition of the respective elements:

$$A + B = \begin{bmatrix} 2+0 & 0+5 & -1+3 \\ 1+2 & 3+1 & 4+4 \end{bmatrix} = \begin{bmatrix} 2 & 5 & 2 \\ 3 & 4 & 8 \end{bmatrix}. \qquad (1.13)$$

□

Product αA of the real number α and the matrix $A = (a_{ij})$ is the matrix $C = (c_{ij})$, consisting of the elements $c_{ij} = \alpha \cdot a_{ij}$.

Example 1.6 Assume that the real numbers $\alpha = 2$, $\beta = -3$ are set, and the matrix $A = \begin{bmatrix} 0 & -1 & 2 \\ -2 & 3 & 4 \end{bmatrix}$. Then $\alpha A = 2A = \begin{bmatrix} 0 & -2 & 4 \\ -4 & 6 & 8 \end{bmatrix}$, $\beta A = (-3)A = \begin{bmatrix} 0 & 3 & -6 \\ 6 & -9 & -12 \end{bmatrix}$.

□

Based on the introduced operations, we may make up a difference of matrices according to the definition: $A - B = A + (-1)B$. Thus, the matrices difference is nothing but the sum of the first summand and the second summand, multiplied by the number (-1).

Note that for antisymmetric matrix A the equality $A^T = -A$ is true.

Example 1.7 Find the difference of the matrices defined in Example 1.5:

$$A - B = \begin{bmatrix} 2 & 0 & -1 \\ 1 & 3 & 4 \end{bmatrix} - \begin{bmatrix} 0 & 5 & 3 \\ 2 & 1 & 4 \end{bmatrix} = \begin{bmatrix} 2 & 0 & -1 \\ 1 & 3 & 4 \end{bmatrix} + (-1) \begin{bmatrix} 0 & 5 & 3 \\ 2 & 1 & 4 \end{bmatrix}$$

$$= \begin{bmatrix} 2+(-1)0 & 0+(-1)5 & -1+(-1)3 \\ 1+(-1)2 & 3+(-1)1 & 4+(-1)4 \end{bmatrix} = \begin{bmatrix} 2 & -5 & -4 \\ -1 & 2 & 0 \end{bmatrix}.$$

□

The introduced operations have the following properties that are true for arbitrary matrices A, B and C and all $\lambda, \mu \in \mathbb{R}$:

1. $A + B = B + A$ (commutativity of addition);
2. $(A + B) + C = A + (B + C)$ (associativity of addition);
3. $\lambda(\mu A) = (\lambda \cdot \mu)A$;
4. $\lambda(A \pm B) = \lambda A \pm \lambda B$;
5. $(\lambda \pm \mu)A = \lambda A \pm \mu A$;
6. $A + O = O + A = A$.

The primary operation of linear algebra is the product of matrices. It, based on the two initial matrices, allows constructing a new matrix.

In order to introduce this notion, let us first consider one special case. The product of a row of n elements by a column of n elements is the element, equal to the sum of the products of the respective elements of the raw and the column:

$$\begin{bmatrix} a_1 & a_2 & \ldots & a_n \end{bmatrix} \begin{bmatrix} b_1 \\ b_2 \\ \vdots \\ b_n \end{bmatrix} = a_1 b_1 + a_2 b_2 + \cdots + a_n b_n. \tag{1.14}$$

Example 1.8 Calculate the product of the row $[1, 2, 4, 8, 16]$ by the column $[16, 8, 4, 2, 1]^T$:

$$\begin{bmatrix} 1 & 2 & 4 & 8 & 16 \end{bmatrix} \begin{bmatrix} 16 \\ 8 \\ 4 \\ 2 \\ 1 \end{bmatrix} = 1 \cdot 16 + 2 \cdot 8 + 4 \cdot 4 + 2 \cdot 8 + 1 \cdot 16 = 80. \tag{1.15}$$

□

Now let us consider the general case of matrices of an arbitrary size.

Product of the matrix $A = (a_{ij})$ of size $m \times n$ and the matrix $B = (b_{ij})$ of size $n \times p$ is the matrix $C = (c_{ij})$ of size $m \times p$, whose elements are expressed in accordance with the rule:

$$c_{ij} = \sum_{k=1}^{n} a_{ik} b_{kj}. \tag{1.16}$$

The product of matrices is written as $C = A \cdot B$ or $C = AB$.

Thus, the element c_{ij} of the matrix $C = AB$ is the sum of the products of the elements of the i-th row of the matrix A by the respective elements of the j-th column of the matrix B (Fig. 1.2).

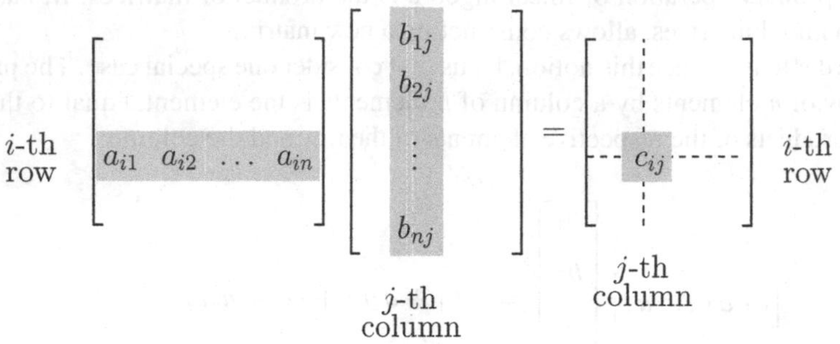

$$i\text{-th row} \begin{bmatrix} & & & \\ a_{i1} & a_{i2} & \dots & a_{in} \\ & & & \end{bmatrix} \begin{bmatrix} b_{1j} \\ b_{2j} \\ \vdots \\ b_{nj} \end{bmatrix} = \begin{bmatrix} & & \\ \cdots & c_{ij} & \cdots \\ & & \end{bmatrix} i\text{-th row}$$

$$j\text{-th column} \qquad\qquad\qquad j\text{-th column}$$

$$c_{ij} = \sum_{k=1}^{n} a_{ik}b_{kj}$$

Fig. 1.2 Multiplication of matrices (a_{ij}) and (b_{ij})

Example 1.9 Execute the operation of multiplication of matrices $\begin{bmatrix} 1 & 2 \\ -3 & 4 \end{bmatrix}$ and $\begin{bmatrix} -3 & 6 \\ 5 & -4 \end{bmatrix}$:

$$\begin{bmatrix} 1 & 2 \\ -3 & 4 \end{bmatrix} \begin{bmatrix} -3 & 6 \\ 5 & -4 \end{bmatrix} = \begin{bmatrix} 1\cdot(-3)+2\cdot 5 & 1\cdot 6+2\cdot(-4) \\ (-3)\cdot(-3)+4\cdot 5 & (-3)\cdot 6+4\cdot(-4) \end{bmatrix}$$

$$= \begin{bmatrix} 7 & -2 \\ 29 & -34 \end{bmatrix}. \tag{1.17}$$

$\square$

Note The definition of the product of matrices introduced above looks less natural than the definition of the sum. However, exactly this method of introducing the operation of multiplication allows, in matrix algebra, preserving many properties typical for the product of real numbers.

The following properties are met:

1. $A(B+C) = AB+AC, (B+C)A = BA+CA$ (distributivity of multiplication with respect to addition);
2. $(AB)C = A(BC)$ (associativity of multiplication);
3. $OA = AO = O$ (property of zero matrix);
4. $IA = AI = A$ (property of identity matrix).

In the general case, in the product of matrices, their order is essential, which is demonstrated by the following example.

Example 1.10 Let $A = \begin{bmatrix} 2 & -1 \\ 1 & 0 \end{bmatrix}$ and $B = \begin{bmatrix} 3 & 0 \\ 1 & -1 \end{bmatrix}$.

Then we have

$$A B = \begin{bmatrix} 2 & -1 \\ 1 & 0 \end{bmatrix} \begin{bmatrix} 3 & 0 \\ 1 & -1 \end{bmatrix} = \begin{bmatrix} 2 \cdot 3 + (-1) \cdot 1 & 2 \cdot 0 + (-1) \cdot (-1) \\ 1 \cdot 3 + 0 \cdot 1 & 1 \cdot 0 + 0 \cdot (-1) \end{bmatrix} = \begin{bmatrix} 5 & 1 \\ 3 & 0 \end{bmatrix},$$

(1.18)

at the same time, the product of matrices, executed in a different order, is equal to

$$B A = \begin{bmatrix} 3 & 0 \\ 1 & -1 \end{bmatrix} \begin{bmatrix} 2 & -1 \\ 1 & 0 \end{bmatrix} = \begin{bmatrix} 3 \cdot 2 + 0 \cdot 1 & 3 \cdot (-1) + 0 \cdot 0 \\ 1 \cdot 2 + (-1) \cdot 1 & 1 \cdot (-1) + (-1) \cdot 0 \end{bmatrix} = \begin{bmatrix} 6 & -3 \\ 1 & -1 \end{bmatrix}.$$

(1.19)

□

So, matrix multiplication is **non-commutative**, i.e. when the multipliers are permuted, the result may change.

As it directly follows from the definition of matrix product, they can be multiplied when and only when the number of rows of the first multiplier—matrix A, coincides with the number of rows of the second multiplier—matrix B. We should also note that the existence of the product $A B$ does not imply the existence of the product $B A$.

Matrix Commutator and Matrix Trace

Matrices A and B are called **commuting** (or **permutation**), if $A B = B A$. The commuting matrices are necessarily square and have the same order.

Commutator of two square matrices of the same order is the value

$$[A, B] = A B - B A.$$

(1.20)

By definition for commuting matrices, the condition $[A, B] = O$ is met.

Example 1.11 Calculate $[A, B]$, if

$$A = \begin{bmatrix} -1 & 2 & -2 \\ 2 & 1 & -1 \\ -1 & -1 & -1 \end{bmatrix}, \quad B = \begin{bmatrix} 1 & 0 & -1 \\ -1 & 1 & 1 \\ 2 & 0 & 0 \end{bmatrix}.$$

(1.21)

Solution

$$AB = \begin{bmatrix} -1 & 2 & -2 \\ 2 & 1 & -1 \\ -1 & -1 & -1 \end{bmatrix} \begin{bmatrix} 1 & 0 & -1 \\ -1 & 1 & 1 \\ 2 & 0 & 0 \end{bmatrix} = \begin{bmatrix} -7 & 2 & 3 \\ -1 & 1 & -1 \\ -2 & -1 & 0 \end{bmatrix}, \qquad (1.22)$$

$$BA = \begin{bmatrix} 1 & 0 & -1 \\ -1 & 1 & 1 \\ 2 & 0 & 0 \end{bmatrix} \begin{bmatrix} -1 & 2 & -2 \\ 2 & 1 & -1 \\ -1 & -1 & -1 \end{bmatrix} = \begin{bmatrix} 0 & 3 & -1 \\ 2 & -2 & 0 \\ -2 & 4 & -4 \end{bmatrix}, \qquad (1.23)$$

$$[A, B] = AB - BA = \begin{bmatrix} -7 & -1 & 4 \\ -3 & 3 & -1 \\ 0 & -5 & 4 \end{bmatrix}. \qquad (1.24)$$

$\square$

Example 1.12 Prove **Jacobi**[2] **identity**, true for the commutators of any matrices of size $n \times n$:

$$[[P, Q], R] + [[Q, R], P] + [[R, P], Q] \equiv O. \qquad (1.25)$$

Proof Use the definition of commutator $[P, Q] = PQ - QP$, then

$$[[P, Q], R] = [PQ - QP, R] = (PQ - QP)R - R(PQ - QP)$$

$$= PQR - QPR - RPQ + RQP. \qquad (1.26)$$

Then, in a similar manner, we will present the remaining summands in the sum:

$$[[Q, R], P] = QRP - RQP - PQR + PRQ, \qquad (1.27)$$

$$[[R, P], Q] = RPQ - PRQ - QRP + QPR. \qquad (1.28)$$

The sum of the values (1.26), (1.27) and (1.28), as is easy to see after reducing such summands, is equal to zero. Thus, the Jacobi identity is proved.

$\square$

---

[2]Carl Gustav Jacob Jacobi (1804–1851), German mathematician.

Trace tr A of the square matrix $A = (a_{ij})$, where $1 \leqslant i, j \leqslant n$, is the sum of its diagonal elements:

$$\text{tr } A = \sum_{i=1}^{n} a_{ii}. \tag{1.29}$$

Another designation of the matrix A trace is Sp A, from a German word "spur".

Example 1.13 The trace of the identity matrix I of size $n \times n$ is equal to its order: tr $I = n$. □

Estimate of the Number of Multiplication Operations When Multiplying Matrices

In order to estimate the working time of the computing algorithms it is necessary to know the number of the multiplication operations executed in the program. Let us determine this number for the matrix multiplication operation.

Let both product matrices be square and have the same order n. Then AB is the matrix $n \times n$. For calculation of all the result elements we will need n^2 multiplications of row by column. The multiplication of row by column contains exactly n multiplications of real numbers. Hence, in order to determine the product AB we will need n^3 real multiplications.

Note There exist non-elementary algorithms that allow multiplying matrices in a smaller number of operations. Among the most well known of such algorithms is **Strassen[3] algorithm**. Note that the advantages of using Strassen algorithm and similar non-elementary methods of matrices multiplication become apparent only for sufficiently large matrix size values [20].

Modern scientific and technical tasks, game industry projects, and technologies of augmented and alternative reality require fast execution of matrix operations on the mass data. This is why such actions with matrices as transposition, multiplication and others are presently executed using the parallel programming methods. Working with matrices on high-performance parallel systems has its own peculiarities associated with the methods of data presentation in the computer memory and the methods of interprocessor communication. In the works [5, 27, 56] basic algorithms of matrix algebra are presented, adapted for application on high-performance computing systems. The examples of implementation of such algorithms are provided in [42].

Note As was noted above, the matrix elements are real and complex numbers. Apart from this, the elements may also be functions on which algebraic operations can be performed. In such a case we say about **functional matrices**. Later on, unless otherwise specified, only numerical matrices are considered.

---

[3] Volker Strassen (born 1936), German mathematician.

1.2   Concept of Algorithm, Correctness of Algorithms

In the Sect. 1.4 will be shown algorithms for working with matrices in Python language. This is why, below we will preliminarily consider the concept of algorithm and algorithm correctness and show how to estimate their efficiency.

Algorithm is an exact prescription defining the computational process leading from the varying source data to the result sought for (the data is the ordered set of symbols) [48, 49]. In other words, an algorithm describes a certain computational procedure with the help of which a computational problem is solved. As a rule, the algorithm is used for solving some class of problems rather than one certain problem [16, 69]. The term "algorithm" derives from the name of a medieval mathematician al-Khwarizmi.[4]

The concept of algorithm belongs to basic fundamental notions of mathematics. Many researchers use various definitions of algorithm that differ from each other. However, all definitions express or imply the following **algorithm properties** [48, 49].

1. *Discreteness.* An algorithm must represent a process of problem solving as a sequential execution of separate steps. Execution of each algorithm step takes some time, and each operation is only executed wholly and cannot be executed partly.
2. *Elementary character of steps.* The method of execution of each command should be known and simple enough.
3. *Determinateness* (from Latin *dētermināre*—determine). Each successive step of the algorithm operation is uniquely determined. The result should be the same for the same source data.
4. *Directedness.* It should be known what to consider as the algorithm operation result.
5. *Mass character.* There must be a possibility to apply the algorithm to all collections of source data from the certain pre-fixed set.

Correctness of Algorithms
Consider the algorithm $\mathcal{A}$ that solves a certain computational problem. The possibility of applying this algorithm in a computer program requires justification of correct problem solution for all input data, i.e. we should carry out the **proof of the algorithm $\mathcal{A}$ correctness**. For this, we need to trace all changes of the variables' values that occur as a result of the algorithm's operation. From the mathematical point of view, we are talking about establishing the true values of some predicates describing the variables.

---

[4]al-Khwarizmi (Muḥammad ibn Mūsā al-Khwārizmī) (about 780–about 850), a distinguished mathematician, astronomer, geographer and philosopher. The term "algebra" derives from the name of his work containing the general techniques for solving problems reduced to several algebraic equations [9].

Assume that P is a predicate true for the input data of the algorithm $\mathcal{A}$, Q is a predicate taking a true value after completion of $\mathcal{A}$. The introduced predicates are called **precondition** and **postcondition**, respectively.

Proposition $\{P\}\mathcal{A}\{Q\}$ means the following: "if operation of the algorithm $\mathcal{A}$ starts from the true value of the predicate P, then it will end at the true value of Q". We obtain that the proof of correctness of the algorithm $\mathcal{A}$ is equivalent to the proof of trueness of $\{P\}\mathcal{A}\{Q\}$. The pre- and postcondition together with the algorithm itself are referred to as the **Hoare[5] triple**. The Hoare triple describes how the execution of the given fragment of the computer program changes the state of computation [59].

Example 1.14 Let us prove the correctness of the algorithm of exchanging the values of two variables.

Listing 1.1

```
1  # Exchanging of values of variables a and b
2  temp = a
3  a = b
4  b = temp
```

Proof Let the variables a and b take the following values: $a = a_0$, $b = b_0$.

Precondition: $P = \{a = a_0, b = b_0\}$, postcondition: $Q = \{a = b_0, b = a_0\}$.

Substitute the values of the variables a and b into the body of the algorithm $\mathcal{A}$, which will result in the following values: $temp = a_0$, $a = b_0$, $b = a_0$. This is why the predicate $\{P\}\mathcal{A}\{Q\}$ takes the true value, and thus the correctness of the algorithm $swap(a, b)$ is proved. □

1.3 Estimation of Algorithm Efficiency

An important task of the algorithm analysis is the estimation of the number of operations executed by the algorithm over a certain class of input data. The exact number of elementary operations does not play any significant role here, since it depends on the software implementation of the algorithm, the computer's architecture and other factors. This is why the algorithm's performance indicator is the growth rate of this value with the growth of the input data volume [16, 51].

In order to analyse the algorithm efficiency, it is necessary to estimate the running time of the computer that solves the set problem, as well as the volume of memory used. The estimate of the running time of the computing system is usually obtained by calculating the number of elementary operations performed during computations (such operations are called **basic operations**). With the supposition

---

[5]Charles Antony Richard Hoare (born 1934), English scientist specializing in computer science.

that one elementary operation is performed in a strictly defined time, the function $f(n)$, defined as the number of operations during computations on input data of size n, is called a **time-complexity function** [51].

In algorithm analysis, the number of **basic operations** is estimated, and it is assumed that execution of each of the listed below operations takes constant and not depending on n time [52].

1. Binary arithmetic operations ($+$, $-$, $*$, $/$) and operations of comparison of real numbers ($<$, $\leqslant$, $>$, $\geqslant$, $=$, $\neq$).
2. Logic operations (**and**, **or**).
3. Branching operations.
4. Calculation of the values of elementary functions for relatively small values of their arguments.

During implementation of matrix algorithms, in most cases the basic operation is considered as the operation of multiplication of two real numbers.

Let us consider the functions $f, g \colon \mathbb{N} \to (0, \infty)$. Assume that $g(n)$ describes the time complexity of the known algorithm.

It is said that a function $f(n)$ belongs to the class $O(g(n))$ (read as "big O of g"), if the growth rate of $f(n)$ does not exceed the growth rate of $g(n)$. We give a strict definition: $f(n) = O(g(n))$, if, for all values of the argument n, starting from a threshold value $n = n_0$, the inequality $f(n) \leqslant cg(n)$ is valid for some positive c:

$$O(g(n)) = \{f(n) \colon \exists c > 0, \ n_0 \in \mathbb{N} \ \text{ such that for all } \ n \geqslant n_0$$

$$f(n) \leqslant cg(n) \ \text{ is valid}\}. \tag{1.30}$$

The notation $f(n) \in O(g(n))$ can be read as "the function g **majorizes** the function f".

Since $O(g(n))$ denotes a set of functions growing no faster than the function $g(n)$, then, in order to indicate that a function belongs to this set, the notation $f(n) \in O(g(n))$ is used. Another notation is rather common in the literature: $f(n) = O(g(n))$, where the equals sign is understood conventionally, namely in the sense of belonging to the set. The class $O(g(n))$ are referred also to as the **"big O notation"**.

Example 1.15 Prove that the asymptotic estimate $3n^3 \in O(n^4)$ is true.

Proof According to the definition (1.30) we should prove that there exists a positive constant c such that starting from some number n_0, the inequality $3n^3 \leqslant cn^4$ is met, or $(cn - 3)n^3 \geqslant 0$.

Assume that $c = 3$, then, starting from $n_0 = 1$, the last inequality is true. Then, $3n^3 \in O(n^4)$. $\qquad\qquad\square$

Note The notation $O(f(t))$ is used not only for $t \to \infty$, but may also be generalized in case of an arbitrary limit value of the argument $t \to t_0$. For example, the expression

$$f(t) = O(g(t)) \ \text{ at } \ t \to t_0 \tag{1.31}$$

means that the limit of the ration limit of the functions $f(t)$ and $g(t)$ is taken at the point $t = t_0$:

$$\lim_{t \to t_0} \frac{f(t)}{g(t)} = \text{const} \geqslant 0. \qquad (1.32)$$

1.4 Primitive Matrix Operations in Python

In the programs in Python language, the matrices are presented in the form of two-dimensional arrays [62]. For the arrays in Python, a special term "*list*" is used. List is an ordered sequence of numbers or other presentable in the computing system's memory objects. Thus, the matrix is specified in the form of a list, whose elements are lists of the same length. In particular, the matrix

$$A = \begin{bmatrix} 11 & 13 & 15 & 17 \\ -9 & -8 & -7 & -6 \\ -1 & -2 & 12 & 14 \end{bmatrix} \qquad (1.33)$$

in a Python program will be presented as
    A=[[11, 13, 15, 17], [-9, -8, -7, -6], [-1, -2, 12, 14]]

As is seen, for formation of a list an enumeration of its elements separated by commas is used. In order to address the matrix elements, square brackets are used, for example, A[i, j].

Note that the indices of arrays in Python begin from zero rather than one. For example, for the matrix (1.33) we have the following equalities:
    A[0, 0] = 11
    A[2, 1] = -2

Note The agreement about zero starting values of indices is also used in such programming languages as C and Java [39]. However, in Fortran and Pascal languages, the indices by default begin from one [12, 74].

Let us show a program code used for inputting the matrix elements from the console and outputting the matrix to the console (see Listing 1.2).

Listing 1.2

```
1  def read_matrix_from_console():
2      n = int(input()) # Number of rows
3      m = int(input()) # Number of columns
4      A = []
5
6      for i in range(n):
```

```
 7        row = input().split()
 8        for j in range(m):
 9            row[j] = int(row[j])
10        A.append(row)
11    return A
12
13
14 def print_matrix_to_console(A):
15    for row in A:
16        for elem in row:
17            print(elem, end=' ')
18        print()
```

Example of call of functions read_matrix_from_console() and
print_matrix_to_console():

```
A = read_matrix_from_console()
print_matrix_to_console(A)
```

The following functions presented in Listing 1.3 perform standard operations on
matrices: addition, multiplication by a number and transposition.

Listing 1.3

```
 1 def matrix_add(A, B):
 2     if len(A) == len(B) and \
 3             len(A[0]) == len(B[0]):
 4         C = [[0 for j in range(len(A[0]))] \
 5               for i in range(len(A))]
 6
 7         for i in range(len(A)):
 8             for j in range(len(A[0])):
 9                 C[i][j] = A[i][j] + B[i][j]
10
11         return C
12
13
14 def matrix_mult_by_scalar(A, alpha):
15     C = [[0 for j in range(len(A[0]))] \
16           for i in range(len(A))]
17
18     for i in range(len(A)):
19         for j in range(len(A[0])):
20             C[i][j] = alpha * A[i][j]
21
22     return C
23
```

```
24
25 def matrix_subtract(A, B):
26     if len(A) == len(B) and \
27             len(A[0]) == len(B[0]):
28         C = [[0 for j in range(len(A[0]))] \
29                 for i in range(len(A))]
30
31         for i in range(len(A)):
32             for j in range(len(A[0])):
33                 C[i][j] = A[i][j] - B[i][j]
34
35         return C
36
37
38 def matrix_transpose(A):
39     C = [[0 for j in range(len(A))] \
40             for i in range(len(A[0]))]
41
42     for i in range(len(A)):
43         for j in range(len(A[0])):
44             C[j][i] = A[i][j]
45
46     return C
```

An important function calculating the product of matrices by formula (1.16) is presented in Listing 1.4.

Listing 1.4

```
1 # Multiplication of matrices A and B
2 def matrix_mult(A, B):
3     C = [[0 for j in range(len(B[0]))] \
4             for i in range(len(A))]
5
6     for i in range(len(A)):
7         for j in range(len(B[0])):
8             s = 0
9
10            for k in range(len(B)):
11                s += A[i][k] * B[k][j]
12
13            C[i][j] = s
14
15     return C
```

Table 1.1 Matrix functions and NumPy procedures

Name	Comment
dot(A,B)	The product of matrices A and B
trace(A)	Trace of matrix
linalg.inv(A)	Inversion of matrix
linalg.det(A)	Determinant of matrix
linalg.matrix_power(A, n)	Raising matrix A to power n
linalg.eigvals(A)	Calculation of eigenvalues of matrix
linalg.eig(A)	Solution of problems on eigenvalues and eigenvectors, the function return all solutions (λ, X) of the system $AX = \lambda X$
linalg.solve(A, B)	Solution of the system of linear equations $AX = B$ with vector B on its right side

1.4.1   NumPy Library

For high-performance calculations, the library NumPy with open source code is widely used [46, 57]. In this package, for presentation of matrices in the memory, the data type *array* is introduced. Apart from this, when including NumPy using the command

```
from numpy import*
```
a great number of matrix functions and procedures become available. The most important ones are listed in Table 1.1.

In particular, transposition of arbitrary rectangular matrices is performed using the method ".T":

```
A=array([[11, 13, 15, 17], [-9, -8, -7, -6], [-1, -2,
12, 14]])
A.T
```
To the console (more specifically, into the standard output stream) will be sent

```
A=array([[11, -9, -1],
         [13, -8, -2],
         [15, -7, 12],
         [17, -6, 14]])
```

1.5   Matrix Algorithms in the Graph Theory

As an example of the algorithm's operation with matrices, let us consider one of the important algorithms of the graph theory—Warshall[6] algorithm [1, 61], which is used for calculating the reachability matrix of the specified directed graph $D(V, E)$.

---

[6]Stephen Warshall (1935–2006), American researcher in the field of computer sciences.

First we will recall the basic notions of the graph theory. Everywhere below, the multiplication operation signs "·" and "×" will be considered as equivalent. In some cases, when it is clear that we are dealing with multiplication, they may be omitted.

Graph is a pair $G = (V, E)$, where V is the set of vertices, while E is the set of edges, connecting some pairs of vertices [22, 31, 55, 73]. In **directed graphs**, the edges are the *ordered* pair of vertices, i.e. it is of importance which vertex is the beginning of the edge and which one is the end. Directed graphs are also referred to as **digraphs**.

A drawing where the graph vertex is shown as points and the edges are shown as segments or arcs is called a **graph diagram**.

Two vertices u and v of the graph are **adjacent**, if they are connected by the edge $r = uv$. In this case it is said that the vertices u and v are the **endpoints** of the edge r. If the vertex v is the endpoint of the edge r, then v and r are considered to be **incident** (from Latin *incēdere*—to distribute).

The number of elements (**cardinality**) of any set, for example V, is denoted as $|V|$.

Adjacency matrix M is a binary matrix of a relation over the set of vertices of the graph $G(V, E)$, which is specified by its edges. The adjacency matrix as the size $|V| \times |V|$, and its elements are determined in accordance with the rule

$$M(i, j) = \begin{cases} 1, & \text{if edge } ij \in E, \\ 0, & \text{if edge } ij \notin E. \end{cases} \tag{1.34}$$

A **path** of length k in the graph G is a sequence of vertices $v_0, v_1, \ldots, v_k$ such that $\forall i = 1, \ldots, k$ the vertices v_{i-1} and v_i are adjacent. There are also considered **trivial** paths of the form v_i, v_i. For undirected graphs, paths are also called **routes**.

The **length** of the path is the number of edges in it, taking into account the iterations.

Example 1.16 Consider a digraph $D(V, E)$, the set of vertices V and the set of edges E of which are specified as follows:

$$V = \{a, b, c, d, e\}, \quad E = \{ab, ae, bc, bd, dc, de, ec\}.$$

The graph $D(V, E)$ is presented in Fig. 1.3.

The adjacency matrix M of the digraph D has the form:

$$M = \begin{array}{c} \\ a \\ b \\ c \\ d \\ e \end{array} \begin{array}{c} a\ b\ c\ d\ e \\ \begin{bmatrix} 0 & 1 & 0 & 0 & 1 \\ 0 & 0 & 1 & 1 & 0 \\ 0 & 0 & 0 & 0 & 0 \\ 0 & 0 & 1 & 0 & 1 \\ 0 & 0 & 1 & 0 & 0 \end{bmatrix} \end{array}. \tag{1.35}$$

Fig. 1.3 The digraph
$D(V, E)$ to the Example 1.3

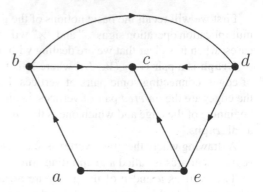

Reachability matrix M^* of the digraph $D(V, E)$ is a logic closing matrix by transitivity relation E. The reachability matrix stores the information about the existence of paths between the digraph vertices: at the intersection of the i-th row and the j-the column stands 1 when and only when there exists a path from the vertex v_i to v_j. M^* may be calculated by a formula using the logical operation **or** [29]

$$M^* = M \text{ or } M^2 \text{ or } \dots \text{ or } M^n, \tag{1.36}$$

where n is the number of vertices of the directed graph, i.e. $n = |V|$. Note, that determining the elements of the matrix M^* by formula (1.36) is associated with a considerable volume of calculations, this is why for the digraphs with a great number of vertices, the **Warshall algorithm** is used, also known as the **algorithm of Roy[7]–Warshall** [61].

The Warshall algorithm is based on formation of a sequence of auxiliary binary matrices $W^{(0)}$, $W^{(1)}$, ..., $W^{(n)}$, where $n = |V|$. The first matrix is set equal to the adjacency matrix M of the digraph. The elements $W_{ij}^{(k)}$, where $1 \leqslant i, j, k \leqslant n$, are calculated by the rule: $W_{ij}^{(k)} = 1$, if there exists a path connecting the vertices v_i and v_j such that all the inner vertices belong to the set $V_k = \{v_1, v_2, \dots, v_k\}$, and $W_{ij}^{(k)} = 0$ otherwise. Note that the inner vertex of the path $P = v_i, \dots, v_l, \dots, v_j$ is any vertex v_l, $1 \leqslant l \leqslant n$, belonging to P, except the first v_i and the last v_j. The resulting matrix $W^{(n)}$ appears to be equal to $W^{(n)} = M^*$, since $M_{ij}^* = 1$ when and only when there exists the path $v_i, \dots, v_j$, all inner vertices of which are contained in $V = \{v_1, v_2, \dots, v_n\}$.

The principal moment is that the matrix $W^{(k)}$ can be obtained from $W^{(k-1)}$ as follows. The path $v_i, \dots, v_j$, containing the inner vertices only from the set V_k, exists when and only when one of the conditions is fulfilled:

1. there exists a path $v_i, \dots, v_j$ with inner vertices only from $V_{k-1} = \{v_1, v_2, \dots, v_{k-1}\}$;

---

[7]Bernard Roy (born 1934), French mathematician.

2. there are paths $v_1, \ldots, v_k$ and $v_k, \ldots, v_j$, also containing inner vertices only from V_{k-1}.

We obtain two cases: either $W_{ij}^{(k-1)} = 1$, if v_k is included into the set of vertices allowed at this stage, or $W_{ik}^{(k-1)} = 1$ and $W_{kj}^{(k-1)} = 1$. Therefore, using the logical operations **or** (disjunction) and **and** (conjunction) we may write

$$W_{ij}^{(k)} = W_{ij}^{(k-1)} \text{ or } \left(W_{ik}^{(k-1)} \text{ and } W_{kj}^{(k-1)} \right). \tag{1.37}$$

Let us show a respective algorithm for constructing M^* by the specified adjacency matrix M of size $n \times n$, where $n > 1$. The intermediate matrices $W^{(k)}$, where $0 \leqslant k \leqslant n - 1$, should not necessarily be stored in memory until the end of the algorithm's operation, this is why, in the suggested realization, the elements $W^{(k-1)}$ are substituted by the elements of the subsequent matrix $W^{(k)}$.

Listing 1.5

```
1  def Warshall_algorithm(M):
2      n = len(M)
3
4      W = [[0 for j in range(n)] \
5              for i in range(n)]
6
7      for i in range(n):
8          for j in range(n):
9              W[i][j] = M[i][j]
10
11     for k in range(n):
12         for i in range(n):
13             for j in range(n):
14                 W[i][j] = W[i][j] or \
15                           (W[i][k] and W[k][j])
16
17     return W
```

Correctness of algorithm WarshallAlgo can be proved by the method of mathematical induction (see the description of this method below on page 53) [61]. Solution of the problem for finding M^* is also investigated in Problem **1.39** and in [61].

Example 1.17 Let digraph D be specified (Fig. 1.4). Construct the reachability matrix M^*, using Warshall algorithm.

Fig. 1.4 Directed graph D

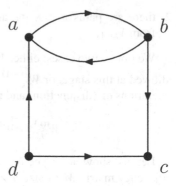

Solution The matrix $W^{(0)}$ coincides with the adjacency matrix of the digraph and has the form

$$W^{(0)} = \begin{array}{c} \\ a \\ b \\ c \\ d \end{array} \overset{\displaystyle a\ b\ c\ d}{\begin{bmatrix} 0 & 1 & 0 & 0 \\ 1 & 0 & 1 & 0 \\ 0 & 0 & 0 & 0 \\ 1 & 0 & 1 & 0 \end{bmatrix}}.$$

Calculate $W^{(1)}$. If $W_{ij}^{(0)} = 1$, then the respective element $W_{ij}^{(1)}$ is also equal to 1: $W_{ij}^{(1)} = 1$. If $W_{ij}^{(0)} = 0$, then attention should be paid to the elements of the first row and the first column, standing at the intersection with the j-th column and the i-the row: if $W_{1j}^{(0)} = W_{i1}^{(0)} = 1$, then $W_{ij}^{(1)} = 1$. The condition $W_{1j}^{(0)} = W_{i1}^{(0)} = 1$ is fulfilled for the two pairs (i, j), namely for $i = j = 2$ and $i = 4$, $j = 2$. Then, $W_{22}^{(1)} = W_{42}^{(1)} = 1$, and all the rest elements $W^{(1)}$ coincide with the respective elements of the matrix $W^{(0)}$. For illustration, in the notation of the matrix we will highlight in bold and underline the elements $W^{(1)}$ that have changed values at this step:

$$W^{(1)} = \begin{array}{c} \\ a \\ b \\ c \\ d \end{array} \overset{\displaystyle a\ b\ c\ d}{\begin{bmatrix} 0 & 1 & 0 & 0 \\ 1 & \mathbf{\underline{1}} & 1 & 0 \\ 0 & 0 & 0 & 0 \\ 1 & \mathbf{\underline{1}} & 1 & 0 \end{bmatrix}}.$$

Then we will calculate $W^{(2)}$. Consider the second row and the second column of the matrix $W^{(1)}$. The elements $W^{(1)}$ that are positioned in the same row with the elements of $W_{i2}^{(1)} = 1$ from the second column and in the same column with the elements of $W_{2j}^{(1)} = 1$ from the second row, will change their values in $W^{(1)}$ for 1. Such will be the elements $W_{11}^{(1)}$ and $W_{13}^{(1)}$. The rest of the elements $W^{(2)}$ coincide with the respective elements of the matrix $W^{(1)}$.

$$
W^{(2)} = \begin{array}{c} \\ a \\ b \\ c \\ d \end{array}
\begin{array}{c} a\ b\ c\ d \\ \left[\begin{array}{cccc} \underline{1}\ 1\ \underline{1}\ 0 \\ 1\ 1\ 1\ 0 \\ 0\ 0\ 0\ 0 \\ 1\ 1\ 1\ 0 \end{array} \right] \end{array}.
$$

At the next step, the vertex c is added to the set of vertices. This does not result in appearance of new elements with value 1.

$$
W^{(3)} = \begin{array}{c} \\ a \\ b \\ c \\ d \end{array}
\begin{array}{c} a\ b\ c\ d \\ \left[\begin{array}{cccc} 1\ 1\ 1\ 0 \\ 1\ 1\ 1\ 0 \\ 0\ 0\ 0\ 0 \\ 1\ 1\ 1\ 0 \end{array} \right] \end{array}.
$$

At the final step, we obtain $W^{(4)} = W^{(3)}$, and the reachability matrix of digraph D will have the form

$$
M^* = \begin{array}{c} \\ a \\ b \\ c \\ d \end{array}
\begin{array}{c} a\ b\ c\ d \\ \left[\begin{array}{cccc} 1\ 1\ 1\ 0 \\ 1\ 1\ 1\ 0 \\ 0\ 0\ 0\ 0 \\ 1\ 1\ 1\ 0 \end{array} \right] \end{array}.
$$

□

Review Questions

1. Define diagonal matrix, upper triangular matrix, lower triangular matrix, symmetric matrix and binary matrix.
2. How is the matrix transposition operation performed?

3. How is the Kronecker delta determined?
4. What matrices are called symmetric? antisymmetric?
5. Formulate the definition of product of two matrices of size $m \times n$ and $n \times p$.
6. What is the commutator of the matrices A and B?
7. Define trace of a square matrix.
8. What is algorithm?
9. Enumerate the properties of an algorithm.
10. How is the algorithm efficiency estimated?
11. Explain the meaning of the notation $O(f(n))$.
12. Describe the presentation of matrices in Python.
13. Enumerate the basic matrix functions and the procedures of NumPy library.
14. How are graphs presented in the computer memory?
15. For solution of what problem is the Warshall algorithm used?

Problems

1.1. Calculate $3A + 2B$, where $A = \begin{bmatrix} 2 & 1 & -1 \\ 0 & 1 & 4 \end{bmatrix}$, $B = \begin{bmatrix} -2 & 1 & 0 \\ -3 & 2 & 2 \end{bmatrix}$.

1.2. Calculate AB, where $A = \begin{bmatrix} 1 & -1 & 0 \\ 2 & 3 & 4 \end{bmatrix}$, $B = \begin{bmatrix} 1 & 1 \\ 2 & -1 \\ 3 & 0 \end{bmatrix}$. Find $B^T A^T$ and $(AB)^T$.

1.3. Let the matrix $A = (a_{ij})$ be of size $n_1 \times n_2$ and the matrix $B = (b_{ij})$ be of size $n_2 \times n_1$. Prove that the following equality is fulfilled:

$$(AB)^T = B^T A^T, \tag{1.38}$$

i.e. the transposed product of two matrices is equal to product of the transposed matrices in reversed order.

1.4. Write the matrices of size 3×3, whose elements are determined by the formulas:

(1) $a_{ij} = (-1)^{i+j-1}$;
(2) $b_{ij} = \dfrac{i + j + |i - j|}{2}$;
(3) $c_{ij} = (i - 2)^2 + (j - 2)^2$;
(4) $d_{ij} = \sin(|i - j|)$.

Calculate the sum of all elements S of each matrix.

1.5. Let $A = (a_{ij})$ be a square matrix of order $n \geqslant 3$. Using the summation symbol $\sum$, write the following values:

(1) the sum of the elements of the third row;
(2) the sum of the elements of the second column;
(3) the sum of squares of the diagonal elements;
(4) the module of the sum of the elements positioned on the secondary diagonal.

1.6. How to write the sum of the elements of the square matrix positioned above the main diagonal using the summation sign? How to do this for the elements positioned below the main diagonal?

1.7. A student carrying out an experiment in a chemical laboratory has accidentally spilled a reagent on an algebra notes page, where antisymmetric matrix was written. As a result, it was impossible to read some of its elements. If we denote such elements by symbol "?", the notation will look as

$$\begin{bmatrix} 0 & 1 & -1 & ? \\ ? & 0 & 2 & 2 \\ ? & ? & ? & ? \\ 7 & ? & 0 & ? \end{bmatrix}.$$

Restore the unknown elements and write the original matrix.

1.8. Determine the number of binary matrices of n rows and n columns that are

(1) symmetric and
(2) antisymmetric

relative to the main diagonal.

1.9. Calculate

$$\text{a)} \quad \begin{bmatrix} 0 & 0 & 1 \\ 1 & 1 & 2 \\ 2 & 2 & 3 \\ 3 & 3 & 4 \end{bmatrix} \begin{bmatrix} -1 & -1 \\ 2 & 2 \\ 1 & 1 \end{bmatrix} \begin{bmatrix} 4 \\ 1 \end{bmatrix} ; \quad \text{b)} \quad \begin{bmatrix} 1 & -2 \\ 3 & -4 \end{bmatrix}^3 .$$

1.10. Consider the binary matrices

$$Q_1 = \begin{bmatrix} 0 & 1 & 0 \\ 1 & 0 & 0 \\ 0 & 0 & 1 \end{bmatrix}, \quad Q_2 = \begin{bmatrix} 0 & 0 & 1 \\ 0 & 1 & 0 \\ 1 & 0 & 0 \end{bmatrix}.$$

Calculate $Q_1 Q_2$, $Q_2 Q_1$, Q_1^2 and Q_2^2.

1.11. For what matrices D of the second order the square of D^2 is equal to the zero matrix?

1.12. Let the matrices $L = [-2, -1, 0, 1, 2]$, $M = [0, 2, 4, 6, 8]$ be specified. Calculate the products of LM^T and $M^T L$.

1.13. The elements of the matrix $G = (g_{ij})$ are determined in accordance with the rule

$$g_{ij} = \begin{cases} 1, & \text{if } i \geqslant j, \\ 0, & \text{if } i < j. \end{cases}$$

What are the elements of the matrix G^2 equal to?

1.14. At the examination in linear algebra, a student says that multiplication of two non-zero matrices will necessarily result in a non-zero matrix. Is the student right?

*1.15. Let us denote by x_i the number of processors produced by some plant starting from the beginning of the year. In particular, the plant's operation during the first month is described by the vector $X = [x_1, x_2, \ldots, x_{31}]^T$. Determine the matrix D, which should influence the X, in order to obtain the vector $Y = [x_2 - x_1, x_3 - x_2, \ldots, x_n - x_{n-1}]^T$: $Y = D \cdot X$. The vector Y reflects the daily production capacity gain of the plant.

1.16. Calculate the product of the functional matrices $A(\psi)B(\theta)A(\varphi)$, if

$$A(\varphi) = \begin{bmatrix} \cos\varphi & \sin\varphi & 0 \\ -\sin\varphi & \cos\varphi & 0 \\ 0 & 0 & 1 \end{bmatrix}, \quad B(\theta) = \begin{bmatrix} 1 & 0 & 0 \\ 0 & \cos\theta & \sin\theta \\ 0 & -\sin\theta & \cos\theta \end{bmatrix}.$$

1.17. Calculate the commutator $[A, B]$, if

$$A = \begin{bmatrix} 7 & 5 & 3 \\ 1 & 3 & 2 \\ 2 & 2 & 7 \end{bmatrix}, B = \begin{bmatrix} 6 & 2 & 3 \\ 5 & 2 & 1 \\ 1 & 1 & 6 \end{bmatrix}.$$

1.18. Calculate the commutator $[A, B]$, if A is an arbitrary matrix, $B = I$ is identity matrix of the same order as A.

1.19. Consider the matrices

$$P_1 = \begin{bmatrix} 0 & -1 & 0 \\ 1 & 0 & 0 \\ 0 & 0 & 0 \end{bmatrix}, \quad P_2 = \begin{bmatrix} 0 & 0 & 0 \\ 0 & 0 & -1 \\ 0 & 1 & 0 \end{bmatrix}, \quad P_3 = \begin{bmatrix} 0 & 0 & 1 \\ 0 & 0 & 0 \\ -1 & 0 & 0 \end{bmatrix}. \qquad (1.39)$$

Calculate the commutators $[P_1, P_2]$, $[P_2, P_3]$ and $[P_3, P_1]$.

1.20. Is it true that for any square matrices A, B and C of the same size the equality $[A + B, C] = [A, C] + [B, C]$ is fulfilled?

1.21. Prove that for any matrices A, B and C of the same size, the identity is true

$$[AB, C] \equiv A[B, C] + [A, C]B. \tag{1.40}$$

1.22. Is it true that for any square matrices A, B and C of the same size the equality $[A, [B, C]] = [[A, B], C]$ is fulfilled?

1.23. Given the square matrices A and B of the same order. In what case the equality $(A + B)^2 = A^2 + 2AB + B^2$ is true?

1.24. Is it true that if $[A, B] = O$ and $[A, C] = O$, then the matrices B and C are commuting?

1.25. Suppose that the sizes of the matrices A, B and C are equal to $n_1 \times n_2$, $n_2 \times n_3$ and $n_3 \times n_4$, respectively. In order to calculate the product of ABC, the multiplication operations can be executed in two ways: $(A \cdot B) \cdot C$ or $A \cdot (B \cdot C)$. With what relation between the variables n_1, n_2, n_3 and n_4 the calculation using the first method—as $(A \cdot B) \cdot C$—will require less operations of multiplication of real numbers in comparison with the second method?

1.26. Prove the correctness of the algorithm of change of values of two variables without using the auxiliary variable.

Listing 1.6

```
1  # Exchanging of values of variables a and b
2  # without using the auxiliary variable
3  a = a + b
4  b = a - b
5  a = a - b
```

1.27. Prove the correctness of the algorithm of addition of square matrices.

Listing 1.7

```
1   # Addition of matrices A and B
2   def matrix_add(A, B):
3       if len(A) == len(B) and \
4               len(A[0]) == len(B[0]):
5           C = [[0 for j in range(len(A[0]))] \
6                   for i in range(len(A))]
7
8           for i in range(len(A)):
9               for j in range(len(A[0])):
10                  C[i][j] = A[i][j] + B[i][j]
11
12          return C
```

1.28. Prove the correctness of the square matrix multiplication algorithm.

Listing 1.8

```
1  # Multiplication of matrices A and B
2  def matrix_mult(A, B):
3      C = [[0 for j in range(len(B[0]))] \
4           for i in range(len(A))]
5
6      for i in range(len(A)):
7          for j in range(len(B[0])):
8              s = 0
9
10             for k in range(len(B)):
11                 s += A[i][k] * B[k][j]
12
13             C[i][j] = s
14
15     return C
```

1.29. Prove the correctness of the optimized matrix multiplication algorithm.

Listing 1.9

```
1  # Optimized multiplication
2  # of matrices a and b
3  def matrix_mult2(A, B):
4      n = len(A)
5
6      C = [[0 for i in range(n)] \
7           for j in range(n)]
8
9      D = [0 for i in range(n)]
10
11     for i in range(n):
12         for j in range(n):
13             s = 0
14
15             for k in range(n):
16                 D[k] = B[k][j]
17
18             for k in range(n):
19                 s += A[i][k] * D[k]
20
21             C[i][j] = s
22
23     return C
```

Note The considered option of matrix multiplication is optimized in such a manner as to preliminarily select the elements of the column $b(k, j)$ into the intermediate array d, which may fully be placed in a fast cache memory.

1.30. Determine the number of operations of addition of two numbers executed by the algorithm `matrix_add` for the matrices of size $N \times N$.

1.31. Determine the number of operations of addition and multiplication, executed by the algorithm `matrix_mult` for the matrices of size $N \times N$.

1.32. Suggest a method to decrease the number of addition operations executed by the algorithm `matrix_mult`.

1.33. Let A_1, A_2 and A_3 be numerical matrices of size 50×25, 25×30 and 30×10, respectively. Determine the minimal number of multiplication operations required to calculate the product of $A_1 A_2 A_3$ by the standard algorithm `matrix_mult`, whose realization for square matrices is presented in Problem **1.28**.

1.34. Let A_1, A_2, A_3 and A_4 be numerical matrices of size 25×10, 10×50, 50×5 and 5×30, respectively. Find the minimal number of multiplication operations, required for calculation of the product of $A_1 A_2 A_3 A_4$ by the standard algorithm `matrix_mult`.

1.35. Let A_1, A_2, A_3 and A_4 be numerical matrices of size 100×20, 20×15, 15×50 and 50×100, respectively. Find the minimal number of multiplication operations required for calculation of the product of $A_1 A_2 A_3 A_4$ by the standard algorithm `matrix_mult`.

∗1.36. Prove that the number of methods for calculation of the product of the matrices $A_1 A_2 \ldots A_{m+1}$, $m \geqslant 1$, or, in other words, the number of ways to place brackets in this product, where A_1, A_2, $\ldots$, A_{m+1} are numerical matrices of size $n_1 \times n_2, n_2 \times n_3, \ldots, n_{m+1} \times n_{m+2}$, respectively, is equal to the **Catalan**[8] **number** $\mathcal{C}_m$, determined by formula

$$\mathcal{C}_m = \frac{1}{m+1} C(2m, m) \text{ for all } m \geqslant 1, \tag{1.41}$$

where $C(2m, m)$ is a binomial coefficient.

1.37. Estimate the number of operations executed by Warshall algorithm for obtaining the reachability matrix of a digraph.

1.38. Using the Warshall algorithm, calculate the reachability matrix of the digraph D, presented in Fig. 1.5.

1.39. One of the ways to modify the Warshall algorithm consists in presentation of the rows of binary matrices as bit strings. In this case, for calculation of the elements of the matrices $W^{(k)}$ for $1 \leqslant i \leqslant N$, the bitwise operation **or** is used. Find the number of operation in the bit strings executed by the given realization of the Warshall algorithm.

---

[8]Eugène Charles Catalan (1814–1894), Belgian mathematician.

Fig. 1.5 To Problem **1.38**

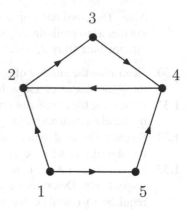

1.40. Write the analytical expression for the function $f(n)$, represented by the algorithm:

Listing 1.10

```python
1  def f(n: int):
2      temp = 0
3
4      for i in range(1, n + 1):
5          for j in range(1, n + 1):
6              for k in range(j, n + 1):
7                  temp += 1
8
9      return temp
```

∗**1.41.** Write the analytical expression for the function $g(n)$, represented by the algorithm:

Listing 1.11

```python
1  def g(n: int):
2      temp = 0
3
4      for i in range(1, n + 1):
5          for j in range(n, i - 1, -1):
6              for k in range(1, j + 1):
7                  temp += 1
8
9      return temp
```

*1.42. Write the analytical expression for the function $h(n)$, represented by the algorithm:

Listing 1.12

```
1  def h(n: int):
2      temp = 0
3
4      for i in range(1, n + 1):
5          for j in range(1, i + 1):
6              for k in range(1, j + 1):
7                  for l in range(1, k + 1):
8                      temp += 1
9
10     return temp
```

1.43. Calculate the trace of the matrix $A = (a_{ij})$, where $1 \leqslant i, j \leqslant n$, whose elements are specified by the formulas:

(1) $a_{ij} = i + j$;
(2) $a_{ij} = i - j$;
(3) $a_{ij} = \ln(i^2 + j^2)$;
(4) $a_{ij} = \max(i, n - j)$;
(5) $a_{ij} = \min\left(\dfrac{1}{i}, \dfrac{1}{j}\right)$;
(6) $a_{ij} = \sin(\pi(i + 2j))$;
(7) $a_{ij} = \dfrac{n}{2} - (i + j)$;
(8) $a_{ij} = i + \dfrac{1}{j}$.

1.44. Prove that the trace of the product of the square matrices does not depend on the order of multipliers: $\operatorname{tr}(AB) = \operatorname{tr}(BA)$.

Answers and Solutions

1.1 *Solution.*

We calculate the summands in the sum: $3A = \begin{bmatrix} 6 & 3 & -3 \\ 0 & 3 & 12 \end{bmatrix}$, $2B = \begin{bmatrix} -4 & 2 & 0 \\ -6 & 4 & 4 \end{bmatrix}$.

We perform the matrix summation operation: $3A + 2B = \begin{bmatrix} 2 & 5 & -3 \\ -6 & 7 & 16 \end{bmatrix}$.

1.2 *Solution.*

We will calculate the product AB relying on the formula (1.16) on page 7:

$$
A \cdot B = \begin{bmatrix} 1 & -1 & 0 \\ 2 & 3 & 4 \end{bmatrix} \begin{bmatrix} 1 & 1 \\ 2 & -1 \\ 3 & 0 \end{bmatrix}
$$

$$
= \begin{bmatrix} 1 \cdot 1 + (-1) \cdot 2 + 0 \cdot 3 & 1 \cdot 1 + (-1) \cdot (-1) + 0 \cdot 0 \\ 2 \cdot 1 + 3 \cdot 2 + 4 \cdot 3 & 2 \cdot 1 + 3 \cdot (-1) + 4 \cdot 0 \end{bmatrix} = \begin{bmatrix} -1 & 2 \\ 20 & -1 \end{bmatrix}.
$$

We execute the matrix transposition operation:

$$
A^T = \begin{bmatrix} 1 & 2 \\ -1 & 3 \\ 0 & 4 \end{bmatrix}, \quad B^T = \begin{bmatrix} 1 & 2 & 3 \\ 1 & -1 & 0 \end{bmatrix};
$$

$$
B^T \cdot A^T = \begin{bmatrix} 1 \cdot 1 + 2 \cdot (-1) + 3 \cdot 0 & 1 \cdot 2 + 2 \cdot 3 + 3 \cdot 4 \\ 1 \cdot 1 + (-1) \cdot (-1) + 0 \cdot 0 & 1 \cdot 2 + (-1) \cdot 3 + 0 \cdot 4 \end{bmatrix}
$$

$$
= \begin{bmatrix} -1 & 20 \\ 2 & -1 \end{bmatrix}.
$$

$$
(A \cdot B)^T = \begin{bmatrix} -1 & 20 \\ 2 & -1 \end{bmatrix}.
$$

Note that the equality $(AB)^T = B^T A^T$ is fulfilled for any matrices A and B, for which the product AB is determined.

1.3 *Proof.*

Based on the definition of the matrix transposition and on the formula (1.16), the left part of the equality (1.38) consists of elements:

$$
\left((AB)^T \right)_{ij} = (AB)_{ji} = \sum_{k=1}^{n_1} a_{jk} b_{ki},
$$

where $1 \leqslant i \leqslant n_3$, $1 \leqslant j \leqslant n_1$.

Further, we represent elements of the right side of Eq. (1.38) in the form

$$
(B^T A^T)_{ij} = \sum_{k=1}^{n_2} (B^T)_{ik} (A^T)_{kj} = \sum_{k=1}^{n_2} b_{ki} a_{jk} = \sum_{k=1}^{n_2} a_{jk} b_{ki},
$$

where $1 \leqslant i \leqslant n_3$, $1 \leqslant j \leqslant n_1$.

It is proved that for all possible numbers i and j elements of matrices $(AB)^T$ and $B^T A^T$ coincide. In accordance with the definition of equality of matrices on page 4 we get

$$(AB)^T = B^T A^T.$$

1.4 *Solution.*

Having calculated the matrix elements by the specified formulas, we obtain

(1) $\begin{bmatrix} -1 & 1 & -1 \\ 1 & -1 & 1 \\ -1 & 1 & -1 \end{bmatrix}$, the sum of all elements $S = -1$;

(2) $\begin{bmatrix} 1 & 2 & 3 \\ 2 & 2 & 3 \\ 3 & 3 & 3 \end{bmatrix}$, $S = 22$;

(3) $\begin{bmatrix} 2 & 1 & 2 \\ 1 & 0 & 1 \\ 2 & 1 & 2 \end{bmatrix}$, $S = 12$;

(4) $\begin{bmatrix} 0 & \sin 1 & \sin 2 \\ \sin 1 & 0 & \sin 1 \\ \sin 2 & \sin 1 & 0 \end{bmatrix}$, $S = 4\sin 1 + 2\sin 2$.

1.5 *Solution.*

(1) For the elements of the third row we have $i = 3$, $j = 1, 2, \ldots, n$. Therefore, the sum of the elements of the third row is presented in the form $\sum\limits_{j=1}^{n} a_{3j}$.

(2) The elements of the second column may be written as a_{i2}, where $i = 1, 2, \ldots, n$. Then, their sum is equal to $\sum\limits_{i=1}^{n} a_{i2}$.

(3) For the diagonal elements, the indices i and j coincide: a_{ii}. The sum of squares of such elements is $\sum\limits_{i=1}^{n} a_{ii}^2$.

(4) The secondary diagonal is formed by the elements whose sum of indices is greater by one to the order of the matrix: $i+j = n+1$. Therefore, $j = (n+1)-i$, the module of the sum of these elements is also equal to $\mathrm{abs}\left(\sum\limits_{i=1}^{n} a_{i(n+1-i)} \right)$.

1.6 *Solution.*

Let us write the sum of the elements of the square matrix positioned above the main diagonal. The number of the elements to be summed up in the rows above the main diagonal decreases by one as the row number increases. Due to this, summation of the elements in the i-th row begins with $i + 1$, then the sum in one row will be $\sum\limits_{j=i+1}^{n} a_{ij}$. Performing the summation operation over all rows, we obtain

$$\sum_{i=1}^{n} \sum_{j=i+1}^{n} a_{ij}.$$

Now let us write the sum of the elements positioned below the main diagonal. The number of the elements to be summed up below the main diagonal increases by one as the row number increases. Then, the upper summation threshold should be equal to $i - 1$, and the sum for one row is calculated as $\sum\limits_{j=1}^{i-1} a_{ij}$. For the sum of the elements positioned below the main diagonal, we obtain $\sum\limits_{i=1}^{n} \sum\limits_{j=1}^{i-1} a_{ij}$.

1.7 *Solution.*

By definition of an antisymmetric matrix, $A^T = -A$, then

$$A^T = \begin{bmatrix} 0 & a_{21} & a_{31} & 7 \\ 1 & 0 & a_{32} & a_{42} \\ -1 & 2 & a_{33} & 0 \\ a_{14} & 2 & a_{34} & a_{44} \end{bmatrix}, \quad -A = \begin{bmatrix} 0 & -1 & 1 & -a_{14} \\ -a_{21} & 0 & -2 & -2 \\ -a_{31} & -a_{32} & -a_{33} & -a_{34} \\ -7 & -a_{42} & 0 & -a_{44} \end{bmatrix}.$$

Equating the respective matrix elements, we obtain

$$A = \begin{bmatrix} 0 & 1 & -1 & -7 \\ -1 & 0 & 2 & 2 \\ 1 & -2 & 0 & 0 \\ 7 & -2 & 0 & 0 \end{bmatrix}.$$

1.8 *Solution.*

First of all we should note that the number of the elements on the main diagonal of the matrix is n, while the number of the elements lying above it is $n(n-1)/2$.

(1) The elements of the symmetric matrix lying below the main diagonal are uniquely determined by the upper triangular part of the matrix; this can be done by $2^{n(n-1)/2}$ methods. There exist 2^n methods for selection of diagonal elements. We obtain that the number of symmetric matrices of n rows and n columns is equal to $2^{n(n-1)/2} \cdot 2^n = 2^{n(n+1)/2}$.

(2) The main diagonal of the antisymmetric matrix is filled with zeros. In order to determine such a matrix, it is sufficient to specify the elements above the main diagonal; this can be done by $2^{n(n-1)/2}$ methods. Therefore, all in all there exist $2^{n(n-1)/2}$ antisymmetric matrices of size $n \times n$.

1.9 *Solution.*

a) $\begin{bmatrix} 0\ 0\ 1 \\ 1\ 1\ 2 \\ 2\ 2\ 3 \\ 3\ 3\ 4 \end{bmatrix} \begin{bmatrix} -1\ -1 \\ 2\ \ 2 \\ 1\ \ 1 \end{bmatrix} \begin{bmatrix} 4 \\ 1 \end{bmatrix} = \begin{bmatrix} 1\ 1 \\ 3\ 3 \\ 5\ 5 \\ 7\ 7 \end{bmatrix} \begin{bmatrix} 4 \\ 1 \end{bmatrix} = \begin{bmatrix} 5 \\ 15 \\ 25 \\ 35 \end{bmatrix}$;

b) $\begin{bmatrix} 1\ -2 \\ 3\ -4 \end{bmatrix}^3 = \begin{bmatrix} 1\ -2 \\ 3\ -4 \end{bmatrix} \begin{bmatrix} 1\ -2 \\ 3\ -4 \end{bmatrix} \begin{bmatrix} 1\ -2 \\ 3\ -4 \end{bmatrix} = \begin{bmatrix} -5\ \ 6 \\ -9\ 10 \end{bmatrix} \begin{bmatrix} 1\ -2 \\ 3\ -4 \end{bmatrix}$

$= \begin{bmatrix} 13\ -14 \\ 21\ -22 \end{bmatrix}$.

1.10 *Solution.*

$Q_1 Q_2 = \begin{bmatrix} 0\ 1\ 0 \\ 1\ 0\ 0 \\ 0\ 0\ 1 \end{bmatrix} \begin{bmatrix} 0\ 0\ 1 \\ 0\ 1\ 0 \\ 1\ 0\ 0 \end{bmatrix} = \begin{bmatrix} 0\ 1\ 0 \\ 0\ 0\ 1 \\ 1\ 0\ 0 \end{bmatrix}$;

$Q_2 Q_1 = \begin{bmatrix} 0\ 0\ 1 \\ 0\ 1\ 0 \\ 1\ 0\ 0 \end{bmatrix} \begin{bmatrix} 0\ 1\ 0 \\ 1\ 0\ 0 \\ 0\ 0\ 1 \end{bmatrix} = \begin{bmatrix} 0\ 0\ 1 \\ 1\ 0\ 0 \\ 0\ 1\ 0 \end{bmatrix}$;

$Q_1^2 = \begin{bmatrix} 0\ 1\ 0 \\ 1\ 0\ 0 \\ 0\ 0\ 1 \end{bmatrix} \begin{bmatrix} 0\ 1\ 0 \\ 1\ 0\ 0 \\ 0\ 0\ 1 \end{bmatrix} = \begin{bmatrix} 1\ 0\ 0 \\ 0\ 1\ 0 \\ 0\ 0\ 1 \end{bmatrix}$;

$Q_2^2 = \begin{bmatrix} 0\ 0\ 1 \\ 0\ 1\ 0 \\ 1\ 0\ 0 \end{bmatrix} \begin{bmatrix} 0\ 0\ 1 \\ 0\ 1\ 0 \\ 1\ 0\ 0 \end{bmatrix} = \begin{bmatrix} 1\ 0\ 0 \\ 0\ 1\ 0 \\ 0\ 0\ 1 \end{bmatrix}$.

1.11 *Solution.*

Let the matrix D have the form $\begin{bmatrix} a\ b \\ c\ d \end{bmatrix}$, where a, b, c, d are unknown real numbers.

Let us square the D and equate the obtained result to the zero matrix.

$$\begin{bmatrix} a & b \\ c & d \end{bmatrix}\begin{bmatrix} a & b \\ c & d \end{bmatrix} = \begin{bmatrix} a^2 + bc & ab + bd \\ ac + dc & bc + d^2 \end{bmatrix} = \begin{bmatrix} 0 & 0 \\ 0 & 0 \end{bmatrix}.$$

Let us write the system of relatively unknown a, b, c, d:

$$\begin{cases} a^2 + bc = 0, \\ c(a + d) = 0, \\ b(a + d) = 0, \\ bc + d^2 = 0. \end{cases}$$

From the obtained system it follows that the matrix D should have the form

$\begin{bmatrix} a & b \\ c & -a \end{bmatrix}$, and the variables a, b and c are bound by the condition $a^2 + bc = 0$.

1.12 *Answer:*

The size of the matrix L is equal to 1×5, the size of the matrix M^T is equal to 5×1. Therefore, the matrices LM^T and $M^T L$ have the sizes 1×1 (this is a scalar) and 5×5, respectively. Having executed the multiplication operations, we obtain

$$LM^T = [-2, -1, 0, 1, 2]\begin{bmatrix} 0 \\ 2 \\ 4 \\ 6 \\ 8 \end{bmatrix} = (-2) \cdot 0 + (-1) \cdot 2 + 0 \cdot 4 + 1 \cdot 6 + 2 \cdot 8 = 20;$$

$$M^T L = \begin{bmatrix} 0 \\ 2 \\ 4 \\ 6 \\ 8 \end{bmatrix}[-2, -1, 0, 1, 2] = \begin{bmatrix} 0 & 0 & 0 & 0 & 0 \\ -4 & -2 & 0 & 2 & 4 \\ -8 & -4 & 0 & 4 & 8 \\ -12 & -6 & 0 & 6 & 12 \\ -16 & -8 & 0 & 8 & 16 \end{bmatrix}.$$

1.13 *Answer:*

Let $H = G^2$, then

$$h_{ij} = \begin{cases} i - j + 1, & \text{if } i \geqslant j, \\ 0, & \text{if } i < j. \end{cases}$$

1.14 *Solution.*

Let $A = \begin{bmatrix} t & 0 \\ 0 & 0 \end{bmatrix}$, $B = \begin{bmatrix} 0 & 0 \\ t & 0 \end{bmatrix}$, where $t \in \mathbb{R}$.

As is easy to see, the equality $AB = O$ is fulfilled here, therefore, the product of two non-zero matrices may be equal to a zero matrix. The student is wrong.

1.16 *Solution.*

$$A(\psi)B(\theta) = \begin{bmatrix} \cos\psi & \sin\psi & 0 \\ -\sin\psi & \cos\psi & 0 \\ 0 & 0 & 1 \end{bmatrix} \begin{bmatrix} 1 & 0 & 0 \\ 0 & \cos\theta & \sin\theta \\ 0 & -\sin\theta & \cos\theta \end{bmatrix}$$

$$= \begin{bmatrix} \cos\psi & \sin\psi\cos\theta & \sin\psi\sin\theta \\ -\sin\psi & \cos\psi\cos\theta & \cos\psi\sin\theta \\ 0 & -\sin\theta & \cos\theta \end{bmatrix}.$$

$A(\psi)B(\theta)A(\varphi)$

$$= \begin{bmatrix} \cos\psi & \sin\psi\cos\theta & \sin\psi\sin\theta \\ -\sin\psi & \cos\psi\cos\theta & \cos\psi\sin\theta \\ 0 & -\sin\theta & \cos\theta \end{bmatrix} \begin{bmatrix} \cos\varphi & \sin\varphi & 0 \\ -\sin\varphi & \cos\varphi & 0 \\ 0 & 0 & 1 \end{bmatrix}$$

$$= \begin{bmatrix} \cos\psi\cos\varphi - \sin\psi\cos\theta\sin\varphi & \cos\psi\sin\varphi + \sin\psi\cos\theta\cos\varphi & \sin\psi\sin\theta \\ -\sin\psi\cos\varphi - \cos\psi\cos\theta\sin\varphi & -\sin\psi\sin\varphi + \cos\psi\cos\theta\cos\varphi & \cos\psi\sin\theta \\ \sin\theta\sin\varphi & -\sin\theta\cos\varphi & \cos\theta \end{bmatrix}.$$

1.17 *Solution.*

$$AB = \begin{bmatrix} 7 & 5 & 3 \\ 1 & 3 & 2 \\ 2 & 2 & 7 \end{bmatrix} \begin{bmatrix} 6 & 2 & 3 \\ 5 & 2 & 1 \\ 1 & 1 & 6 \end{bmatrix} = \begin{bmatrix} 70 & 27 & 44 \\ 23 & 10 & 18 \\ 29 & 15 & 50 \end{bmatrix},$$

$$BA = \begin{bmatrix} 6 & 2 & 3 \\ 5 & 2 & 1 \\ 1 & 1 & 6 \end{bmatrix} \begin{bmatrix} 7 & 5 & 3 \\ 1 & 3 & 2 \\ 2 & 2 & 7 \end{bmatrix} = \begin{bmatrix} 50 & 42 & 43 \\ 39 & 33 & 26 \\ 20 & 20 & 47 \end{bmatrix},$$

$$[A, B] = AB - BA = \begin{bmatrix} 20 & -15 & 1 \\ -16 & -23 & -8 \\ 9 & -5 & 3 \end{bmatrix}.$$

1.18 *Answer:* $[A, I] = O$.

1.19 *Answer:*

$[P_1, P_2] = P_3$, $[P_2, P_3] = P_1$ and $[P_3, P_1] = P_2$.

1.20 *Solution.*

Yes, the equality $[A + B, C] = [A, C] + [B, C]$ is true for any square matrices A, B and C of the same size. Simple calculations show that

$$[(A+B), C] = (A+B)C - C(A+B) = AC - CA + BC - CB = [A, C] + [B, C].$$

1.21 *Proof.*

Transform the right side of the equality (1.40), relying on the definition of (1.20):

$$A[B, C] + [A, C]B = A(BC - CB) + (AC - CA)B.$$

Then, remove the brackets and indicate the similar summands, following which use the definition of the commutator once again:

$$A[B, C] + [A, C]B = ABC - ACB + ACB - CAB = ABC - CAB = [AB, C].$$

Thus the identity (1.40) is proved.

1.22 *Solution.*

No, the equality $[A, [B, C]] = [[A, B], C]$ is fulfilled not for all A, B and C, as the following counterexample shows:

$$A = \begin{bmatrix} 1 & 0 \\ 0 & 0 \end{bmatrix}, \quad B = \begin{bmatrix} 0 & 1 \\ 0 & 0 \end{bmatrix}, \quad C = \begin{bmatrix} 0 & 0 \\ 1 & 0 \end{bmatrix},$$

$$[A, [B, C]] = \begin{bmatrix} 0 & 0 \\ 0 & 0 \end{bmatrix}, \quad [[A, B], C] = \begin{bmatrix} 1 & 0 \\ 0 & -1 \end{bmatrix}.$$

Therefore, for the arbitrary matrices $[A, [B, C]] \neq [[A, B], C]$.

1.23 *Answer:* the equality is true in the case $AB = BA$, i.e. if the matrices are commuting.

1.24 *Solution.*

This is not true. The following counterexample can be given to the proposition from the problem situation:

$$A = \begin{bmatrix} 1 & 0 \\ 0 & 1 \end{bmatrix}, \quad B = \begin{bmatrix} 1 & 0 \\ 0 & -1 \end{bmatrix}, \quad C = \begin{bmatrix} 0 & 1 \\ 1 & 0 \end{bmatrix}.$$

In this case, the equalities $[A, B] = [A, C] = O$ are fulfilled, while

$$[B, C] = \begin{bmatrix} 1 & 0 \\ 0 & -1 \end{bmatrix}\begin{bmatrix} 0 & 1 \\ 1 & 0 \end{bmatrix} - \begin{bmatrix} 0 & 1 \\ 1 & 0 \end{bmatrix}\begin{bmatrix} 1 & 0 \\ 0 & -1 \end{bmatrix} = \begin{bmatrix} 0 & -2 \\ -2 & 0 \end{bmatrix} \neq O.$$

Therefore, the matrices B and C are not necessarily commuting.

1.25 *Solution.*

Calculate the number of multiplication operations for each of the two methods.

(1) During multiplication of each row of the matrix A by each column of the matrix B, n_2 multiplication operations are performed. Since the number of rows is n_1, and the number of columns is n_3, as a result we obtain $n_1 n_2 n_3$ operations. The matrix AB has the size $n_1 \times n_3$; then for obtaining the matrix ABC from AB and C we need $n_1 n_3 n_4$ operations of multiplication of real numbers. Thus, the calculation of $(AB)C$ requires $n_1 n_3 (n_2 + n_4)$ multiplication operations.

(2) Reasoning similarly, we obtain $n_2 n_4 (n_1 + n_3)$ multiplication operations required for calculation by the scheme $A(BC)$.

As a result, calculation by the first method—as $(AB)C$—requires less operations of multiplication of real numbers when fulfilling the condition $n_1 n_3 (n_2 + n_4) < n_2 n_4 (n_1 + n_3)$.

1.30 *Answer:* for calculation of the sum of two matrices A and B we will need N^2 additions for determining each of N^2 elements $A + B$.

1.31 *Solution.*

Each of N^2 elements of the matrix AB is calculated as a scalar product of two vectors of size N, which, respectively, requires N additions and N multiplications. The total number of both additions and multiplications appears to be equal to $N \cdot N^2 = N^3$.

1.32 *Solution.*

The cycle body by the variable k may be rewritten in the form

```
C[i][j] = A[i][0] * B[0][j]

for k in range(1, len(B)):
    C[i][j] = C[i][j] + A[i][k] * B[k][j]
```

Then the number of additions reduces to $N^3 - N^2$, while the number of multiplications remains unchanged.

1.33 *Solution.*

As is known, the number of multiplication operations required to calculate the product of matrices of size $n_1 \times n_2$ and $n_2 \times n_3$ equals to $n_1 n_2 n_3$. (Recall that the product of two matrices is defined if the number of columns of the first one coincides with the number of rows of the second one.) Due to associativity of the multiplication operation, the product $A_1 A_2 A_3$ can be calculated in two ways: $(A_1 A_2)A_3$ and $A_1(A_2 A_3)$.

In the first case, we will need $50 \cdot 25 \cdot 30 + 50 \cdot 30 \cdot 10 = 52\,500$ multiplications and in the second case— $25 \cdot 30 \cdot 10 + 50 \cdot 25 \cdot 10 = 20\,000$. So, the minimal number of multiplication operations required to calculate the elements of the matrix $A_1 A_2 A_3$ by the standard algorithm equals to $20\,000$.

Note There exists an efficient algorithm for finding the order of multiplications in the product $A_1 A_2 \ldots A_n$, $n > 2$, with the minimal number of operations [16].

1.34 *Answer:* 7500.

1.35 *Answer:* 255,000.

1.37 *Solution.*

In order to calculate $W[i, j]$ in the rows with numbers 14–15 (see the algorithm on page 21), two logical operations are required. Since this row is executed $N \times N \times N = N^3$ times, where N is the size of the adjacency matrix of the digraph, the full number of operations for obtaining M^* equals to $2N^3$.

1.38 *Answer:*

$$M^* = \begin{bmatrix} 0 & 1 & 1 & 1 & 1 \\ 0 & 1 & 1 & 1 & 0 \\ 0 & 1 & 1 & 1 & 0 \\ 0 & 1 & 1 & 1 & 0 \\ 0 & 1 & 1 & 1 & 0 \end{bmatrix}.$$

1.40 *Answer:* $f(N) = \dfrac{N^2(N+1)}{2}$.

1.41 *Answer:* $g(N) = \dfrac{N(N+1)(2N+1)}{6}$.

1.42 *Answer:* $h(N) = \dfrac{N(N+1)(N+2)(N+3)}{24}$.

1.43 *Answer:*

(1) $\operatorname{tr} A = n(n+1)$;

(2) $\operatorname{tr} A = 0$;

(3) $\operatorname{tr} A = n \ln 2 + 2 \ln(n!)$, where $n! = \displaystyle\prod_{i=1}^{n} i$ is factorial of the number n;

(4) $\operatorname{tr} A = \begin{cases} 3k^2, & \text{if } n = 2k, \\ 3k^2 + 3k + 1, & \text{if } n = 2k+1, \end{cases}$ where $k \in \mathbb{N}$;

(5) $\operatorname{tr} A = H_n$, where $H_n = \displaystyle\sum_{i=1}^{n} \frac{1}{i}$ is the n-th harmonic number;

(6) $\operatorname{tr} A = 0$;

(7) $\operatorname{tr} A = -\dfrac{1}{2}n(n+2)$;

(8) $\operatorname{tr} A = \dfrac{1}{2}n(n+1) + H_n$, where H_n is harmonic number (see above).

1.44 *Proof.*

The trueness of the statement $\operatorname{tr}(AB) = \operatorname{tr}(BA)$ follows from the chain of equalities:

$$\operatorname{tr}(AB) = \sum_{i,j} a_{ij}b_{ji} = \sum_{i,j} b_{ji}a_{ij} \overset{i \leftrightarrow j}{=} \sum_{i,j} b_{ij}a_{ji} = \operatorname{tr}(BA).$$

Chapter 2
Matrix Algebra

2.1 Determinant of a Matrix: Determinants of the Second and Third Order

One of the fundamental notions of linear algebra is the **determinant of a square matrix**. Let us begin considering this notion with the determinants of the second and third order.

We will associate the matrix $A = \begin{bmatrix} a_{11} & a_{12} \\ a_{21} & a_{22} \end{bmatrix}$ of size 2×2 with the number

$$a_{11}a_{22} - a_{21}a_{12}, \tag{2.1}$$

which is called the **second order determinant** of the matrix A and is denoted as

$$\Delta \equiv \det A \equiv |A| \equiv \begin{vmatrix} a_{11} & a_{12} \\ a_{21} & a_{22} \end{vmatrix}. \tag{2.2}$$

Example 2.1 If $A = \begin{bmatrix} 3 & 2 \\ 1 & 5 \end{bmatrix}$, then $\Delta = \begin{vmatrix} 3 & 2 \\ 1 & 5 \end{vmatrix} = 3 \cdot 5 - 1 \cdot 2 = 13$. $\qquad \square$

© Springer Nature Switzerland AG 2021
S. Kurgalin, S. Borzunov, *Algebra and Geometry with Python*,
https://doi.org/10.1007/978-3-030-61541-3_2

Third order determinant of the matrix A of size 3×3 is a number obtained by the following formula:

$$\Delta \equiv \det A \equiv |A| \equiv \begin{vmatrix} a_{11} & a_{12} & a_{13} \\ a_{21} & a_{22} & a_{23} \\ a_{31} & a_{32} & a_{33} \end{vmatrix}$$

$$= a_{11}a_{22}a_{33} + a_{12}a_{23}a_{31} + a_{13}a_{21}a_{32}$$

$$-a_{13}a_{22}a_{31} - a_{12}a_{21}a_{33} - a_{11}a_{23}a_{32}. \tag{2.3}$$

The formula (2.3) can be easily remembered with the help of the **triangle rule**: the value of the third order determinant is equal to the algebraic sum of six terms, each being a product of three elements, one from each row and each column of the matrix A. The sign "+" is taken by the product of the elements lying on the main diagonal, and two products of the elements forming within the matrix triangles with bases parallel to the main diagonal. The sign "−" is taken by the products of the elements lying on the secondary diagonal, and two products of the elements forming triangles with bases parallel to the secondary diagonal (see Fig. 2.1). Unfortunately, triangle rule is only applicable to calculation of the determinants of matrices of size 3×3. The next Sect. 2.2 describes what to do with the matrices of greater size.

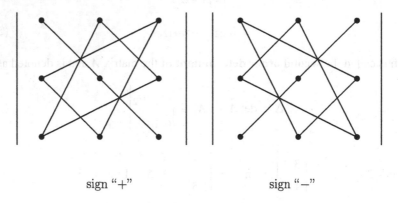

sign "+"   sign "−"

Fig. 2.1 The scheme "triangle rule" for calculation of the third order determinant

Example 2.2

$$\det A = |A| = \begin{vmatrix} 4 & 2 & 2 \\ -4 & -3 & 1 \\ 3 & 0 & 3 \end{vmatrix}$$

$$= 4 \cdot (-3) \cdot 3 + 2 \cdot 1 \cdot 3 + 2 \cdot (-4) \cdot 0$$

$$- 2 \cdot (-3) \cdot 3 - 2 \cdot (-4) \cdot 3 - 4 \cdot 1 \cdot 0$$

$$= -36 + 6 + 0 + 18 + 24 - 0 = 12.$$

□

Note that the first order determinant for the matrix $A = [a_{11}]$, consisting of one element, is equal to the value of this element: $\det[a_{11}] = a_{11}$.

2.2 Determinants of the n-th order: Minors

The determinant of the n-th order, where $n \geqslant 2$, has the form

$$\Delta = \begin{vmatrix} a_{11} & a_{12} & \ldots & a_{1n} \\ a_{21} & a_{22} & \ldots & a_{2n} \\ \vdots & \vdots & \ddots & \vdots \\ a_{n1} & a_{n2} & \ldots & a_{nn} \end{vmatrix}. \tag{2.4}$$

By analogy with (2.3) it is a polynomial each summand of which is a product of exactly n elements of the matrix (a_{ij}), and only one multiplier is included into the product from each row and each column of this matrix.

Now let us turn to strict definitions.

Let the ordered set of indices $(k_1, k_2, \ldots, k_n)$ form some permutation of the numbers $1, 2, \ldots, n$. For n first natural numbers, as is easy to see, there exist $n!$ pairwise different permutations.

Inversion in permutation or simply **inversion** is the pair (k_1, k_2), where the greater number stands before the smaller one: $k_1 > k_2$.

Example 2.3 In the collection $(4, 3, 2, 1)$ there are six inversions: $(4, 3)$, $(4, 2)$, $(4, 1)$, $(3, 2)$, $(3, 1)$, $(2, 1)$; there are no other inversions in this collection. □

Note The only permutation that contains no inversion is the identity permutation $(1, 2, \ldots, n)$.

Determinant of the n-th order of the matrix (a_{ij}) is the sum

$$\det A = \sum_{\text{perm}} (-1)^\sigma a_{1k_1} a_{2k_2} \dots a_{nk_n}, \qquad (2.5)$$

where summation is performed over all permutations, σ is the number of inversions in the permutation $(k_1, k_2, \dots, k_n)$. The determinant contains $n!$ terms, half of which is taken with a positive sign and half with a negative sign.

Example 2.4 Let us expand the definition (2.5) for the case $n = 2$.

Solution For $n = 2$ we have $2! = 1 \cdot 2 = 2$ permutations, namely $k_1 = 1, k_2 = 2$ and $k_1 = 2, k_2 = 1$. The sum in $\det A$ will consist of two terms: $a_{11}a_{22}$ and $a_{12}a_{21}$, taken with positive and negative signs, respectively:

$$\det A = \sum_{\text{perm}} (-1)^\sigma a_{1k_1} a_{2k_2} = (-1)^0 a_{11}a_{22} + (-1)^1 a_{12}a_{21}. \qquad (2.6)$$

The obtained formula exactly correlates with the definition (2.1), provided in Sect. 2.1.   □

If we remove from the matrix A of the n-th order the i-th row and the j-th column, then we will obtain the matrix of the $(n-1)$-th order, whose determinant is called the **complementary minor of the element** a_{ij} of the matrix A and is denoted by M_{ij}.

The variable $A_{ij} = (-1)^{i+j} M_{ij}$ is called the **algebraic complement (cofactor)** of the element a_{ij} of the matrix A.

Example 2.5 Given the determinant

$$\Delta = \begin{vmatrix} 2 & 1 & -2 \\ 3 & -1 & 4 \\ -3 & 5 & 0 \end{vmatrix}, \qquad (2.7)$$

then the complementary minors of the elements of the matrix a_{23} and a_{31} and their cofactors are equal to

$$M_{23} = \begin{vmatrix} 2 & 1 \\ -3 & 5 \end{vmatrix} = 10 + 3 = 13, \quad A_{23} = (-1)^{2+3} M_{23} = (-1) \cdot 13 = -13, \qquad (2.8)$$

$$M_{31} = \begin{vmatrix} 1 & -2 \\ -1 & 4 \end{vmatrix} = 4 - 2 = 2, \quad A_{31} = (-1)^{3+1} M_{31} = 2. \qquad (2.9)$$

□

Theorem 2.1 (Laplace[1]) *For the determinant of the matrix $A = (a_{ij})$ the following formulas are true:*

$$\det A = \sum_{j=1}^{n} a_{ij} A_{ij}, \quad i = 1, 2, \ldots, n, \tag{2.10}$$

$$\det A = \sum_{i=1}^{n} a_{ij} A_{ij}, \quad j = 1, 2, \ldots, n, \tag{2.11}$$

i.e. the determinant can be expanded over any row or any column, using the cofactors of the matrix elements.

The formulas (2.10) and (2.11) allow reducing the calculation of the determinant of the n-th order to calculation of the determinant of the $(n - 1)$-th order. The procedure of reduction of order continues until we arrive at the determinants of the second and third order, which are relatively easy to calculate.

Relations (2.10) and (2.11) are referred to as the **Laplace expansions**.

Example 2.6 Let us calculate the determinant of the fourth order, expanding it in the first column:

$$\Delta = \begin{vmatrix} 2 & 3 & -3 & 4 \\ 2 & 1 & -1 & 2 \\ 6 & 2 & 1 & 0 \\ 2 & 3 & 0 & -5 \end{vmatrix} = 2\begin{vmatrix} 1 & -1 & 2 \\ 2 & 1 & 0 \\ 3 & 0 & -5 \end{vmatrix} - 2\begin{vmatrix} 3 & -3 & 4 \\ 2 & 1 & 0 \\ 3 & 0 & -5 \end{vmatrix} + 6\begin{vmatrix} 3 & -3 & 4 \\ 1 & -1 & 2 \\ 3 & 0 & -5 \end{vmatrix}$$

$$- 2\begin{vmatrix} 3 & -3 & 4 \\ 1 & -1 & 2 \\ 2 & 1 & 0 \end{vmatrix} = 2(-5 + 0 + 0 - 6 - 0 - 10)$$

$$- 2(-15 + 0 + 0 - 12 - 0 - 30) + 6(15 - 18 + 0 + 12 - 0 - 15)$$

$$- 2(0 - 12 + 4 + 8 - 6 - 0) = -42 + 114 - 36 + 12 = 48.$$

$\square$

---

[1] Pierre-Simon, marquis de Laplace (1749–1827), French mathematician, physicist and astronomer.

2.3   General Properties of Determinants: Elementary Transformations of a Matrix

Let us enumerate the general properties of determinants [43, 63].

Property 1   During transposition of a matrix, its determinant does not change:

$$\det A^T = \det A. \tag{2.12}$$

Property 2   After exchange of two rows (columns) of a matrix, the determinant changes its sign.

Property 3   The determinant of a matrix with two identical rows (columns) is equal to zero.

Property 4   The determinant of a matrix with two proportional rows (columns) is equal to zero.

Property 5   The determinant of a matrix will not change if to all elements of its row (column) are added the respective elements of another row (column), multiplied by the same number.

Property 6   The determinant of a product of matrices is equal to the product of determinants, i.e.

$$\det(A \cdot B) = \det A \cdot \det B. \tag{2.13}$$

Property 7   The determinant of a triangular matrix coincides with the product of the elements standing on the main diagonal.

Property 8   If all the elements of some row (column) of a matrix are multiplied by the same number, its determinant will be multiplied by this number.

Property 9   If all the elements of the i-th row of a determinant are given as a sum of two terms $a_{ij} = u_j + v_j$ for $j = 1, \ldots, n$, then the determinant is equal to the sum of two determinants of the following form:

$$\begin{vmatrix} a_{11} & a_{12} & \ldots & a_{1n} \\ \hdotsfor{4} \\ u_1 + v_1 & u_2 + v_2 & \ldots & u_n + v_n \\ \hdotsfor{4} \\ a_{n1} & a_{n2} & \ldots & a_{nn} \end{vmatrix} = \begin{vmatrix} a_{11} & a_{12} & \ldots & a_{1n} \\ \hdotsfor{4} \\ u_1 & u_2 & \ldots & u_n \\ \hdotsfor{4} \\ a_{n1} & a_{n2} & \ldots & a_{nn} \end{vmatrix} + \begin{vmatrix} a_{11} & a_{12} & \ldots & a_{1n} \\ \hdotsfor{4} \\ v_1 & v_2 & \ldots & v_n \\ \hdotsfor{4} \\ a_{n1} & a_{n2} & \ldots & a_{nn} \end{vmatrix}. \tag{2.14}$$

Property 10   If all the elements of some row or column of a matrix are equal to zero, then its determinant is equal to zero.

Elementary transformations of a matrix are such transformations that are associated with Properties 2, 5, 8: exchange of two rows (columns); multiplication of a row (column) by a non-zero number; addition to one of the matrix row of another one, multiplied by any non-zero number (the same for the columns).

With the help of the elementary transformations, the matrix may be reduced to a triangular form, while its determinant may then be easily obtained using Property 7.

Example 2.7 Calculate the determinant

$$
\begin{vmatrix}
2 & 3 & -3 & 4 \\
2 & 1 & -1 & 2 \\
6 & 2 & 1 & 0 \\
2 & 3 & 0 & -5
\end{vmatrix}
\tag{2.15}
$$

using the elementary transformations of the matrix corresponding to this determinant.

Solution

(1) We subtract from the second and fourth rows the first row, and from the third row—the first row multiplied by 3.

As a result we obtain

$$
\begin{vmatrix}
2 & 3 & -3 & 4 \\
2 & 1 & -1 & 2 \\
6 & 2 & 1 & 0 \\
2 & 3 & 0 & -5
\end{vmatrix}
=
\begin{vmatrix}
2 & 3 & -3 & 4 \\
0 & -2 & 2 & -2 \\
0 & -7 & 10 & -12 \\
0 & 0 & 3 & -9
\end{vmatrix}
;
\tag{2.16}
$$

(2) add to the third row the second row multiplied by $-\dfrac{7}{2}$:

$$
\begin{vmatrix}
2 & 3 & -3 & 4 \\
0 & -2 & 2 & -2 \\
0 & 0 & 3 & -5 \\
0 & 0 & 3 & -9
\end{vmatrix}
;
\tag{2.17}
$$

(3) subtract from the fourth row the third row:

$$
\begin{vmatrix}
2 & 3 & -3 & 4 \\
0 & -2 & 2 & -2 \\
0 & 0 & 3 & -5 \\
0 & 0 & 0 & -4
\end{vmatrix}
.
\tag{2.18}
$$

As a result we obtain the determinant of the upper triangular matrix, which is equal to the product of the elements positioned on the main diagonal:

$$2 \cdot (-2) \cdot 3 \cdot (-4) = 48. \tag{2.19}$$

□

Note Note that linear algebra knows alternative methods to find the variable det A, in particular, the axiomatic definition [63].

2.4 Inverse Matrix

Let A be a square matrix of size $n \times n$, and I is an identity matrix of the same size.

The matrix B is called the **inverse** of the matrix A, if the following equalities are fulfilled

$$A \cdot B = B \cdot A = I.$$

The matrix inverse of A is denoted as A^{-1}.

Note that the inverse matrix exists not for every matrix.

Theorem 2.2 *If the determinant of the matrix A is equal to zero, i.e.* det $A = 0$, *then the inverse matrix of A^{-1} does not exist.*

A square matrix is referred to as **nonsingular** (or **nondegenerate**), if an inverse matrix is determined for it. Otherwise, A is a **singular** (**degenerate**) matrix. It is known that the matrix inverse of the nonsingular matrix is the only one.

Theorem 2.3 *If the determinant of the matrix $A = (a_{ij})$ is other than zero, i.e.* $\Delta = $ det $A \neq 0$, *then the inverse matrix exists:*

$$A^{-1} = \frac{1}{\Delta} \left(A_{ij} \right)^T, \tag{2.20}$$

where (A_{ij}) is a matrix formed by cofactors of the elements a_{ij} of the matrix A (cofactor matrix).

Note A matrix transposed to (A_{ij}) is called **adjugate**, or **classical adjoint**, relative to the original one [15, 35, 64, 65].

The following statements are true:

1. If the matrix A is invertible, then A^T is also invertible, and $(A^T)^{-1} = (A^{-1})^T$.
2. If the matrices A and B are invertible, then $(A\,B)^{-1} = B^{-1}\,A^{-1}$.

3. The matrix inverse of the upper (lower) triangular matrix is also the upper (lower) triangular one.

4. If A^{-1} exists, then $\det(A^{-1}) = \dfrac{1}{\det A}$.

Example 2.8 Find the inverse matrix of the matrix:

$$A = \begin{bmatrix} 2 & 5 & 7 \\ 6 & 3 & 4 \\ 5 & -2 & -3 \end{bmatrix}. \tag{2.21}$$

Solution The determinant of the matrix is equal to

$$\det A = 2 \begin{vmatrix} 3 & 4 \\ -2 & -3 \end{vmatrix} - 5 \begin{vmatrix} 6 & 4 \\ 5 & -3 \end{vmatrix} + 7 \begin{vmatrix} 6 & 3 \\ 5 & -2 \end{vmatrix}$$

$$= 2(-9 + 8) - 5(-18 - 20) + 7(-12 - 15) = -1. \tag{2.22}$$

Calculate the cofactors:

$$A_{11} = (-1)^{1+1} \begin{vmatrix} 3 & 4 \\ -2 & -3 \end{vmatrix} = -1, \quad A_{12} = (-1)^{1+2} \begin{vmatrix} 6 & 4 \\ 5 & -3 \end{vmatrix} = 38, \tag{2.23}$$

$$A_{13} = (-1)^{1+3} \begin{vmatrix} 6 & 3 \\ 5 & -2 \end{vmatrix} = -27, \quad A_{21} = (-1)^{2+1} \begin{vmatrix} 5 & 7 \\ -2 & -3 \end{vmatrix} = 1, \tag{2.24}$$

$$A_{22} = (-1)^{2+2} \begin{vmatrix} 2 & 7 \\ 5 & -3 \end{vmatrix} = -41, \quad A_{23} = (-1)^{2+3} \begin{vmatrix} 2 & 5 \\ 5 & -2 \end{vmatrix} = 29, \tag{2.25}$$

$$A_{31} = (-1)^{3+1} \begin{vmatrix} 5 & 7 \\ 3 & 4 \end{vmatrix} = -1, \quad A_{32} = (-1)^{3+2} \begin{vmatrix} 2 & 7 \\ 6 & 4 \end{vmatrix} = 34, \tag{2.26}$$

$$A_{33} = (-1)^{3+3} \begin{vmatrix} 2 & 5 \\ 6 & 3 \end{vmatrix} = -24. \tag{2.27}$$

Write the matrix formed by the cofactors:

$$\begin{bmatrix} -1 & 38 & -27 \\ 1 & -41 & 29 \\ -1 & 34 & -24 \end{bmatrix}. \tag{2.28}$$

As a result, the desired inverse matrix will have the form:

$$A^{-1} = \frac{1}{(-1)} \begin{bmatrix} -1 & 1 & -1 \\ 38 & -41 & 34 \\ -27 & 29 & -24 \end{bmatrix} = \begin{bmatrix} 1 & -1 & 1 \\ -38 & 41 & -34 \\ 27 & -29 & 24 \end{bmatrix}. \tag{2.29}$$

$\square$

2.5 Integer Powers of a Matrix

The notion of raising a number to an integer power in matrix algebra is easily generalized. By definition we have

$$A^0 = I, \quad A^1 = A, \quad A^2 = AA, \quad A^3 = AAA, \quad \ldots, \tag{2.30}$$

and if A is a nondegenerate matrix, then

$$A^{-p} = (A^{-1})^p = (A^p)^{-1}. \tag{2.31}$$

For diagonal matrices, their p-th power preserves the property of diagonality:

$$\begin{bmatrix} d_1 & 0 & \ldots & 0 \\ 0 & d_2 & \ldots & 0 \\ \multicolumn{4}{c}{\ldots\ldots\ldots\ldots} \\ 0 & 0 & \ldots & d_n \end{bmatrix}^p = \begin{bmatrix} d_1^p & 0 & \ldots & 0 \\ 0 & d_2^p & \ldots & 0 \\ \multicolumn{4}{c}{\ldots\ldots\ldots\ldots} \\ 0 & 0 & \ldots & d_n^p \end{bmatrix} \tag{2.32}$$

for all integer p.

In order to solve the following problems, we will need the mathematical induction method.

2.5.1 *Mathematical Induction Method*

For proving the statements depending on a natural parameter, mathematics widely uses the **mathematical induction method** [1, 17, 66] (from Latin *inductio*— derivation).

Mathematical Induction Principle Let $P(n)$ be a statement defined for all natural numbers n, and let the following conditions be fulfilled:

(1) $P(1)$ is true;
(2) $\forall k \geqslant 1$ the logical implication is true $P(k) \Rightarrow P(k+1)$.
     Then $P(n)$ is true for any natural n.

The proposition *1* is usually referred to as the **basis step**, and the proposition *2*—**inductive step**.

In order to prove identities by the mathematical induction method, the following is done. Let the statement $P(k)$ take the true value when the considered identity is true for some natural number k. Then, two statements are proved:

(1) basis step, i.e. $P(1)$;

(2) inductive step, i.e. $P(k) \Rightarrow P(k+1)$ for an arbitrary $k \geqslant 1$.

According to the mathematical induction method, a conclusion is made about trueness of the considered identity for all natural values of n.

Note that in mathematical logic a statement of the form $P(n)$ is called **predicate**.

Example 2.9 Using the mathematical induction method, we will prove the statement:

$$1 + 3 + 5 + \cdots + (2n - 1) = n^2 \text{ for all natural numbers } n.$$

Proof Let $P(n)$ be the predicate "$1 + 3 + 5 + \cdots + (2n - 1) = n^2$".
    B a s i s   s t e p
For $n = 1$ we obtain $1 = 1^2$, i. e. $P(1)$ is true.
    I n d u c t i v e   s t e p
Assume that for $n = k$ the statement $1 + 3 + 5 + \cdots + (2k - 1) = k^2$ is true. Prove the trueness of $P(k + 1)$:

$$1 + 3 + 5 + \cdots + (2k - 1) + (2(k + 1) - 1)$$
$$= k^2 + (2(k + 1) - 1) = k^2 + 2k + 1 = (k + 1)^2.$$

Thus, for any natural k the implication $P(k) \Rightarrow P(k + 1)$ is true. Then, by the principle of mathematical induction the predicate $P(n)$ has a true value for all natural n.   □

Example 2.10 Relying on the mathematical induction method, prove that $n^2 + n$ is even for all natural n.

Proof Denote $f(n) = n^2 + n$ and $P(n)$—the predicate "$f(n)$ is divisible by 2".
Basis step
For $n = 1$ we obtain $f(1) = 1^2 + 1 = 2$—even number, this is why $P(1)$ is true.
Inductive step
Assume that $f(k)$ is even for natural $k \geqslant 1$. Let us prove that it implies evenness
of $f(k + 1)$:

$$f(k + 1) = (k + 1)^2 + (k + 1) = k^2 + 2k + 1 + k + 1$$
$$= (k^2 + k) + 2k + 2 = f(k) + 2(k + 1).$$

Since in the right side of the obtained relation stands the sum of two even numbers,
then $f(k + 1)$ is divisible by 2.

Note The statement of the example becomes apparent if we represent the expression
$k^2 + k$ in the form $k^2 + k = k(k + 1)$. Out of two consecutive natural numbers, one
is necessarily even, and their product is divisible by two.   □

Example 2.11 Calculate the hundredth power of the matrix $A = \begin{bmatrix} a & c \\ 0 & a \end{bmatrix}$.

Let us find several lower powers of this matrix:

$$A^1 = \begin{bmatrix} a & c \\ 0 & a \end{bmatrix}, \quad A^2 = \begin{bmatrix} a & c \\ 0 & a \end{bmatrix}\begin{bmatrix} a & c \\ 0 & a \end{bmatrix} = \begin{bmatrix} a^2 & 2ac \\ 0 & a^2 \end{bmatrix}, \quad A^3 = \begin{bmatrix} a^3 & 3a^2c \\ 0 & a^3 \end{bmatrix}, \dots$$

$$(2.33)$$

An assumption arises that for all natural values of n the following equality is true:

$$\begin{bmatrix} a & c \\ 0 & a \end{bmatrix}^n = \begin{bmatrix} a^n & na^{n-1}c \\ 0 & a^n \end{bmatrix}.$$

In order to verify this assumption, let us use the mathematical induction method.
Prove that for all natural n the equality is fulfilled:

$$A^n = \begin{bmatrix} a^n & na^{n-1}c \\ 0 & a^n \end{bmatrix}. \tag{2.34}$$

Proof Denote the predicate "$A^n = \begin{bmatrix} a^n & na^{n-1}c \\ 0 & a^n \end{bmatrix}$" through $P(n)$.

Basis step
Consider the case $n = 1$. The equality takes the form $A^1 = A$, which is the true
statement.
Inductive step

Assume that $P(k)$ for some $k = 1, 2, \ldots$ takes the true value, i.e. $A^k = \begin{bmatrix} a^k & ka^{k-1}c \\ 0 & a^k \end{bmatrix}$. Prove that $P(k+1)$ is true.

Write the predicate $P(k+1)$ in the form:

$$A^{k+1} = \begin{bmatrix} a^{k+1} & (k+1)a^k c \\ 0 & a^{k+1} \end{bmatrix}. \tag{2.35}$$

Let us use the inductive supposition and rewrite the sum A^{k+1} as $A^k A$:

$$A^{k+1} = \begin{bmatrix} a^k & ka^{k-1}c \\ 0 & a^k \end{bmatrix} \begin{bmatrix} a & c \\ 0 & a \end{bmatrix} = \begin{bmatrix} a^k \cdot a & a^k c + ka^{k-1}c \cdot a \\ 0 & a^k \cdot a \end{bmatrix}.$$

After algebraic transformations we obtain

$$A^{k+1} = \begin{bmatrix} a^{k+1} & (k+1)a^k c \\ 0 & a^{k+1} \end{bmatrix}.$$

This expression coincides with (2.35). Then, for any $k = 1, 2, \ldots$ the implication $P(k) \Rightarrow P(k+1)$ is true, and the mathematical induction method has proved that $A^n = \begin{bmatrix} a^n & na^{n-1}c \\ 0 & a^n \end{bmatrix}$ for all natural n.

Substituting in (2.34) the value $n = 100$, we finally obtain $A^{100} = \begin{bmatrix} a^{100} & 100a^{99}c \\ 0 & a^{100} \end{bmatrix}.$ □

The operation of raising the matrix to power can be executed by a relatively small number of calculations, if the original matrix is presentable in the form

$$B = U^{-1}DU, \tag{2.36}$$

where D is a diagonal matrix, U is a nonsingular matrix: $UU^{-1} = I$. We will carry of the calculation of $B^p = (U^{-1}DU)^p$ relying on the associativity property of the multiplication operation (see page 8):

$$B^p = (U^{-1}DU)^p = \underbrace{(U^{-1}DU)(U^{-1}DU)(U^{-1}DU)\ldots(U^{-1}DU)(U^{-1}DU)}_{p \text{ times}}$$

$$= U^{-1}D(UU^{-1})D(UU^{-1})D(U\ldots U^{-1})D(UU^{-1})DU$$

$$= U^{-1}DIDID\ldots DIDU = U^{-1}\underbrace{(DD\ldots D)}_{p \text{ times}}U = U^{-1}D^p U. \tag{2.37}$$

The rule of raising the matrix B to power is generalized in case of integer negative $p = -\text{abs}(p)$:

$$
\begin{aligned}
B^{-\text{abs}(p)} &= (U^{-1}BU)^{-\text{abs}(p)} = ((U^{-1}BU)^{-1})^{\text{abs}(p)} \\
&= (U^{-1}B^{-1}(U^{-1})^{-1})^{\text{abs}(p)} = (U^{-1}B^{-1}U)^{\text{abs}(p)} \\
&= U^{-1}B^{-\text{abs}(p)}U = B^p.
\end{aligned}
\tag{2.38}
$$

Therefore, the above reasoning proves the theorem of raising to an integer power of matrices B of a special form (2.36).

Theorem 2.4 (Matrix Power Theorem) *For an arbitrary matrix B, presentable in the form $B = U^{-1}DU$, and the integer number p the following equality is true:*

$$
B^p = U^{-1}D^pU.
\tag{2.39}
$$

Example 2.12 Prove that various integer powers of the matrix commute

$$
A^{p_1}A^{p_2} = A^{p_2}A^{p_1}
\tag{2.40}
$$

Proof Indeed, based on the properties of the operation of raising to power, we have $A^{p_1}A^{p_2} = A^{p_1+p_2} = A^{p_2+p_1} = A^{p_2}A^{p_1}$. $\square$

2.6 Functions of Matrices

A matrix may act as an argument of some function. Let us begin with consideration of a polynomial of a matrix. As is known, a polynomial of degree p in variable x is the sum $f(x)$ of the form:

$$
f(x) = c_0 + c_1 x + c_2 x^2 + \cdots + c_p x^p,
\tag{2.41}
$$

where c_i ($i = 0, 1, \ldots, p$) are arbitrary numerical coefficients.

By the **polynomial** $f(A)$ of degree p of the matrix A we will understand the following expression:

$$
f(A) = c_0 I + c_1 A + c_2 A^2 + \cdots + c_p A^p.
\tag{2.42}
$$

As is easy to see, the value of the function $f(A)$ is, in its turn, some matrix as well. Its elements are expressed by the formulas:

$$
(f(A))_{ij} = c_0 \delta_{ij} + c_1 (A)_{ij} + c_2 (A^2)_{ij} + \cdots + c_p (A^p)_{ij}.
\tag{2.43}
$$

Using the theorem about the power of a matrix of the special form (2.39), we will write the relation

$$f(U^{-1}AU) = U^{-1}f(A)U, \qquad (2.44)$$

which is true for an arbitrary orthogonal matrix U.

Fractionally rational function of a matrix is the value

$$\frac{f_1(A)}{f_2(A)} = f_1(A)(f_2(A))^{-1} = (f_2(A))^{-1}f_1(A) \qquad (2.45)$$

if $\det f_2(A) \neq 0$.

Let us demonstrate the correctness of the introduced definition, i.e. that two products in (2.45) are always equal to each other. Indeed, having multiplied both parts of the equality $f_1(A)(f_2(A))^{-1} = (f_2(A))^{-1}f_1(A)$ by $f_2(A)$ on the right and then by $f_2(A)$ on the left, we obtain

$$f_2(A)f_1(A) = f_1(A)f_2(A). \qquad (2.46)$$

The polynomials $f_1(A)$ and $f_2(A)$ depend only on the matrix A, and, therefore, commute. Due to this, the fraction $\dfrac{f_1(A)}{f_2(A)}$ is defined correctly [53].

2.6.1 Exponent and Logarithm

The infinite sum of the matrices of size $m \times n$ of the form

$$A + B + C + \cdots + Z + \cdots \qquad (2.47)$$

is called a **series**. It is said that the series **converges**, if for all $i = 1, 2, \ldots, m$ and $j = 1, 2, \ldots, n$ converge the sequences of the respective components of these matrices [53]:

$$a_{ij} + b_{ij} + \cdots + t_{ij} + \cdots \qquad (2.48)$$

By definition, the **exponent** e^A of the square matrix A is set equal to the sum

$$e^A = I + A + \frac{1}{2!}A^2 + \frac{1}{3!}A^3 + \cdots + \frac{1}{p!}A^p + \cdots \qquad (2.49)$$

It is known that this series converges for any real square matrix [37]. The matrix I in the formula (2.49) is taken of the same order as A.

In a compact form, this sum can be presented as

$$e^A = \sum_{p=0}^{\infty} \frac{A^p}{p!}. \tag{2.50}$$

Sometimes, especially when writing formulas with fractions, another notation for the matrix exponent is used, namely exp A.

Example 2.13 Calculate $\exp \begin{bmatrix} 1 & \lambda \\ 0 & 1 \end{bmatrix}$, where $\lambda \in \mathbb{R}$.

Solution As follows from the formula (2.34) in Example 2.11, for all natural p the equality is fulfilled $\begin{bmatrix} 1 & \lambda \\ 0 & 1 \end{bmatrix}^p = \begin{bmatrix} 1 & p\lambda \\ 0 & 1 \end{bmatrix}$. Denote $A = \begin{bmatrix} 1 & \lambda \\ 0 & 1 \end{bmatrix}$.

Using the definition (2.49), we obtain

$$
\begin{aligned}
e^A &= \sum_{p=0}^{\infty} \frac{A^p}{p!} = \begin{bmatrix} 1 & 0 \\ 0 & 1 \end{bmatrix} + \begin{bmatrix} 1 & \lambda \\ 0 & 1 \end{bmatrix} + \frac{1}{2!}\begin{bmatrix} 1 & 2\lambda \\ 0 & 1 \end{bmatrix} + \frac{1}{3!}\begin{bmatrix} 1 & 3\lambda \\ 0 & 1 \end{bmatrix} + \cdots \\
&= \begin{bmatrix} \sum_{p=0}^{\infty}\frac{1}{p!} & \lambda\sum_{p=0}^{\infty}\frac{p}{p!} \\ 0 & \sum_{p=0}^{\infty}\frac{1}{p!} \end{bmatrix} = \begin{bmatrix} e & \lambda\sum_{p=1}^{\infty}\frac{1}{(p-1)!} \\ 0 & e \end{bmatrix} = \begin{bmatrix} e & \lambda e \\ 0 & e \end{bmatrix} = e\begin{bmatrix} 1 & \lambda \\ 0 & 1 \end{bmatrix}.
\end{aligned}
\tag{2.51}
$$

During calculation of the elements of the matrix e^A we used the definition of the number $e = 2.71818\ldots$ (also known as, **bases of the natural logarithms**), known from the course the mathematical analysis [76]:

$$e = \sum_{p=0}^{\infty} \frac{1}{p!} = 1 + \frac{1}{1!} + \frac{1}{2!} + \cdots + \frac{1}{p!} + \cdots \tag{2.52}$$

□

Note The notion of the matrix exponent is widely used in the theory of systems of differential equations [3].

Logarithm of the square matrix A is the sum

$$\ln A = (A-I) - \frac{1}{2}(A-I)^2 + \frac{1}{3}(A-I)^3 - \cdots + (-1)^{p-1}\frac{1}{p}(A-I)^p + \cdots, \tag{2.53}$$

if this series converges. As in the case with the matrix exponent, the identity matrix I in the formula (2.53) must have the same order as A. In a compact form, the sum (2.53) can be presented as follows:

$$\ln A = \sum_{p=1}^{\infty} \frac{(-1)^p}{p} (A - I)^p.$$

Example 2.14 Calculate $\ln \begin{bmatrix} 1 & \lambda \\ 0 & 1 \end{bmatrix}$, where $\lambda \in \mathbb{R}$.

Solution We will need the natural powers of the difference of the matrices $(A - I) = \begin{bmatrix} 0 & \lambda \\ 0 & 0 \end{bmatrix}$.

Since $(A - I)^2 = \begin{bmatrix} 0 & \lambda \\ 0 & 0 \end{bmatrix} \begin{bmatrix} 0 & \lambda \\ 0 & 0 \end{bmatrix} = \begin{bmatrix} 0 & 0 \\ 0 & 0 \end{bmatrix}$, then all the summands of the series in (2.53), except the first one, are in this case equal to the zero matrix. Therefore, the value $\ln A$ is fully defined by the first summand of the series $\ln A = \begin{bmatrix} 0 & \lambda \\ 0 & 0 \end{bmatrix}$. □

Theorem 2.5 *For the determinant of an arbitrary matrix A, the formula*

$$\det A = \exp(tr \ln A) \tag{2.54}$$

is valid.

Note Many examples of using various functions of matrices for solving practical problems are considered in [34].

2.7 Matrix Rank

Minor of the k-th order $M_{j_1, j_2, \ldots, j_k}^{i_1, i_2, \ldots, i_k}$ of the matrix A of size $m \times n$ is the determinant of the matrix obtained from the elements, standing at the intersection of the selected rows with numbers $i_1, i_2, \ldots, i_k$ and columns with numbers $j_1, j_2, \ldots, j_k$ on condition that $1 \leqslant i_1 < i_2 < \cdots < i_k \leqslant m$ and $1 \leqslant j_1 < j_2 < \cdots < j_k \leqslant m$.

Example 2.15 Given the square matrix

$$A = \begin{bmatrix} a_{11} & a_{12} & a_{13} \\ a_{21} & a_{22} & a_{23} \\ a_{31} & a_{32} & a_{33} \end{bmatrix}, \tag{2.55}$$

for which the number of rows m and the number of columns n are equal to $m = n = 3$.

Then the second order minors of the matrix A have the form

$$M_{1,2}^{1,2} = \begin{vmatrix} a_{11} & a_{12} \\ a_{21} & a_{22} \end{vmatrix}, \qquad M_{1,3}^{1,2} = \begin{vmatrix} a_{11} & a_{13} \\ a_{21} & a_{23} \end{vmatrix}, \qquad (2.56)$$

$$M_{1,2}^{1,3} = \begin{vmatrix} a_{11} & a_{12} \\ a_{31} & a_{32} \end{vmatrix}, \qquad M_{1,3}^{1,3} = \begin{vmatrix} a_{11} & a_{13} \\ a_{31} & a_{33} \end{vmatrix} \qquad (2.57)$$

and so on until the minor

$$M_{2,3}^{2,3} = \begin{vmatrix} a_{22} & a_{23} \\ a_{32} & a_{33} \end{vmatrix}. \qquad (2.58)$$

All in all, for the matrix of size 3×3 nine second order minors can be formed (the number of arbitrary order matrix minors see in solution of Problem **2.18**).   □

The number r is called the **rank** of the matrix A, if there exists a minor of order r, other than zero, and all minors of greater order are equal to zero.

Any minor of maximal order r, other than zero, is referred to as **basic minor**.

For the rank of the matrix A the designation rk A is used.

In the following presentation, we need the concept of linear dependence of the matrix rows. Let us give k rows of the form

$$\begin{aligned} U_1 &= [u_{11}\, u_{12}\, \ldots\, u_{1n}], \\ U_2 &= [u_{21}\, u_{22}\, \ldots\, u_{2n}], \\ &\quad\ldots\ldots\ldots\ldots\ldots\ldots \\ U_k &= [u_{k1}\, u_{k2}\, \ldots\, u_{kn}], \end{aligned} \qquad (2.59)$$

each of which contains n real numbers.

We multiply every element of the first row by real number α_1, every element of the second row multiply by α_1, etc. Then we add corresponding elements of rows. As a result, a new row W forms

$$\begin{aligned} W &= \alpha_1 U_1 + \alpha_2 U_2 + \cdots + \alpha_k u_k \\ &= \alpha_1 [u_{11}\, u_{12}\, \ldots\, u_{1n}] \\ &\quad + \alpha_2 [u_{21}\, u_{22}\, \ldots\, u_{2n}] \\ &\quad\ldots\ldots\ldots\ldots\ldots\ldots \\ &\quad + \alpha_k [u_{k1}\, u_{k2}\, \ldots\, u_{kn}]. \end{aligned} \qquad (2.60)$$

The row W is called a **linear combination** of rows U ($i = 1, 2, \ldots, k$). The numbers α_i, where $i = 1, 2, \ldots, k$, are said to be the **coefficients of a linear combination** of rows.

If there is a collection of real numbers $\alpha_1, \alpha_2, \ldots, \alpha_k$, among which at least one is not equal to zero, such that W is the zero row, then $U_1, U_2, \ldots, U_k$ are called **linearly dependent**. Otherwise, these rows are said to be **linearly independent**.

Similarly, the definition of linear dependence/independence of matrix columns is introduced.

Example 2.16 The rows $U_1 = [1 \ 0 \ -3]$, $U_2 = [3 \ -2 \ 1]$, $U_3 = [5 \ -2 \ -5]$ are linearly dependent, since there is a linear combination of them equal to the zero row:

$$2U_1 + U_2 - U_3 = [0 \ 0 \ 0]. \tag{2.61}$$

$\square$

Theorem 2.6 (Basic Minor Theorem) *The number of linearly independent rows and columns of a matrix is the same and equals to the order of the basic minor. And the rows (columns), included into the basic minor, are linearly independent, and the rest are linearly expressed through them.*

Note A zero matrix has no linearly independent rows. This is why the rank of the matrix formed by zero elements will be considered as equal to zero by definition [36, 47].

The elementary transformations do not change the rank of a matrix.

Recall that the **elementary transformations** of a matrix are (see Sect. 2.3):

(a) exchange of two rows (columns);
(b) multiplication of a row (column) by a non-zero number;
(c) addition to one of the matrix row of another one, multiplied by any non-zero number (the same for the columns).

When calculating the rank of the matrix, deletion of a zero row (column) or one of the two proportional rows (columns) is used, it does not change the rank of the matrix.

The matrix is referred to as **echelon matrix**, if each its row begins with a strictly greater number of zeroes, than the previous row.

One of the basic methods of finding the rank of the matrix is the method of elementary transformations.

Elementary transformations method allows bringing the matrices to the echelon form with the help of the following algorithm:

(1) Select the row at the beginning of which stands a non-zero element. This row is written first and is called the **pivot row**.

(2) To all the remaining rows the pivot row is added, multiplied by

$$-\frac{a_{i1}}{a_{11}} \qquad (2.62)$$

(here i is the number of the row to which the pivot row is added). As a result, in all rows except the pivot one the first elements will be equal to zero.

(3) Out of the remaining rows the one is selected whose second element is not equal to zero. It is written second and is considered to be the pivot one.

(4) To the remaining rows the pivot row is added, multiplied by

$$-\frac{a_{i2}}{a_{22}}. \qquad (2.63)$$

As a result in the second column the zero elements have formed, except the first and the second rows.

This process continues until obtaining the echelon matrix. The number of non-zero rows in this matrix will be its rank.

Example 2.17 Find the rank of the matrix using the method of elementary transformations:

$$A = \begin{bmatrix} 3 & 5 & 7 \\ 1 & 2 & 3 \\ 1 & 3 & 5 \end{bmatrix}. \qquad (2.64)$$

The third order minor is the determinant of the matrix: $M_{1,2,3}^{1,2,3} = \det A$.
Calculate the determinant of the matrix A by the first column expansion method:

$$\det A = 3 \cdot (-1)^{1+1} \begin{vmatrix} 2 & 3 \\ 3 & 5 \end{vmatrix} + 1 \cdot (-1)^{2+1} \begin{vmatrix} 5 & 7 \\ 3 & 5 \end{vmatrix} + 1 \cdot (-1)^{3+1} \begin{vmatrix} 5 & 7 \\ 2 & 3 \end{vmatrix}$$

$$= 3 \cdot (10 - 9) - 1 \cdot (25 - 21) + 1 \cdot (15 - 14) = 3 - 4 + 1 = 0.$$

Consider the second order minor

$$M_{1,2}^{1,2} = \begin{vmatrix} 3 & 5 \\ 1 & 2 \end{vmatrix} = 1 \neq 0. \qquad (2.65)$$

Since it is not equal to zero, then rk $A = 2$.   □

It is often difficult to calculate the matrix rank based on its definition because one has to search through a great number of minors. As a rule, in order to simplify the calculation of the matrix rank, it is reduced to a simpler form using the elementary transformations (see Sect. 2.3).

So, the rank of the matrix presented in the echelon form is equal to the number of the non-zero rows.

In what follows, we will use the following designations for the equivalent transformations:

- $A \rightarrow B$—the matrix B is obtained as a result of the elementary transformation of the matrix A;
- $(i) + a(j)$—addition to the i-th row of the matrix of the row with number j, multiplied by the constant a;
- $(i) - a(j)$—subtraction from the i-th row of the matrix of the row with number j, multiplied by the constant a;
- $(i) \leftrightarrow (j)$—exchange of two rows.

Similar designations will also be used for operations with columns, and the column with number j will be denoted by $[j]$.

Example 2.18 Find the rank of the matrix

$$\begin{bmatrix} 3 & -1 & 3 & 2 & 5 \\ 5 & -3 & 2 & 3 & 4 \\ 1 & -3 & -5 & 0 & -7 \\ 7 & -5 & 1 & 4 & 1 \end{bmatrix}. \tag{2.66}$$

Solution Swap the first and the third rows:

$$\begin{bmatrix} 1 & -3 & -5 & 0 & -7 \\ 5 & -3 & 2 & 3 & 4 \\ 3 & -1 & 3 & 2 & 5 \\ 7 & -5 & 1 & 4 & 1 \end{bmatrix}. \tag{2.67}$$

To the second row, add the elements of the first row, multiplied by (-5), to the third—by (-3) and to the fourth—by (-7). As a result we obtain

$$\begin{bmatrix} 1 & -3 & -5 & 0 & -7 \\ 0 & 12 & 27 & 3 & 39 \\ 0 & 8 & 18 & 2 & 26 \\ 0 & 16 & 36 & 4 & 50 \end{bmatrix}. \tag{2.68}$$

Divide the second row by 3, and the third and fourth rows by 2:

$$
\begin{bmatrix}
1 & -3 & -5 & 0 & -7 \\
0 & 4 & 9 & 1 & 13 \\
0 & 4 & 9 & 1 & 13 \\
0 & 8 & 18 & 2 & 25
\end{bmatrix}. \tag{2.69}
$$

Then subtract from the third row the second one, and from the fourth row the doubled second one:

$$
\begin{bmatrix}
1 & -3 & -5 & 0 & -7 \\
0 & 4 & 9 & 1 & 13 \\
0 & 0 & 0 & 0 & 0 \\
0 & 0 & 0 & 0 & -1
\end{bmatrix}
\xrightarrow{(3)\leftrightarrow(4)}
\begin{bmatrix}
1 & -3 & -5 & 0 & -7 \\
0 & 4 & 9 & 1 & 13 \\
0 & 0 & 0 & 0 & -1 \\
0 & 0 & 0 & 0 & 0
\end{bmatrix}. \tag{2.70}
$$

We have obtained the echelon form of the matrix. The number of non-zero rows is three and therefore the rank of this matrix is three.   □

There is one more method to calculate the matrix rank. It is referred to as the **bordering minor method** and consists in the following. For computing the variable rk A, consecutively compute the minors, passing from the lower order minors to the higher order minors. If the r-th order minor is already found

$$
M =
\begin{bmatrix}
a_{11} & \ldots & a_{1r} \\
\ldots & \ldots & \ldots \\
a_{r1} & \ldots & a_{rr}
\end{bmatrix}, \tag{2.71}
$$

that is other than zero, then it is enough just to compute the minors of the $(r+1)$-th order, bordering the minor M:

$$
\begin{bmatrix}
a_{11} & \ldots & a_{1r} & a_{1t_2} \\
\ldots & \ldots & \ldots & \ldots \\
a_{r1} & \ldots & a_{rr} & a_{rt_2} \\
a_{t_1 1} & \ldots & a_{t_1 r} & a_{t_1 t_2}
\end{bmatrix}
\quad \text{for all } t_1, t_2 > r. \tag{2.72}
$$

If they all appeared to be equal to zero, we obtain rk $A = r$.

The bordering minor method is especially convenient for the problems where functional matrices are present, or the matrices whose elements depend on the parameters (see, for example, Problem **2.60**).

Review Questions

1. What is a determinant of the second order? of the third order?
2. Formulate the rule of triangle for computing a determinant of the third order.
3. Define the concept of inversion.
4. Write a formula for the determinant of the n-th order.
5. How are the additional minor M_{ij} of the element a_{ij} and its cofactor A_{ij} interconnected?
6. Formulate Laplace theorem.
7. Enumerate the general properties of determinants.
8. What matrix transformations are elementary?
9. What is an inverse matrix?
10. What matrix is called degenerate?
11. What is the method of mathematical induction used for?
12. How are the functions of matrices computed?
13. Define exponential and logarithm of a matrix.
14. Which rows are linear dependent?
15. Formulate the theorem of a basic minor.
16. What is the elementary transformations method based on?
17. Enumerate the methods of finding the matrix rank.

Problems

2.1. Expand the definition (2.5) for the case $n = 3$.

2.2. Find the number of inversions in the permutation $(5, 1, 4, 3, 6, 8, 7, 2)$.

2.3. How many inversions are there in the permutation $(n, n - 1, \ldots, 2, 1)$?

2.4. Find the number of inversions in each permutation of a collection of $2n$ numbers:

(1) $(1, 3, 5, 7, \ldots, 2n - 1, 2, 4, 6, \ldots, 2n)$;
(2) $(2, 4, 6, \ldots, 2n, 1, 3, 5, 7, \ldots, 2n - 1)$.

2.5. With what sign does the summand $a_{n1}a_{n-1,2} \ldots a_{2,n-1}a_{1n}$ appear in the expression for the determinant (2.5)?

2.6. Compute the third order determinants:

$$(1) \begin{vmatrix} 3 & -2 & 1 \\ -2 & 1 & 3 \\ 2 & 0 & -2 \end{vmatrix}; \quad (2) \begin{vmatrix} 1 & 2 & 0 \\ 0 & 1 & 3 \\ 5 & 0 & -1 \end{vmatrix}; \quad (3) \begin{vmatrix} 2 & 0 & 5 \\ 1 & 3 & 16 \\ 0 & -1 & 10 \end{vmatrix};$$

$$(4) \begin{vmatrix} 2 & -1 & 3 \\ -2 & 3 & 2 \\ 0 & 2 & 5 \end{vmatrix}; \quad (5) \begin{vmatrix} 2 & 1 & 0 \\ 1 & 0 & 3 \\ 0 & 5 & -1 \end{vmatrix}; \quad (6) \begin{vmatrix} 2 & 0 & 0 \\ 3 & 3 & 0 \\ 4 & 4 & 4 \end{vmatrix}.$$

2.7. Compute the determinants of the fourth order:

$$(a) \begin{vmatrix} -3 & 0 & 0 & 0 \\ 2 & 2 & 0 & 0 \\ 1 & 3 & -1 & 0 \\ -1 & 5 & 3 & 5 \end{vmatrix}; \quad (b) \begin{vmatrix} 2 & -1 & 3 & 4 \\ 0 & -1 & 5 & -3 \\ 0 & 0 & 5 & -3 \\ 0 & 0 & 0 & 2 \end{vmatrix}; \quad (c) \begin{vmatrix} 2 & -1 & 1 & 0 \\ 0 & 1 & 2 & -1 \\ 3 & -1 & 2 & 3 \\ 3 & 1 & 6 & 1 \end{vmatrix}.$$

2.8. Solve the following equation relative to the variable x:

$$\begin{vmatrix} x & 1 & 1 & 1 \\ 1 & x & 1 & 1 \\ 1 & 1 & x & 1 \\ 1 & 1 & 1 & x \end{vmatrix} = 0.$$

2.9. Compute the determinant

$$\begin{vmatrix} 1 & 1 & 1 & \ldots & 1 & 1 \\ 1 & 2 & 1 & \ldots & 1 & 1 \\ 1 & 1 & 3 & \ldots & 1 & 1 \\ \ldots & \ldots & \ldots & \ldots & \ldots & \ldots \\ 1 & 1 & 1 & \ldots & p & 1 \\ 1 & 1 & 1 & \ldots & 1 & p+1 \end{vmatrix},$$

where p is a natural number.

∗2.10. Compute the determinant of the matrix Q of size $n \times n$, whose elements are equal to

(a) $q_{ij} = \delta_{i,j+1} + \delta_{i+1,j}$,
(b) $q_{ij} = \delta_{i,j+1} - \delta_{i+1,j}$

for all $1 \leqslant i, j \leqslant n$. Here δ_{k_1,k_2} is the Kronecker symbol (see page 4).

2.11. Fibonacci[2] sequence

$$F_n = 1, 1, 2, 3, 5, 8, 13, 21, 34, 55, \ldots \quad \text{for } n = 1, 2, \ldots$$

is determined by the recurrence relation: $F_{n+2} = F_n + F_{n+1}$ with the initial conditions $F_1 = F_2 = 1$ [28, 70].

Prove that the $(n + 1)$-the term of the Fibonacci sequence is equal to the determinant of the n-th order

$$F_{n+1} = \begin{vmatrix} 1 & 1 & 0 & 0 & \ldots & 0 & 0 \\ -1 & 1 & 1 & 0 & \ldots & 0 & 0 \\ 0 & -1 & 1 & 1 & \ldots & 0 & 0 \\ \multicolumn{7}{c}{\ldots\ldots\ldots\ldots\ldots\ldots} \\ 0 & 0 & 0 & 0 & \ldots & 1 & 0 \\ 0 & 0 & 0 & 0 & \ldots & 1 & 1 \\ 0 & 0 & 0 & 0 & \ldots & -1 & 1 \end{vmatrix}. \tag{2.73}$$

*2.12. The **Vandermonde**[3] **determinant** is the determinant of size $n \times n$, composed of the real numbers $a_1, a_2, \ldots, a_n$:

$$V_n = \begin{vmatrix} 1 & 1 & 1 & \ldots & 1 \\ a_1 & a_2 & a_3 & \ldots & a_n \\ a_1^2 & a_2^2 & a_3^2 & \ldots & a_n^2 \\ \multicolumn{5}{c}{\ldots\ldots\ldots\ldots\ldots} \\ a_1^{n-1} & a_2^{n-1} & a_3^{n-1} & \ldots & a_n^{n-1} \end{vmatrix}. \tag{2.74}$$

(1) Verify that for $n = 2$ this determinant is equal to $V_2 = a_2 - a_1$.
(2) By the mathematical induction method, prove that the Vandermonde determinant V_n is equal to the product of all the possible differences $a_j - a_i$ for $1 \leqslant i < j \leqslant n$:

$$V_n = \prod_{i<j} (a_j - a_i). \tag{2.75}$$

---

[2]Under the name Fibonacci is known Middle Age mathematician Leonardo Pisano (about 1170—about 1250).

[3]Alexandre-Théophile Vandermonde (1735–1796), French mathematician and musician.

2.13. Let A be a matrix of size $n \times n$, m is an arbitrary natural number. Is it true that the equality $\det(mA) = m^n \det A$ is fulfilled?

2.14. At the examination in linear algebra, the student says that the determinant of the sum of two matrices is always equal to the sum of the determinants of these matrices: $\det(A + B) = \det A + \det B$. Is the student right?

2.15. Is the determinant

$$\begin{vmatrix} 1 & 1 & 1 & 1 & 10000 \\ 10000 & 1 & 1 & 1 & 1 \\ 1 & 1 & 1 & 10000 & 1 \\ 1 & 10000 & 1 & 1 & 1 \\ 1 & 1 & 10000 & 1 & 1 \end{vmatrix} \tag{2.76}$$

a positive number, a negative number or zero?

2.16. Is the determinant

$$\begin{vmatrix} 1 & 10000 & 3 & 4 & 5 \\ 5 & 1 & 2 & 3 & 10000 \\ 4 & 5 & 1 & 10000 & 3 \\ 3 & 4 & 10000 & 1 & 2 \\ 10000 & 3 & 4 & 5 & 1 \end{vmatrix} \tag{2.77}$$

a positive number, a negative number or zero?

2.17. Using Python, compute the determinants in Problems **2.15** and **2.16** and verify the correctness of the solutions of these problems.

2.18. Find the number of minors of the k-th order in a matrix consisting of m rows and n columns.

2.19. How many minors of the k-th order in a matrix of size $n \times n$ do not contain diagonal elements of the original matrix?

***2.20.** What greatest value can take the determinant of a matrix of size 3×3, consisting of the elements $+1$ and -1?

***2.21.** What greatest value can take the determinant of a matrix of size 3×3, consisting of the elements 0 and 1?

2.22. Will the determinant of the matrix change if its columns are permuted in the reverse order?

2.23. **Computing the determinant of the matrix using Python**

In the text file `input.txt` are successively written in rows the elements of the integer square matrix A. Using the recursion, compute the determinant $\det A$. Enter the result into the text file `output.txt`.

2.24. Evaluate the number of multiplications executed by the recursive algorithm for computing the determinant of the matrix of size $n \times n$ (see previous Problem).

∗2.25. Prove that the exact number of multiplications executed by the recursive algorithm for computing the determinant of the matrix of size $n \times n$ (see Problems **2.23** and **2.24**) is equal to

$$T(n) = en\Gamma(n, 1) - n!,$$

where e is the base of natural logarithms, $\Gamma(n, x) = \int\limits_x^\infty e^{-t}t^{n-1}dt$ — **incomplete gamma function** $\left(\text{for the natural } n \in \mathbb{N} \text{ the following equality is true } \Gamma(n, x) = (n - 1)! e^{-x} \sum\limits_{k=0}^{n-1} \dfrac{x^k}{k!}\right)$.

∗2.26. Obtain the asymptotic estimate of the mean value of the number of inversions $A(N)$ in the array consisting of N elements, for $N \to \infty$.

∗2.27. Prove the validity of the identity

$$\det(I + \varepsilon B) = 1 + \varepsilon \operatorname{tr} B + O(\varepsilon^2) \text{ for } \varepsilon \to 0, \tag{2.78}$$

where B is some square matrix, I is an identity matrix of the same size.

∗2.28. Obtain the asymptotic estimate of the variable

$$\det\left(\frac{I}{(I + \varepsilon B)^p}\right) \text{ for } \varepsilon \to 0, \tag{2.79}$$

if B is an arbitrary square matrix, I is the identity matrix of the same size, $p \in \mathbb{N}$.

∗2.29. Assume that $M = G + \varepsilon H$, where ε is a real number, G is an invertible matrix. Relying upon the equality $\det A = \exp(\operatorname{tr} \ln A)$ (see the Theorem 2.5 on page 59), prove that

$\det M$

$$= \det G\left(1 + \varepsilon\operatorname{tr}(G^{-1}H) + \frac{1}{2}\varepsilon^2(\operatorname{tr}^2(G^{-1}H) - \operatorname{tr}(G^{-1}H)^2) + O(\varepsilon^3)\right) \tag{2.80}$$

for $\varepsilon \to 0$.

2.30. Assume that the real parameters a, b, c and d are selected, such that the inequality $ad - bc \neq 0$ is valid. Compute the matrix inverse of $A = \begin{bmatrix} a & b \\ c & d \end{bmatrix}$.

2.31. Find the inverse matrices of the following matrices:

$$(1) \begin{bmatrix} 1 & 2 \\ 3 & 4 \end{bmatrix}; \quad (2) \begin{bmatrix} 3 & 4 \\ 5 & 7 \end{bmatrix};$$

$$(3) \begin{bmatrix} 3 & -4 & 5 \\ 2 & -3 & 1 \\ 3 & -5 & -1 \end{bmatrix}; \quad (4) \begin{bmatrix} 2 & 7 & 3 \\ 3 & 9 & 4 \\ 1 & 5 & 3 \end{bmatrix}; \quad (5) \begin{bmatrix} 1 & 2 & 2 \\ 2 & 1 & -2 \\ 2 & -2 & 1 \end{bmatrix}.$$

2.32. Find the inverse matrices of the following matrices:

$$(1) \begin{bmatrix} 1 & 1 \\ 0 & 2 \end{bmatrix}; \quad (2) \begin{bmatrix} 1 & 1 & 1 \\ 0 & 2 & 2 \\ 0 & 0 & 3 \end{bmatrix}; \quad (3) \begin{bmatrix} 1 & 1 & 1 & 1 \\ 0 & 2 & 2 & 2 \\ 0 & 0 & 3 & 3 \\ 0 & 0 & 0 & 4 \end{bmatrix}.$$

2.33. Calculate the commutators $[A, A^{-1}]$ and $[A, A^{-1}]$, if A is an arbitrary nonsingular matrix.

2.34. For what values of the real parameter λ the matrix does not have the inverse one?

$$(1) \begin{bmatrix} -1 & \lambda & \lambda \\ \lambda & \lambda & 0 \\ 6 & 4 & \lambda \end{bmatrix};$$

$$(2) \begin{bmatrix} 3 & \lambda & \lambda \\ \lambda & \lambda & -1 \\ -1 & -1 & \lambda \end{bmatrix}.$$

2.35. Compute the matrix inverse of A:

$$A = \begin{bmatrix} 1 & \alpha & 0 & 0 \\ 0 & 1 & \beta & 0 \\ 0 & 0 & 1 & \gamma \\ 0 & 0 & 0 & 1 \end{bmatrix},$$

where $\alpha, \beta, \gamma \in \mathbb{R}$. Does the matrix A^{-1} exist for all possible values of the parameters?

*2.36. Compute the matrix inverse of G_n of size $n \times n$, where $n \geqslant 2$:

$$
G_n = \begin{bmatrix}
0 & 0 & 0 & \dots & 0 & 1 \\
0 & 0 & 0 & \dots & 1 & 0 \\
& & \dotfill & & & \\
0 & 1 & 0 & \dots & 0 & 0 \\
1 & 0 & 0 & \dots & 0 & 0
\end{bmatrix}.
\tag{2.81}
$$

2.37. The elements of a **Hilbert**[4] **matrix** $H = (h_{ij})$ are set by the rule $h_{ij} = \dfrac{1}{i+j-1}$, where $i, j = 1, 2, \dots, n$.

With the help of Python, compute H^{-1}, $H \cdot H^{-1}$ and $H^{-1} \cdot H$ for $n = 6, 7, 8$.

*2.38. Find the inverse of the following matrix:

$$
\begin{bmatrix}
1 & \lambda & \lambda^2 & \lambda^3 & \dots & \lambda^n \\
0 & 1 & \lambda & \lambda^2 & \dots & \lambda^{n-1} \\
0 & 0 & 1 & \lambda & \dots & \lambda^{n-2} \\
& & & \dotfill & & \\
0 & 0 & 0 & 0 & \dots & 1
\end{bmatrix},
$$

where λ is some real number.

2.39. With the help of the mathematical induction method, prove the **matrix product inversion formula**

$$
(A_1 A_2 \dots A_{n-1} A_n)^{-1} = A_n^{-1} A_{n-1}^{-1} \dots A_2^{-1} A_1^{-1}.
\tag{2.82}
$$

2.40. Solve the matrix equations:

(1) $\begin{bmatrix} 2 & 1 \\ 0 & 2 \end{bmatrix} \cdot X = \begin{bmatrix} -6 & 4 \\ 2 & 1 \end{bmatrix}$;

(2) $\begin{bmatrix} -1 & 1 & 1 \\ 0 & 2 & 2 \\ 0 & 2 & 3 \end{bmatrix} \cdot X = \begin{bmatrix} -2 & 1 & 1 \\ -1 & 0 & 2 \\ -1 & -2 & 0 \end{bmatrix}$;

---

[4]David Hilbert (1862–1943), German mathematician.

$$(3) \ X \cdot \begin{bmatrix} 1 & 10 & -3 \\ -3 & 6 & 2 \\ 2 & 6 & -3 \end{bmatrix} = \begin{bmatrix} 0 & -1 & 2 \\ -1 & -2 & -2 \\ 0 & 1 & 10 \end{bmatrix};$$

$$(4) \ \begin{bmatrix} 2 & 3 \\ -3 & 2 \end{bmatrix} \cdot X \cdot \begin{bmatrix} 1 & 7 \\ 5 & 4 \end{bmatrix} = \begin{bmatrix} 1 & -1 \\ 0 & 1 \end{bmatrix}.$$

2.41. Assuming that a is an arbitrary real number, raise the matrix of power:

$$\begin{bmatrix} 1 & 0 & 0 \\ 0 & 1 & 0 \\ 0 & a & 1 \end{bmatrix}^{256}. \tag{2.83}$$

∗2.42. Raise the matrix of power:

$$\begin{bmatrix} 1 & 0 & 0 \\ g & 1 & 0 \\ h & 0 & 1 \end{bmatrix}^{512}, \tag{2.84}$$

where constants $g, h \in \mathbb{R}$.

2.43. Using the mathematical induction method, prove that the n-th power of the matrix $\mathcal{F} = \begin{bmatrix} 0 & 1 \\ 1 & 1 \end{bmatrix}$ has the form

$$\mathcal{F}^n = \begin{bmatrix} F_{n-1} & F_n \\ F_n & F_{n+1} \end{bmatrix} \tag{2.85}$$

for all natural values $n > 1$, where F_n are the Fibonacci numbers (see definition in Problem **2.11** on page 67).

2.44. Compute the n-th power of the upper triangular matrix $A = \begin{bmatrix} 1 & \alpha & \gamma \\ 0 & 1 & \beta \\ 0 & 0 & 1 \end{bmatrix}$,

where $\alpha, \beta, \gamma \in \mathbb{R}$.

2.45. Compute the q-th power of the functional matrix

$$U(\varphi) = \begin{bmatrix} \cos\varphi & \sin\varphi \\ -\sin\varphi & \cos\varphi \end{bmatrix}$$

for all integer values of $q \in \mathbb{Z}$.

2.46. It is known about the matrices A and B that their commutator is the identity matrix: $[A, B] = I$. Compute $[A, B^q]$ for all integer values of the parameter $q \in \mathbb{Z}$.

2.47. Compute the value of the function $f(x) = x^2 - 3x + 2$, if as the argument is taken the matrix A, where

(1) $A = \begin{bmatrix} 1 & 0 \\ 0 & 1 \end{bmatrix}$;

(2) $A = \begin{bmatrix} 1 & -2 \\ -3 & 1 \end{bmatrix}$.

2.48. Compute the value of the function $g(x) = x^3 + x - 3$, if as the argument is taken the matrix A, where

(1) $A = \begin{bmatrix} 1 & 0 & 0 \\ 1 & 1 & 0 \\ 1 & 1 & 1 \end{bmatrix}$;

(2) $A = \begin{bmatrix} -2 & 1 & 0 \\ 3 & -1 & 1 \\ 2 & 1 & 0 \end{bmatrix}$.

2.49. Compute the value of the fractionally rational function $g(A)$, if $g(x) = \dfrac{x^3 + x^2 - 3x - 5}{x^3 - 5x - 2}$ and

(1) $A = \begin{bmatrix} 2 & 0 & 0 \\ 0 & -4 & 0 \\ 0 & 0 & 1 \end{bmatrix}$;

(2) $A = \begin{bmatrix} 1 & 1 & 0 \\ 0 & 1 & 0 \\ 0 & 1 & 1 \end{bmatrix}$.

2.50. Show that the equality

$$\exp \begin{bmatrix} 0 & -1 \\ 1 & 0 \end{bmatrix} = \begin{bmatrix} \cos 1 & -\sin 1 \\ \sin 1 & \cos 1 \end{bmatrix}$$

is valid.

2.51. Find e^A for

(1) $A = \begin{bmatrix} 1 & 0 & 0 \\ 0 & 1 & 0 \\ 0 & 0 & 0 \end{bmatrix}$;

(2) $A = \begin{bmatrix} 0 & 1 & 0 \\ 0 & 0 & 1 \\ 0 & 0 & 0 \end{bmatrix}$.

∗2.52. Find e^A for

(1) $A = \begin{bmatrix} 1 & 0 & 1 \\ 0 & 1 & 0 \\ 0 & 0 & 1 \end{bmatrix}$;

(2) $A = \begin{bmatrix} 1 & 0 & 1 \\ 1 & 1 & 1 \\ 0 & 0 & 1 \end{bmatrix}$.

2.53. For the matrices specified in the previous problem, compute $\ln A$.

∗2.54. Prove that for the commuting matrices the exponential of the sum is equal to the product of the exponentials of each of the summands:

$$\exp(A + B) = \exp(A)\exp(B).$$

2.55. Prove that $\exp(tA) = I + tA + O(t^2)$ for $t \to 0$.

2.56. Consider all possible matrices of size 3×3 that contain no zero column. Enumerate pairwise various echelon forms of such matrices.

2.57. Which elementary operations with the elements of the echelon matrix preserve its echelon form?

2.58. Find the ranks of the matrices:

$$(1)\ A = \begin{bmatrix} 1 & 2 & -4 & 3 & -2 \\ -1 & 3 & -6 & -2 & 4 \\ 2 & -1 & 2 & 5 & 6 \end{bmatrix};\quad (2)\ A = \begin{bmatrix} 2 & -4 & 3 & 1 & 0 \\ 1 & -2 & 1 & -4 & 2 \\ 0 & 1 & -1 & 3 & 1 \\ 4 & -7 & 4 & -4 & 5 \end{bmatrix};$$

$$(3)\ A = \begin{bmatrix} 1 & 2 & 1 & 3 \\ 4 & -1 & -5 & -6 \\ 1 & -3 & -4 & -7 \\ 2 & 1 & -1 & 0 \end{bmatrix};\quad (4)\ A = \begin{bmatrix} 0 & 2 & -4 \\ -1 & -4 & -5 \\ 3 & 1 & 7 \\ 0 & 5 & -10 \\ 2 & 3 & 0 \end{bmatrix};$$

$$(5)\ A = \begin{bmatrix} 3 & 5 & 7 \\ 1 & 2 & 3 \\ 1 & 3 & 5 \end{bmatrix};\quad (6)\ A = \begin{bmatrix} 4 & 3 & 2 & 2 \\ 0 & 2 & 1 & 1 \\ 0 & 0 & 3 & 3 \end{bmatrix};$$

$$(7)\ A = \begin{bmatrix} 1 & -1 & 2 & 4 & 3 \\ -2 & 1 & 5 & 2 & 6 \\ 2 & -1 & 4 & 7 & 2 \end{bmatrix};\quad (8)\ A = \begin{bmatrix} 1 & 3 & 5 & -1 \\ 2 & -1 & -3 & 4 \\ 5 & 1 & -1 & 7 \\ 7 & 7 & 9 & 1 \end{bmatrix}.$$

2.59. With the help of the bordering minor method, find the maximum value that the rank of the matrix Λ can take

$$\Lambda = \begin{bmatrix} \beta - \gamma & 0 & \gamma & -\beta \\ \gamma - \alpha & -\gamma & 0 & \alpha \\ \alpha - \beta & \beta & -\alpha & 0 \end{bmatrix},$$

if α, β and γ are real constants.

∗2.60. Find the ranks of the following matrices for all possible values of the real parameter λ:

(a) $\begin{bmatrix} 1 & \lambda & 0 \\ 1 & 1 & 0 \\ 0 & 0 & 1 \end{bmatrix};$

(b) $\begin{bmatrix} 1-\lambda & 0 & 0 \\ 0 & 2-\lambda & 0 \\ 0 & 0 & 3-\lambda \end{bmatrix}$;

(c) $\begin{bmatrix} 1 & 2 & -3 \\ -1 & 2-\lambda & 10 \\ -1 & 0 & 3-\lambda \end{bmatrix}$;

(d) $\begin{bmatrix} -\lambda & 1 & 1 \\ 0 & 1-\lambda & 1 \\ 0 & 0 & 2-\lambda \end{bmatrix}$;

(e) $\begin{bmatrix} 1 & -6 & -5 \\ -1 & 2-\lambda & 5 \\ -1 & 6 & 1-\lambda \end{bmatrix}$;

(f) $\begin{bmatrix} 1-\lambda & 2 & 0 & 0 \\ 1 & 2-\lambda & 0 & 0 \\ 0 & 0 & 3-\lambda & 0 \\ 0 & 0 & 0 & 4-\lambda \end{bmatrix}$;

(g) $\begin{bmatrix} 1-\lambda & 2 & 0 & 0 \\ 1 & 1+\lambda & 0 & 0 \\ 0 & 0 & 2-\lambda & 0 \\ 0 & 0 & 0 & 2+\lambda \end{bmatrix}$;

(h) $\begin{bmatrix} 1+\lambda & 2 & 0 & 4 \\ 1 & 1 & 0 & 0 \\ 0 & 0 & 2+\lambda & -1 \\ 1 & 0 & 2 & 1 \end{bmatrix}$;

(i) $\begin{bmatrix} -\lambda & 1 & 1 & 1 \dots & 1 & 1 \\ 0 & 1-\lambda & 1 & 1 \dots & 1 & 1 \\ 0 & 0 & 2-\lambda & 1 \dots & 1 & 1 \\ \hdotsfor{6} \\ 0 & 0 & 0 & 0 \dots & (n-1)-\lambda & 1 \\ 0 & 0 & 0 & 0 \dots & 0 & n-\lambda \end{bmatrix}$.

Answers and Solutions

2.1 *Solution.*

For $n = 3$ we have $3! = 1 \cdot 2 \cdot 3 = 6$ permutations:

$$k_1 = 1, k_2 = 2, k_3 = 3; \quad k_1 = 1, k_2 = 3, k_3 = 2; \quad k_1 = 2, k_2 = 1, k_3 = 3;$$
$$k_1 = 2, k_2 = 3, k_3 = 1; \quad k_1 = 3, k_2 = 1, k_3 = 2; \quad k_1 = 3, k_2 = 2, k_3 = 1.$$

Therefore, the sum in $\det A$ will consist of 6 terms, half of which is taken with a positive sign, and another half—with a negative sign:

$$\det A = \sum_{\text{perm}} (-1)^{\sigma} a_{1k_1} a_{2k_2} a_{3k_3}$$

$$= (-1)^0 a_{11} a_{22} a_{33} + (-1)^1 a_{11} a_{23} a_{32} + (-1)^1 a_{12} a_{21} a_{33}$$
$$+ (-1)^2 a_{12} a_{23} a_{31} + (-1)^2 a_{13} a_{21} a_{32} + (-1)^3 a_{13} a_{22} a_{31}$$
$$= a_{11} a_{22} a_{33} + a_{12} a_{23} a_{31} + a_{13} a_{21} a_{32}$$
$$- a_{13} a_{22} a_{31} - a_{12} a_{21} a_{33} - a_{11} a_{23} a_{32}.$$

The obtained formula, of course, conforms with the definition (2.3).

2.2 *Answer:* the number of inversions is equal to 10.

2.3 *Solution.*

The first element of the permutation, equal to n, forms $n - 1$ inversions paired with each of the elements $n - 1, n - 2, \ldots, 1$. The second element of the permutation $n - 1$ forms with the remaining elements $n - 2$ inversions. As is easy to see, the permutation element, equal to k, where $1 \leqslant k \leqslant n$, forms $k - 1$ inversions. Therefore, the total number of inversions in the permutation $(n, n - 1, \ldots, 2, 1)$ is equal to the sum $\sum\limits_{k=1}^{n} k = n(n-1)/2$.

2.4 *Answer:*

(1) $\dfrac{n(n-1)}{2}$;

(2) $\dfrac{n(n+1)}{2}$.

2.5 *Solution.*

The summand $a_{n1} a_{n-1,2} \ldots a_{2,n-1} a_{1n}$ in the formula (2.5) is assigned the sign $(-1)^{\sigma} = (-1)^{n(n-1)/2}$, since the number of inversions in the permutation $(n, n - 1, \ldots, 2, 1)$ is equal to $\sigma = n(n-1)/2$ (see Problem **2.3**).

Note. With the help of the function "floor" $\lfloor x \rfloor$—the greatest integer, which is less than or equal to the argument x, i.e. $\lfloor x \rfloor = \max(n \in \mathbb{Z}, \; n \leqslant x)$, the answer can be written in a more compact form: $(-1)^\sigma = (-1)^{\lfloor n/2 \rfloor}$.

2.6 *Solution.*

(1) Calculate the determinant, using the rule of triangle for calculation by the formula (2.3).

$$\begin{vmatrix} 3 & -2 & 1 \\ -2 & 1 & 3 \\ 2 & 0 & -2 \end{vmatrix} = 3 \cdot 1 \cdot (-2) + (-2) \cdot 3 \cdot 2 + 1 \cdot (-2) \cdot 0$$

$$- 1 \cdot 1 \cdot 2 - (-2) \cdot (-2) \cdot (-2) - 3 \cdot 3 \cdot 0 = -12.$$

(2) Expand the determinant in the first row:

$$\begin{vmatrix} 1 & 2 & 0 \\ 0 & 1 & 3 \\ 5 & 0 & -1 \end{vmatrix} = 1 \begin{vmatrix} 1 & 3 \\ 0 & -1 \end{vmatrix} - 2 \begin{vmatrix} 0 & 3 \\ 5 & -1 \end{vmatrix} + 0 \begin{vmatrix} 0 & 1 \\ 5 & 0 \end{vmatrix}$$

$$= 1 \cdot (1 \cdot (-1) - 0 \cdot 3) - 2 \cdot (0 \cdot (-1) - 3 \cdot 5) + 0 \cdot (0 \cdot 0 - 5 \cdot 1)$$

$$= -1 + 30 = 29.$$

(3) Expand the determinant in the first row:

$$\begin{vmatrix} 2 & 0 & 5 \\ 1 & 3 & 16 \\ 0 & -1 & 10 \end{vmatrix} = 2 \cdot (3 \cdot 10 - (-1) \cdot 16) - 0 \cdot (1 \cdot 10 - 0 \cdot 16)$$

$$+ 5 \cdot ((-1) \cdot 1 - 0 \cdot 3)$$

$$= 92 - 5 = 87;$$

(4) Expand the determinant, for example, in the third row:

$$\begin{vmatrix} 2 & -1 & 3 \\ -2 & 3 & 2 \\ 0 & 2 & 5 \end{vmatrix} = 0 \cdot (3 \cdot 3 - (-1) \cdot 2) - 2 \cdot (2 \cdot 2 - (-2) \cdot 3)$$

$$+ 5 \cdot (2 \cdot 3 - (-2) \cdot (-1)) = -20 + 20 = 0.$$

(5) Expand the determinant in the first row:

$$
\begin{vmatrix} 2 & 1 & 0 \\ 1 & 0 & 3 \\ 0 & 5 & -1 \end{vmatrix} = 2 \cdot (0 \cdot (-1) - 5 \cdot 3) - 1 \cdot (1 \cdot (-1) - 0 \cdot 3) + 0 \cdot (1 \cdot 5 - 0 \cdot 0)
$$

$$
= -30 + 1 = -29.
$$

(6) The determinant of the lower triangular matrix is calculated as the product of its diagonal elements:

$$
\begin{vmatrix} 2 & 0 & 0 \\ 3 & 3 & 0 \\ 4 & 4 & 4 \end{vmatrix} = 2 \cdot 3 \cdot 4 = 24.
$$

2.7 Solution.

(a) Expand the determinant in the first row:

$$
\begin{vmatrix} -3 & 0 & 0 & 0 \\ 2 & 2 & 0 & 0 \\ 1 & 3 & -1 & 0 \\ -1 & 5 & 3 & 5 \end{vmatrix} = -3 \cdot \begin{vmatrix} 2 & 0 & 0 \\ 3 & -1 & 0 \\ 5 & 3 & 5 \end{vmatrix} - 0 \cdot \begin{vmatrix} 2 & 0 & 0 \\ 1 & -1 & 0 \\ -1 & 3 & 5 \end{vmatrix} + 0 \cdot \begin{vmatrix} 2 & 2 & 0 \\ 1 & 3 & 0 \\ -1 & 5 & 5 \end{vmatrix}
$$

$$
- 0 \cdot \begin{vmatrix} 2 & 2 & 0 \\ 1 & 3 & -1 \\ -1 & 5 & 3 \end{vmatrix} = -3 \cdot (-10 + 0 + 0 - 0 - 0 - 0)
$$

$$
- 0 + 0 - 0 = 30.
$$

Note that the original matrix is the lower triangular one, hence its determinant can be calculated by a simpler method as the product of diagonal elements:
$\Delta = (-3) \cdot 2 \cdot (-1) \cdot 5 = 30$.

(b) Expand the determinant in the first column:

$$
\begin{vmatrix} 2 & -1 & 3 & 4 \\ 0 & -1 & 5 & -3 \\ 0 & 0 & 5 & -3 \\ 0 & 0 & 0 & 2 \end{vmatrix} = 2 \cdot \begin{vmatrix} -1 & 5 & -3 \\ 0 & 5 & -3 \\ 0 & 0 & 2 \end{vmatrix} - 0 \cdot \begin{vmatrix} -1 & 3 & 4 \\ 0 & 5 & -3 \\ 0 & 0 & 2 \end{vmatrix} + 0 \cdot \begin{vmatrix} -1 & 3 & 4 \\ -1 & 5 & -3 \\ 0 & 0 & 2 \end{vmatrix}
$$

$$-0 \cdot \begin{vmatrix} -1 & 3 & 4 \\ -1 & 5 & -3 \\ 0 & 5 & -3 \end{vmatrix} = 2 \cdot (-10 + 0 + 0 - 0 - 0 - 0) - 0$$

$$+0 - 0 = -20.$$

The result can be obtained faster if we note that the matrix is the upper triangular one. Then $\Delta = 2 \cdot (-1) \cdot 5 \cdot 2 = -20$.

(c) Expand the determinant in the first row:

$$\begin{vmatrix} 2 & -1 & 1 & 0 \\ 0 & 1 & 2 & -1 \\ 3 & -1 & 2 & 3 \\ 3 & 1 & 6 & 1 \end{vmatrix} = 2 \cdot \begin{vmatrix} 1 & 2 & -1 \\ -1 & 2 & 3 \\ 1 & 6 & 1 \end{vmatrix} - (-1) \cdot \begin{vmatrix} 0 & 2 & -1 \\ 3 & 2 & 3 \\ 3 & 6 & 1 \end{vmatrix} + 1 \cdot \begin{vmatrix} 0 & 1 & -1 \\ 3 & -1 & 3 \\ 3 & 1 & 1 \end{vmatrix}$$

$$-0 \cdot \begin{vmatrix} 0 & 1 & 2 \\ 3 & -1 & 2 \\ 3 & 1 & 6 \end{vmatrix} = 2 \cdot (2 + 6 + 6 + 2 - 18 + 2)$$

$$- (-1) \cdot (0 + 18 - 18 + 6 - 0 - 6)$$

$$+ 1 \cdot (0 + 9 - 3 - 3 - 0 - 3) - 0 = 0.$$

2.8 *Solution.*

Denote the determinant by Δ and expand it in the first row. We will expand the obtained third order determinants in the first row or in the row consisting of ones, if any:

$$\Delta = x \begin{vmatrix} x & 1 & 1 \\ 1 & x & 1 \\ 1 & 1 & x \end{vmatrix} - \begin{vmatrix} 1 & 1 & 1 \\ 1 & x & 1 \\ 1 & 1 & x \end{vmatrix} + \begin{vmatrix} 1 & x & 1 \\ 1 & 1 & 1 \\ 1 & 1 & x \end{vmatrix} - \begin{vmatrix} 1 & x & 1 \\ 1 & 1 & x \\ 1 & 1 & 1 \end{vmatrix}$$

$$= x\left(x(x^2 - 1) - (x - 1) + (1 - x)\right) - \left((x^2 - 1) - (x - 1) + (1 - x)\right)$$

$$+ \left(-(x^2 - 1) + (x - 1) - (1 - x)\right) - \left((x^2 - 1) - (x - 1) + (1 - x)\right)$$

$$= x(x^3 - 3x + 2) - 3(x^2 - 2x + 1) = x(x - 1)(x^2 + x - 2) - 3(x - 1)^2$$

$$= (x - 1)(x^3 + x^2 - 5x + 3) = (x - 1)^3(x + 3).$$

As a result, the roots of the equation $\Delta = 0$ are equal to $x = 1$ and $x = -3$.

2.9 *Solution.*

Apply to the determinant the following equivalent transformations: from each row, starting with the second one, subtract the first one. As a result we obtain the determinant of the upper triangular matrix:

$$
\begin{vmatrix}
1 & 1 & 1 & \ldots & 1 & 1 \\
0 & 1 & 0 & \ldots & 0 & 0 \\
0 & 0 & 2 & \ldots & 0 & 0 \\
0 & 0 & 0 & \ldots & 0 & 0 \\
0 & 0 & 0 & \ldots & p-1 & 0 \\
0 & 0 & 0 & \ldots & 0 & p
\end{vmatrix}.
$$

Such a determinant is equal to the product of the diagonal elements of the matrix:
$1 \cdot 1 \cdot 2 \ldots (p-1) \cdot p = p!$.

2.10 *Solution.*

(a) Denote the determinant of the matrix Q by Q_n and write it in an explicit form:

$$
Q_n = \begin{vmatrix}
0 & 1 & 0 & 0 & \ldots & 0 & 0 \\
1 & 0 & 1 & 0 & \ldots & 0 & 0 \\
0 & 1 & 0 & 1 & \ldots & 0 & 0 \\
& & & \ldots & & & \\
0 & 0 & 0 & 0 & \ldots & 0 & 1 \\
0 & 0 & 0 & 0 & \ldots & 1 & 0
\end{vmatrix}.
$$

Let us use the expansion in its first row, following which expand the obtained the determinant of order $(n-1) \times (n-1)$ in the first column:

$$
Q_n = (-1)^{1+2} \cdot
\begin{vmatrix}
1 & 1 & 0 & \ldots & 0 & 0 \\
0 & 0 & 1 & \ldots & 0 & 0 \\
0 & 1 & 0 & \ldots & 0 & 0 \\
& & \ldots & & & \\
0 & 0 & 0 & \ldots & 0 & 1 \\
0 & 0 & 0 & \ldots & 1 & 0
\end{vmatrix}
= (-1)^{1+2}(-1)^{1+1} \cdot
\begin{vmatrix}
0 & 1 & \ldots & 0 & 0 \\
1 & 0 & \ldots & 0 & 0 \\
& & \ldots & & \\
0 & 0 & \ldots & 0 & 1 \\
0 & 0 & \ldots & 1 & 0
\end{vmatrix}.
$$

Note that the problem reduced to computing the variable Q_{n-2}. For the smallest possible values of the order of the matrix $n = 1$ and $n = 2$ we have

$$Q_1 = \det[0] = 0, \quad Q_2 = \det \begin{bmatrix} 0 & 1 \\ 1 & 0 \end{bmatrix} = -1.$$

Thus, the recurrence relation is obtained:

$$\begin{cases} Q_n = -Q_{n-2}, \\ Q_1 = 0, \ Q_2 = -1. \end{cases}$$

By the mathematical induction method we can show that its solution Q_n will have the form:

$$Q_n = \begin{cases} (-1)^{n/2}, & \text{if } n \text{ is even,} \\ 0, & \text{if } n \text{ is odd.} \end{cases}$$

(b) Using Laplace's method, similarly to item (a) we obtain the recurrence relation:

$$\begin{cases} Q_n = Q_{n-2}, \\ Q_1 = 0, \ Q_2 = 1. \end{cases}$$

Its solution, as is easy to show with the help of the mathematical induction method, has the form:

$$Q_n = \frac{1}{2}\left(1 + (-1)^n\right).$$

2.11 *Solution.*

Denote by $P(n)$ the predicate "$F_{n+1} = D(n)$", where $D(n)$ is the determinant (2.73). Let us use the mathematical induction method.

B a s i s   s t e p

For $n = 1$ and $n = 2$ we have

$$F_2 = \begin{vmatrix} 1 & 1 \\ -1 & 1 \end{vmatrix} \text{---true,} \quad F_3 = \begin{vmatrix} 1 & 1 & 0 \\ -1 & 1 & 1 \\ 0 & -1 & 1 \end{vmatrix} \text{---true.}$$

I n d u c t i v e   s t e p

Assume the trueness of the statements $P(k)$ for $k = 1, 2, \ldots$.

Prove that this entails the trueness of $P(k + 1)$.

Indeed, expanding the determinant $D(n + 1)$ in the first row, we obtain

$$D_{n+1} = D_n + D_{n-1}.$$

According to the inductive supposition $D(n) = F_{n+1}$, $D(n - 1) = F_n$.
Therefore, we have obtained the true equality $F_{n+2} = F_{n+1} + F_n$.

Thus, the mathematical induction method has proved that $F_{n+1} = D(n)$ for all natural n.

2.12 *Solution.*

(1) According to the definition of (2.74) for $n = 2$ we have

$$V_n = \begin{vmatrix} 1 & 1 \\ a_1 & a_2 \end{vmatrix} = a_2 - a_1.$$

(2) Let us introduce for consideration the predicate "$V_n = \prod_{\substack{i,j \in [1,n] \\ i<j}} (a_j - a_i)$" and

denote it by $P(n)$.

Basis step
The case of the least $n = 2$ is proved in item (1) of this problem.
Inductive step
Let for some natural $k \geqslant 2$ the equality be fulfilled $V_k = \prod_{\substack{i,j \in [1,k] \\ i<j}} (a_j - a_i)$.

Prove that $V_{k+1} = \prod_{\substack{i,j \in [1,k+1] \\ i<j}} (a_j - a_i)$.

Transform V_{k+1} as follows: from the $(k + 1)$-th row subtract the k-th one, multiplied by a_1, then from the k-th one subtract the $(k - 1)$-th one, also multiplied by a_1 and so until the second row inclusive:

$$V_{k+1} = \begin{vmatrix} 1 & 1 & 1 & \cdots & 1 \\ 0 & a_2 - a_1 & a_3 - a_1 & \cdots & a_{k+1} - a_1 \\ 0 & a_2^2 - a_1 a_2 & a_3^2 - a_1 a_3 & \cdots & a_{k+1}^2 - a_1 a_{k+1} \\ & & \cdots\cdots\cdots\cdots\cdots & & \\ 0 & a_2^k - a_1 a_2^{k-1} & a_3^k - a_1 a_3^{k-1} & \cdots & a_{k+1}^k - a_1 a_{k+1}^{k-1} \end{vmatrix}.$$

The first column of the obtained determinant is formed by zeroes, except the element in the upper left corner, equal to one. Using this fact, it is easy to perform the expansion in the first column. After taking the common multipliers outside the sign of determinant, we have

$$V_{k+1} = (a_2 - a_1)(a_3 - a_1)\dots(a_{k+1} - a_1) \begin{vmatrix} 1 & 1 & \dots & 1 \\ a_2 & a_3 & \dots & a_{k+1} \\ a_2^2 & a_3^2 & \dots & a_{k+1}^2 \\ \dots\dots\dots\dots\dots\dots \\ a_2^{k-1} & a_3^{k-1} & \dots & a_{k+1}^{k-1} \end{vmatrix}.$$

As a result we have obtained the determinant V_k, which, according to the inductive supposition, is equal to the product of all the possible differences $a_j - a_i$ for $1 \leqslant i < j \leqslant k$.

Therefore, the mathematical induction method has proved the formula of (2.75).

Note. The Vandermonde matrix, i.e. a matrix of the form

$$\begin{vmatrix} 1 & 1 & \dots & 1 \\ a_1 & a_2 & \dots & a_n \\ a_1^2 & a_2^2 & \dots & a_n^2 \\ \dots\dots\dots\dots\dots\dots \\ a_1^{n-1} & a_2^{n-1} & \dots & a_n^{n-1} \end{vmatrix}$$

is widely met in the theory of approximation of functions by polynomials [11, 58].

2.13 *Solution.*

In case of multiplication of any row by a real number, the determinant of this matrix is multiplied by this number. Therefore, as all the elements of the matrix A are multiplied by m, the determinant is multiplied by the value

$$\underbrace{m \times m \times \cdots \times m}_{n \text{ times}}.$$

Thus, for all $m \in \mathbb{N}$ the equality $\det(mA) = m^n \det A$ is fulfilled.

2.14 *Solution.*

The student is wrong. The determinant of the sum of two matrices is not always equal to the sum of the determinants of these matrices, which is confirmed by the following counterexample.

Consider an identity matrix of size $n \times n$. Then the inequality is valid:

$$\det(I + I) \neq \det I + \det I,$$

since $\det(I + I) = \det(2I) = 2^n$ (see Problem **2.13**). At the same time, $\det I + \det I = 1 + 1 = 2 \neq \det(I + I)$.

2.15 *Solution.*

The expansion of the determinant by the formula (2.5) contains $5! = 120$ summands. One of these summands of the form $a_{15}a_{21}a_{34}a_{42}a_{53}$ includes all the five multipliers of value 10000 and is equal to $(-1)^\sigma 10000^5 = (-1)^\sigma 10^{20}$. Here σ is the number of inversions in the permutation $(5, 1, 4, 2, 3)$. As is easy to see, $\sigma = 6$.

The remaining $5! - 1 = 119$ summands include no more than three multipliers 10000 and, therefore, do not exceed $10000^3 \times 1 \times 1 = 10^{12}$ in absolute magnitude. Hence it may be concluded that the considered determinant is no less than the difference $10^{20} - 119 \cdot 10^{12} > 0$.

As a result, the determinant is positive.

2.16 *Solution.*

Relying of the reasoning provided in the solution of Problem **2.15**, we obtain: the determinant in this case does not exceed the value

$$(-10^{20} + 119 \cdot 4 \cdot 5 \cdot 10^{12}) < 0.$$

Therefore, this determinant is a negative number.

2.17 *Answer:*

Computing of determinants with the help of Python provides the following results:

$$\begin{vmatrix} 1 & 1 & 1 & 1 & 10^4 \\ 10^4 & 1 & 1 & 1 & 1 \\ 1 & 1 & 1 & 10^4 & 1 \\ 1 & 10^4 & 1 & 1 & 1 \\ 1 & 1 & 10^4 & 1 & 1 \end{vmatrix} = 99999990001999850004,$$

$$\begin{vmatrix} 1 & 10^4 & 3 & 4 & 5 \\ 5 & 1 & 2 & 3 & 10^4 \\ 4 & 5 & 1 & 10^4 & 3 \\ 3 & 4 & 10^4 & 1 & 2 \\ 10^4 & 3 & 4 & 5 & 1 \end{vmatrix} = -99999909053183731167.$$

These results, of course, conform with the solution of Problems **2.15** and **2.16**.

2.18 *Solution.*

Recall that in combinatorics the number of combinations of n various elements of k without iterations is denoted as $C(n, k)$ [1, 60].

In order to form the minor of the k-th order, one should select k rows and k columns from the matrix. The rows can be selected by $C(m, k)$ methods, while the columns—by $C(n, k)$ method. Applying the combinatory rule of product, we obtain, that all in all we can have $C(m, k)C(n, k)$ minors of the k-th order.

2.19 *Answer:*

The rows for formation of the minor may be selected using the number of combinations, by $C(n, k)$ methods. Then the columns should be selected so that their numbers should not coincide with the numbers of the selected rows. This can be done in $C(n - k, k)$ ways. In all, according to the combinatory rule of product, we obtain the answer: $C(n, k)C(n - k, k)$ minors.

2.20 *Solution.*

Consider the matrix A of size 3×3

$$A = \begin{bmatrix} a_{11} & a_{12} & a_{13} \\ a_{21} & a_{22} & a_{23} \\ a_{31} & a_{32} & a_{33} \end{bmatrix}$$

and write its determinant in the form

$$\det A = a_{11}a_{22}a_{33} + a_{12}a_{23}a_{31} + a_{13}a_{21}a_{32}$$

$$- a_{11}a_{23}a_{32} - a_{12}a_{21}a_{33} - a_{13}a_{22}a_{31}$$

$$= \alpha + \beta + \gamma$$

$$+ \delta + \varepsilon + \zeta,$$

where designations $\alpha = a_{11}a_{22}a_{33}$, $\beta = a_{12}a_{23}a_{31}$, $\ldots$, $\zeta = -a_{13}a_{22}a_{31}$ are introduced. Each of the variables $\alpha, \beta, \ldots, \zeta$ takes the values from the set $\{-1, 1\}$.

All the six summands of the determinant cannot have the same sign. Indeed, the product $\alpha\beta\gamma$ can be presented in the form of the product of nine elements of the matrix A:

$$\alpha\beta\gamma = \prod_{i,j=1}^{3} a_{ij}.$$

At the same time, there exists the equality $\delta\varepsilon\zeta = (-1)^3 \prod_{i,j=1}^{3} a_{ij} = -\alpha\beta\gamma$.
Therefore, among $\alpha, \beta, \ldots, \zeta$ there exist negative summands, and $\det A < 6$.

If five terms of the determinant have one sign, and the sixth term has a different sign, then $\det A$ as an even number. Then, $\det A < 5$.

As is easy to find by direct calculation, the matrix

$$\begin{bmatrix} -1 & 1 & 1 \\ 1 & -1 & 1 \\ 1 & 1 & -1 \end{bmatrix}$$

has det $A = 4$.

Finally we obtain: the greatest value of the determinant of the matrix of size 3×3, consisting of the elements $+1$ and -1, is equal to 4.

2.21 *Answer:* 2.

2.22 *Answer:* the determinant will be multiplied by $(-1)^{\lfloor n/2 \rfloor}$.

2.23 *Solution.*

```
# The number of multiplications
count = 0


def get_determ(A):
    global count

    size = len(A)

    if size == 1:
        return A[0][0]
    elif size == 2:
        count += 2
        return A[0][0] * A[1][1] - A[0][1] * A[1][0]
    else:
        det = 0

        # Expansion over the first row
        for col in range(size):
            minor = [row[:col] + row[col + 1:] for row in (A[1:])]

            det_sign = 1 if col % 2 == 0 else -1
            det += det_sign * A[0][col] * get_determ(minor)
            count += 1

        return det

mas = [[1, 2, 3], [4, 5, 6], [7, 8, 9]]
print("det =", get_determ(mas))
print("count =", count)
```

2.24 *Solution.*

For the matrix of size $n \times n$, n recursive calls are performed and n multiplications are executed of the form $a_{ij} \times A_{ij}$, $j = 1, \ldots, n$. The exit from the recursion will be at $n = 1$; no multiplications are executed in this case. Due to this, the total number of multiplications satisfies the recurrence relation:

$$\begin{cases} T(n) = nT(n-1) + n, & n > 1, \\ T(1) = 0. \end{cases}$$

Solve the obtained relation by the method of substitution, i.e. successively expressing $T(n-1)$ by $T(n-2)$, then $T(n-2)$ by $T(n-3)$, and so on:

$$\begin{aligned} T(n) &= nT(n-1) + n \\ &= n[(n-1)T(n-2) + n - 1] + n \\ &= n(n-1)T(n-2) + n(n-1) + n \\ &= n(n-1)[(n-2)T(n-3) + n - 2] + n(n-1) + n \\ &= n(n-1)(n-2)T(n-3) + n(n-1)(n-2) + n(n-1) + n. \end{aligned}$$

Similarly continuing this process until $T(1) = 0$, we obtain

$$\begin{aligned} T(n) &= n(n-1)(n-2)\ldots 2 \cdot T(1) \\ &\quad + n(n-1)(n-2)\ldots 2 + n(n-1)(n-2)\ldots 3 + \cdots + n(n-1) + n \\ &= 0 + n! + \frac{n!}{2!} + \cdots + \frac{n!}{(n-2)!} + \frac{n!}{(n-1)!} \\ &= n!\left[1 + \frac{1}{2!} + \cdots + \frac{1}{(n-2)!} + \frac{1}{(n-1)!}\right] = n!\sum_{k=1}^{n-1}\frac{1}{k!}. \end{aligned}$$

It is possible to write an analytical expression for $T(n)$ by non-elementary functions, however, for solution of the posed problem it is enough to evaluate the asymptotic behaviour of the function $T(n)$.

Note that for $n \to \infty$

$$\lim_{n\to\infty}\sum_{k=1}^{n-1}\frac{1}{k!} = \sum_{k=1}^{\infty}\frac{1}{k!} = \sum_{k=0}^{\infty}\frac{1}{k!} - 1 = e - 1,$$

where $e = 2.71828\ldots$ is the base of natural logarithms. Hence we obtain the inequality

$$n! \leqslant n!\sum_{k=1}^{n-1}\frac{1}{k!} \leqslant (e-1)n!,$$

and, finally, $T(n) = O(n!)$.

2.25 *Solution.*

According to the result of the previous problem, $T(n) = n! \sum_{k=1}^{n-1} \frac{1}{k!}$.

Let us expand the expression $en\Gamma(n, 1) - n!$, taking into account the definition of incomplete gamma function:

$$en\Gamma(n, 1) - n! = en \cdot (n-1)! \, e^{-1} \sum_{k=0}^{n-1} \frac{1^k}{k!} - n! = n! \sum_{k=0}^{n-1} \frac{1}{k!} - n! = n! \sum_{k=1}^{n-1} \frac{1}{k!},$$

which coincides with $T(n)$.

2.26 *Solution.*

Let us find the total number of inversions $S(N)$ contained in all permutations of N elements. The relation $\frac{S(N)}{N!}$ will be equal to the mean value of the number of inversions $A(N)$ in the N-element array.

In order to compute the variable $S(N)$ suppose that some permutation $(a_{i_1}, a_{i_2}, \dots, a_{i_{n-1}}, a_{i_n})$ contains exactly σ inversions.

Note that in the permutation $(a_{i_n}, a_{i_{n-1}}, \dots, a_{i_2}, a_{i_1})$ the number of inversions is $\frac{N(N-1)}{2} - \sigma$. This means that the total number of inversions in the pair of arrays

$$(a_{i_1}, a_{i_2}, \dots, a_{i_{n-1}}, a_{i_n}) \text{ and } (a_{i_n}, a_{i_{n-1}}, \dots, a_{i_2}, a_{i_1})$$

is equal to $\sigma + \left(\frac{N(N-1)}{2} - \sigma \right) = \frac{N(N-1)}{2}$.

Since there exist only $N!$ permutations of the N-element array, then $S(N) = \frac{1}{2} N! \times \frac{N(N-1)}{2}$, and therefore there exists the following estimate of the mean value of the number of inversions:

$$A(N) = \frac{S(N)}{N!} = \frac{1}{4} N^2 + O(N) \text{ for } N \to \infty.$$

2.27 *Proof.*

Denote the elements of the matrix B by b_{ij}, $1 \leqslant i, j \leqslant n$. Using the introduced designation, we can write that the matrix $I + \varepsilon B$ is formed by the elements $(\delta_{ij} + \varepsilon b_{ij})$.

According to the definition, the variable $\det(I + \varepsilon B)$ is equal to the sum over all possible permutations:

$$\det(I + \varepsilon B) = \sum_{\text{perm}} (-1)^{\sigma} (\delta_{1i_1} + \varepsilon b_{1i_1})(\delta_{2i_2} + \varepsilon b_{2i_2}) \dots (\delta_{ni_n} + \varepsilon b_{ni_n}).$$

Removing the brackets under the summation sign, we obtain

$$\det(I + \varepsilon B) = \sum_{\text{perm}} (-1)^{\sigma} (\delta_{1i_1} \delta_{2i_2} \ldots \delta_{ni_n}$$

$$+ \varepsilon b_{1i_1} \delta_{2i_2} \delta_{3i_3} \ldots \delta_{ni_n}$$

$$+ \varepsilon b_{2i_2} \delta_{1i_1} \delta_{3i_3} \ldots \delta_{ni_n}$$

$$+ \varepsilon b_{3i_3} \delta_{1i_1} \delta_{2i_2} \ldots \delta_{ni_n} + \cdots$$

$$+ \varepsilon b_{ni_n} \delta_{1i_1} \delta_{2i_2} \ldots \delta_{(n-1)i_{n-1}}$$

$$+ \varepsilon^2 (\ldots) + \cdots).$$

The product $\delta_{1i_1} \delta_{2i_2} \ldots \delta_{ni_n}$ is equal to one, if $i_1 = 1, i_2 = 2, \ldots, i_n = n$, and is equal to zero in other cases.

Further, the products of the form

$$\varepsilon b_{ki_k} \delta_{1i_1} \delta_{2i_2} \ldots \delta_{(k-1)i_{k-1}} \delta_{(k+1)i_{k+1}} \ldots \delta_{ni_n}$$

reduce to the summands εb_{kk}.

This implies that

$$\det(I + \varepsilon B) = (-1)^{\sigma_0} (1 + \varepsilon b_{11} + \varepsilon b_{22} + \cdots + \varepsilon b_{nn} + O(\varepsilon^2)),$$

where σ_0 is the number of inversions in the permutation $(i_1, i_2, \ldots, i_n) = (1, 2, \ldots, n)$.

It is clear that $\sigma_0 = 0$.

As a result we obtain

$$\det(I + \varepsilon B) = (-1)^{\sigma_0} \left(1 + \varepsilon \sum_{k=1}^{n} b_{kk} + O(\varepsilon^2)\right) = 1 + \varepsilon \operatorname{tr} B + O(\varepsilon^2).$$

2.28 *Answer:* $\det \left(\dfrac{I}{(I + \varepsilon B)^p} \right) = 1 - p\varepsilon(\operatorname{tr} B) + O(\varepsilon^2)$.

2.29 *Proof.*

Represent the matrix M in the form $M = G(I + \varepsilon G^{-1}H)$. Calculate the determinant

$$\det M = \det G \det(I + \varepsilon G^{-1}H)$$

$$= \det G \exp(\mathrm{tr}\,\ln(I + \varepsilon G^{-1}H)) = \det G \exp\left(\mathrm{tr}\,\sum_{k=1}^{\infty}\frac{(-1)^{k-1}}{k}(\varepsilon G^{-1}H)^k\right)$$

$$= \det G\left(1 + \varepsilon\,\mathrm{tr}\,(G^{-1}H) + \frac{1}{2}\varepsilon^2(\mathrm{tr}^2\,(G^{-1}H) - \mathrm{tr}\,(G^{-1}H)^2) + O(\varepsilon^3)\right).$$

Thus, the formula (2.80) is proved.

2.30 Solution.

The determinant of the matrix is equal to $\det A = ad - bc$.

Calculate the cofactors for each of the elements of the matrix A:

$$A_{11} = (-1)^{1+1} \cdot d = d, \quad A_{12} = (-1)^{1+2} \cdot c = -c;$$
$$A_{21} = (-1)^{2+1} \cdot b = -b, \quad A_{22} = (-1)^{2+2} \cdot a = a.$$

Write a matrix of cofactors:

$$\begin{bmatrix} d & -c \\ -b & a \end{bmatrix}.$$

Then the sought inverse matrix will have the form:

$$A^{-1} = \frac{1}{ad - bc}\begin{bmatrix} d & -b \\ -c & a \end{bmatrix}.$$

2.31 Solution.

(1) Since $\det A = -2 \neq 0$, then the inverse matrix exists. Find the cofactors of the elements of the matrix A:

$$A_{11} = 4, \qquad\qquad A_{12} = -3,$$
$$A_{21} = -2, \qquad\qquad A_{22} = 1.$$

Therefore, the matrix of cofactors can be written in the form: $\begin{bmatrix} 4 & -3 \\ -2 & 1 \end{bmatrix}.$

Transpose the matrix (A_{ij}): $\begin{bmatrix} 4 & -3 \\ -2 & 1 \end{bmatrix}^T = \begin{bmatrix} 4 & -2 \\ -3 & 1 \end{bmatrix}$.

For computing A^{-1}, divide the obtained adjoint matrix by the determinant:

$$A^{-1} = \begin{bmatrix} 4 & -2 \\ -3 & 1 \end{bmatrix} / (-2) = \begin{bmatrix} -2 & 1 \\ 3/2 & -1/2 \end{bmatrix}.$$

(2) Since $\det A = 1 \neq 0$, then the inverse matrix exists. Find the cofactors of the elements of the matrix A:

$$A_{11} = 7, \quad A_{12} = -5,$$
$$A_{21} = -4, \quad A_{22} = 3.$$

Transpose the matrix (A_{ij}): $\begin{bmatrix} 7 & -5 \\ -4 & 3 \end{bmatrix}^T = \begin{bmatrix} 7 & -4 \\ -5 & 3 \end{bmatrix}$.

Therefore, $A^{-1} = \dfrac{1}{\det A} \begin{bmatrix} 7 & -4 \\ -5 & 3 \end{bmatrix} = \begin{bmatrix} 7 & -4 \\ -5 & 3 \end{bmatrix}$.

(3) Compute the determinant by the method of expansion in the first column:

$$\det A = 3 \begin{vmatrix} -3 & 1 \\ -5 & -1 \end{vmatrix} - 2 \begin{vmatrix} -4 & 5 \\ -5 & -1 \end{vmatrix} + 3 \begin{vmatrix} -4 & 5 \\ -3 & 1 \end{vmatrix} = 24 - 58 + 33 = -1 \neq 0.$$

Find the cofactors:

$$A_{11} = \begin{vmatrix} -3 & 1 \\ -5 & -1 \end{vmatrix} = 8, \quad A_{12} = 5, \quad A_{13} = -1,$$

$$A_{21} = -29, \quad A_{22} = -18, \quad A_{23} = 3,$$
$$A_{31} = 11, \quad A_{32} = 7, \quad A_{33} = -1.$$

We will obtain the matrix of cofactors: $\begin{bmatrix} 8 & 5 & -1 \\ -29 & -18 & 3 \\ 11 & 7 & -1 \end{bmatrix}$.

Perform the transposition operation:

$$\begin{bmatrix} 8 & 5 & -1 \\ -29 & -18 & 3 \\ 11 & 7 & -1 \end{bmatrix}^T = \begin{bmatrix} 8 & -29 & 11 \\ 5 & -18 & 7 \\ -1 & 3 & -1 \end{bmatrix}.$$

With the help of division by det A write the inverse matrix:

$$A^{-1} = \frac{1}{\det A} \begin{bmatrix} 8 & -29 & 11 \\ 5 & -18 & 7 \\ -1 & 3 & -1 \end{bmatrix} = \begin{bmatrix} -8 & 29 & -11 \\ -5 & 18 & -7 \\ 1 & -3 & 1 \end{bmatrix}.$$

(4) Since det $A = -3 \neq 0$, then A^{-1} is determined. Find the cofactors A_{ij}:

$$A_{11} = 7, \quad A_{12} = -5, \quad A_{13} = 6,$$
$$A_{21} = -6, \quad A_{22} = 3, \quad A_{23} = -3,$$
$$A_{31} = 1, \quad A_{32} = 1, \quad A_{33} = -3.$$

The inverse matrix is equal to $A^{-1} = \frac{1}{\det A} \begin{bmatrix} 7 & -6 & 1 \\ -5 & 3 & 1 \\ 6 & -3 & -3 \end{bmatrix} =$

$$\begin{bmatrix} -\dfrac{7}{3} & 2 & -\dfrac{1}{3} \\ \dfrac{5}{3} & -1 & -\dfrac{1}{3} \\ -2 & 1 & 1 \end{bmatrix}.$$

(5) Find the determinant: det $A = -27 \neq 0$. The cofactors are equal to

$$A_{11} = -3, \quad A_{12} = -6, \quad A_{13} = -6,$$
$$A_{21} = -6, \quad A_{22} = -3, \quad A_{23} = 6,$$
$$A_{31} = -6, \quad A_{32} = 6, \quad A_{33} = -3.$$

Compute the elements of the inverse matrix:

$$A^{-1} = \frac{1}{\det A} \begin{bmatrix} -3 & -6 & -6 \\ -6 & -3 & 6 \\ -6 & 6 & -3 \end{bmatrix} = \frac{1}{9} \begin{bmatrix} 1 & 2 & 2 \\ 2 & 1 & -2 \\ 2 & -2 & 1 \end{bmatrix}.$$

2.32 *Answer:*

$$(1) \quad \begin{bmatrix} 1 & -\dfrac{1}{2} \\ 0 & \dfrac{1}{2} \end{bmatrix}; \quad (2) \quad \begin{bmatrix} 1 & -\dfrac{1}{2} & 0 \\ 0 & \dfrac{1}{2} & -\dfrac{1}{3} \\ 0 & 0 & \dfrac{1}{3} \end{bmatrix}; \quad (3) \quad \begin{bmatrix} 1 & -\dfrac{1}{2} & 0 & 0 \\ 0 & \dfrac{1}{2} & -\dfrac{1}{3} & 0 \\ 0 & 0 & \dfrac{1}{3} & -\dfrac{1}{4} \\ 0 & 0 & 0 & \dfrac{1}{4} \end{bmatrix}.$$

2.33 *Answer:* $[A, A^{-1}] = [A, A^{-1}] = O$.

2.34 *Solution.*

(1) As is known, the matrix A does not have the inverse one when the condition $\det A = 0$ is fulfilled.

Calculate the determinant: $\begin{vmatrix} -1 & \lambda & \lambda \\ \lambda & \lambda & 0 \\ 6 & 4 & \lambda \end{vmatrix} = -\lambda^3 - 3\lambda^2 = -\lambda^2(\lambda + 3).$

Therefore, the matrix does not have the inverse one for $\lambda \in \{0, -3\}$.

(2) The determinant is equal to $-\lambda^3 + 3\lambda^2 + \lambda - 3 = -(\lambda - 3)(\lambda - 1)(\lambda + 1)$. The matrix does not have the inverse one for $\lambda \in \{-1, 1, 3\}$.

2.35 *Solution.*

For finding the inverse matrix, compute the determinant $\det A$.

$$\det A = \begin{vmatrix} 1 & \alpha & 0 & 0 \\ 0 & 1 & \beta & 0 \\ 0 & 0 & 1 & \gamma \\ 0 & 0 & 0 & 1 \end{vmatrix},$$

$\det A = 1$, since A is the upper triangular matrix and the determinant is equal to the product of the diagonal elements.

Find the cofactors:

$$A_{11} = 1, \quad A_{12} = 0, \quad A_{13} = 0, \quad A_{14} = 0,$$

$$A_{21} = -\alpha, \quad A_{22} = 1, \quad A_{23} = 0, \quad A_{24} = 0,$$

$$A_{31} = \alpha\beta, \quad A_{32} = -\beta, \quad A_{33} = 1, \quad A_{34} = 0,$$

$$A_{41} = -\alpha\beta\gamma, \quad A_{42} = \beta\gamma, \quad A_{43} = -\gamma, \quad A_{44} = 1.$$

We perform transposition and obtain the adjoint matrix:

$$
\begin{bmatrix}
1 & 0 & 0 & 0 \\
-\alpha & 1 & 0 & 0 \\
\alpha\beta & -\beta & 1 & 0 \\
-\alpha\beta\gamma & \beta\gamma & -\gamma & 1
\end{bmatrix}^{T}
=
\begin{bmatrix}
1 & -\alpha & \alpha\beta & -\alpha\beta\gamma \\
0 & 1 & -\beta & \beta\gamma \\
0 & 0 & 1 & -\gamma \\
0 & 0 & 0 & 1
\end{bmatrix}.
$$

Since the determinant is equal to one, then A^{-1} coincides with the adjoint matrix:

$$
A^{-1} =
\begin{bmatrix}
1 & -\alpha & \alpha\beta & -\alpha\beta\gamma \\
0 & 1 & -\beta & \beta\gamma \\
0 & 0 & 1 & -\gamma \\
0 & 0 & 0 & 1
\end{bmatrix}.
$$

As is shown above, the determinant $\det A$ does not depend on the parameters α, β, γ. Therefore, the matrix A^{-1} is defined for any values of $\alpha, \beta, \gamma \in \mathbb{R}$.

2.36 Solution.

The determinant of the matrix $G_n = (g_{ij})$ is equal to $\det G_n = (-1)^{n(n-1)/2}$, since non-zero product in the sum of the form (2.5) is equal to $g_{1,n} g_{2,n-1} \cdots g_{n,1} = 1$, and the multiplier σ for this product takes the value $\sigma = (-1)^{n(n-1)/2}$ (see Problem **2.5**).

Construct the matrix of cofactors of the elements g_{ij}, where $1 \leqslant i, j \leqslant n$. The cofactor of any element equal to zero will be equal to zero. This follows from the fact that in case of deletion of the non-zero element in the corresponding minor appears a zero row and a zero column, and, in turn, such a minor is equal to zero. The minor of any element $g_{i,(n+1)-i}$, located on the secondary diagonal, will be equal to $\det G_{n-1} = (-1)^{(n-1)(n-2)/2}$, since after deletion of $g_{i,(n+1)-i}$ we will obtain the matrix G_{n-1} of size $(n-1) \times (n-1)$.

Therefore, cofactors of such elements $g_{i,n-i}$ are equal to

$$
(-1)^{i+(n+1-i)} \cdot \det G_{n-1} = (-1)^{n+1} \cdot (-1)^{(n-1)(n-2)/2}
$$

$$
= (-1)^{n(n-1)/2+2} = (-1)^{n(n-1)/2}.
$$

Thus, the matrix of cofactors is equal to the original G_n, multiplied by $(-1)^{n(n-1)/2}$.

As is easy to see, transposition does not change the obtained matrix. The last step—divide the adjoint matrix by the determinant $\det G_n = (-1)^{n(n-1)/2}$. Finally we obtain that the matrix inverse of G coincides with it itself: $G^{-1} \equiv G$ for all values of $n \geqslant 2$.

2.37 *Solution.*

Let us provide a code in Python for solution of the problem.

```python
import numpy as np


def get_Hilbert_matrix(n):
    return np.matrix([[ 1 / (i + j - 1)
        for j in range(1, n + 1)] for i in range(1, n + 1)])


matrix = get_Hilbert_matrix(6)
inversed = np.linalg.inv(matrix)
hh1 = np.matmul(matrix, inversed)
h1h = np.matmul(inversed, matrix)

print(matrix)
print(inversed)
print(hh1)
print(h1h)
```

The difference of the elements of the matrices $H \cdot H^{-1}$ and $H^{-1} \cdot H$ computed with the help of Python from the identity matrix is $\sim 10^{-10}$ for $n = 6$, $\sim 10^{-7}$ for $n = 8$ and $\sim 10^{-8}$ for $n = 7$. (Here, the symbol $\sim$ means equality by the order of value.)

Thus, Hilbert matrices demonstrate accumulation of machine errors when making computations with real numbers [58]. These matrices are very often used for testing of numerical algorithms.

The matrix H^{-1} can be found in an explicit form; the analytical representations for h_{ij} are shown in [40]. An interesting peculiarity of this problem is also the fact that the elements of the inverse matrix are integer numbers.

2.38 *Solution.*

This matrix is an upper triangular one, and its determinant is equal to the product of the elements positioned on the main diagonal, i.e. is equal to one.

Construct the matrix of cofactors. The cofactors of all unit elements positioned on the main diagonal will be equal to one. Upon deleting all zeroes except those standing on the diagonal below the main one, there appear matrices with proportional rows, therefore, their minors will be equal to zero. Those zeroes that are positioned on the diagonal below the main one, as cofactors will have the values $(-\lambda)$. The sum $i + j$ for such elements is always odd, since they are positioned below the main diagonal, and upon their deletion we obtain an upper triangular matrix with one element λ and other unities on the main diagonal. Upon deletion of λ of any degree, except zero, we obtain an upper triangular matrix with zeroes and unities on the main diagonal. Therefore, both the minor and the cofactor are in this case equal to zero.

Finally, having executed the transposition operation, we obtain the inverse matrix:

$$\begin{bmatrix} 1 & -\lambda & 0 & 0 & \ldots & 0 \\ 0 & 1 & -\lambda & 0 & \ldots & 0 \\ 0 & 0 & 1 & -\lambda & \ldots & 0 \\ \multicolumn{6}{c}{\ldots\ldots\ldots\ldots\ldots\ldots} \\ 0 & 0 & 0 & 0 & \ldots & 1 \end{bmatrix}.$$

2.39 *Solution.*

Let us use mathematical induction method.

B a s i s   s t e p

For the least natural $n = 1$ we have

$$(A_1)^{-1} = A_1^{-1} \text{ is true.}$$

I n d u c t i v e   s t e p

Assume that for $n = k$ the equality

$$(A_1 A_2 \ldots A_{k-1} A_k)^{-1} = A_k^{-1} A_{k-1}^{-1} \ldots A_2^{-1} A_1^{-1}$$

is valid. Then we should prove that for $n = k + 1$ the following is true:

$$(A_1 A_2 \ldots A_{k+1-1} A_{k+1})^{-1} = A_{k+1}^{-1} A_{k+1-1}^{-1} \ldots A_2^{-1} A_1^{-1}.$$

Denote the expression for $A_k^{-1} \ldots A_1^{-1}$ by B, Then:

$$(B A_{k+1})^{-1} = A_{k+1}^{-1} B^{-1} = A_{k+1}^{-1} (A_k^{-1} A_{k-1}^{-1} \ldots A_1^{-1}).$$

Therefore, according to the mathematical induction method, $\forall n \in \mathbb{N}$ the identity

$$(A_1 A_2 \ldots A_{n-1} A_n)^{-1} = A_n^{-1} A_{n-1}^{-1} \ldots A_2^{-1} A_1^{-1}$$

is valid.

2.40 *Solution.*

(1) Find the matrix inverse of the matrix $A = \begin{bmatrix} 2 & 1 \\ 0 & 2 \end{bmatrix}$.

Its determinant is equal to $\det A = 4$, the matrix of cofactors has the components $\begin{bmatrix} 2 & 0 \\ -1 & 2 \end{bmatrix}$, inverse of the matrix: $A^{-1} = \begin{bmatrix} 1/2 & -1/4 \\ 0 & 1/2 \end{bmatrix}$.

We obtain the matrix X by multiplying A^{-1} by the matrix $B = \begin{bmatrix} -6 & 4 \\ 2 & 1 \end{bmatrix}$:

$$X = \begin{bmatrix} 1/2 & -1/4 \\ 0 & 1/2 \end{bmatrix} \cdot \begin{bmatrix} -6 & 4 \\ 2 & 1 \end{bmatrix} = \begin{bmatrix} -7/2 & 7/4 \\ 1 & 1/2 \end{bmatrix} = \frac{1}{4} \begin{bmatrix} -14 & 7 \\ 4 & 2 \end{bmatrix}.$$

(2) Let us find the matrix inverse of the matrix $A = \begin{bmatrix} -1 & 1 & 1 \\ 0 & 2 & 2 \\ 0 & 2 & 3 \end{bmatrix}$.

The determinant is equal to $\det A = -2$.
The inverse of the matrix A^{-1}:

$$A^{-1} = \begin{bmatrix} -1 & 1/2 & 0 \\ 0 & 3/2 & -1 \\ 0 & -1 & 1 \end{bmatrix}.$$

Multiply the matrix inverse of the matrix A by the matrix $B = \begin{bmatrix} -2 & 1 & 1 \\ -1 & 0 & 2 \\ -1 & -2 & 0 \end{bmatrix}$:

$$X = A^{-1}B = \begin{bmatrix} -1 & 1/2 & 0 \\ 0 & 3/2 & -1 \\ 0 & -1 & 1 \end{bmatrix} \begin{bmatrix} -2 & 1 & 1 \\ -1 & 0 & 2 \\ -1 & -2 & 0 \end{bmatrix} = \frac{1}{2} \begin{bmatrix} 3 & -2 & 0 \\ -1 & 4 & 6 \\ 0 & -4 & -4 \end{bmatrix}.$$

(3) We obtain the solution of the equation multiplying both sides of the equation $X \cdot A = B$ by A^{-1} on the right.
To do this, find the matrix inverse of the matrix A:

$$A^{-1} = \frac{1}{10} \begin{bmatrix} -30 & 12 & 38 \\ -5 & 3 & 7 \\ -30 & 14 & 36 \end{bmatrix}.$$

Compute the elements of an unknown matrix $X = B \cdot A^{-1}$:

$$X = \begin{bmatrix} 0 & -1 & 2 \\ -1 & -2 & -2 \\ 0 & 1 & 10 \end{bmatrix} \cdot \frac{1}{10} \begin{bmatrix} -30 & 12 & 38 \\ -5 & 3 & 7 \\ -30 & 14 & 36 \end{bmatrix} = \frac{1}{10} \begin{bmatrix} -55 & 25 & 65 \\ 100 & -46 & -124 \\ -305 & 143 & 367 \end{bmatrix}.$$

(4) From the equation $A \cdot X \cdot B = C$, find the unknown matrix X by the formula $X = A^{-1} \cdot C \cdot B^{-1}$.

We have

$$A^{-1} = \frac{1}{13} \begin{bmatrix} 2 & -3 \\ 3 & 2 \end{bmatrix}, \quad B^{-1} = \frac{1}{31} \begin{bmatrix} -4 & 7 \\ 5 & -1 \end{bmatrix}.$$

Consecutively perform multiplications in the following order:

$$(A^{-1} \cdot C) \cdot B^{-1}.$$

Multiply the matrix inverse of the matrix A by the matrix C:

$$A^{-1}C = \frac{1}{13} \begin{bmatrix} 2 & -3 \\ 3 & 2 \end{bmatrix} \begin{bmatrix} 1 & -1 \\ 0 & 1 \end{bmatrix} = \frac{1}{13} \begin{bmatrix} 2 & -5 \\ 3 & -1 \end{bmatrix}.$$

Finally, multiply the product $A^{-1}C$ by the matrix, inverse of the matrix B:

$$X = \frac{1}{13} \begin{bmatrix} 2 & -5 \\ 3 & -1 \end{bmatrix} \cdot \frac{1}{31} \begin{bmatrix} -4 & 7 \\ 5 & -1 \end{bmatrix} = \frac{1}{403} \begin{bmatrix} -33 & 19 \\ -17 & 22 \end{bmatrix}.$$

2.41 *Answer:* $\begin{bmatrix} 1 & 0 & 0 \\ 0 & 1 & 0 \\ 0 & 256a & 1 \end{bmatrix}.$

2.42 *Solution.*
Having calculated the several first powers A^k, namely the second, third, fourth
and fifth powers of the matrix $A = \begin{bmatrix} 1 & 0 & 0 \\ g & 1 & 1 \\ h & 0 & 1 \end{bmatrix}$, we have

$$A^2 = \begin{bmatrix} 1 & 0 & 0 \\ 2g+h & 1 & 2 \\ 2h & 0 & 1 \end{bmatrix}, \quad A^3 = \begin{bmatrix} 1 & 0 & 0 \\ 3g+3h & 1 & 3 \\ 3h & 0 & 1 \end{bmatrix},$$

$$A^4 = \begin{bmatrix} 1 & 0 & 0 \\ 4g+6h & 1 & 4 \\ 4h & 0 & 1 \end{bmatrix}, \quad A^5 = \begin{bmatrix} 1 & 0 & 0 \\ 5g+10h & 1 & 5 \\ 5h & 0 & 1 \end{bmatrix}.$$

Based on the obtained equalities, suppose that for all natural values of n the
identity is fulfilled:

$$A^n = \begin{bmatrix} 1 & 0 & 0 \\ ng+n(n-1)h/2 & 1 & n \\ nh & 0 & 1 \end{bmatrix},$$

denote the respective predicate by $P(n)$.
Let us use the mathematical induction method.
B a s i s   s t e p

$$A^1 = \begin{bmatrix} 1 & 0 & 0 \\ 1 \cdot g + 1(1-1)h/2 & 1 & 1 \\ 1 \cdot h & 0 & 1 \end{bmatrix} = A \text{ is true.}$$

I n d u c t i v e   s t e p
Assume that for $n = k$ the predicate $P(n)$ takes the true value, Then:

$$A^k = \begin{bmatrix} 1 & 0 & 0 \\ kg+k(k-1)h/2 & 1 & k \\ kh & 0 & 1 \end{bmatrix}.$$

Prove that for $n = k + 1$ the equality is valid:

$$A^{k+1} = \begin{bmatrix} 1 & 0 & 0 \\ (k+1)g + k(k+1)h/2 & 1 & k+1 \\ (k+1)h & 0 & 1 \end{bmatrix}.$$

Indeed,

$$A^{k+1} = \begin{bmatrix} 1 & 0 & 0 \\ kg + k(k-1)h/2 & 1 & k \\ kh & 0 & 1 \end{bmatrix} \begin{bmatrix} 1 & 0 & 0 \\ g & 1 & 1 \\ h & 0 & 1 \end{bmatrix}$$

$$= \begin{bmatrix} 1 & 0 & 0 \\ (k+1)g + k(k+1)h/2 & 1 & k+1 \\ (k+1)h & 0 & 1 \end{bmatrix}.$$

Thus, the predicate $P(n)$ is proved for all $n \in \mathbb{N}$.

Substituting as the exponent of the matrix the number $n = 512$, we obtain the answer: $A^{512} = \begin{bmatrix} 1 & 0 & 0 \\ 256(2g + 511h) & 1 & 512 \\ 512h & 0 & 1 \end{bmatrix}.$

2.43 *Solution.*

Denote by $P(n)$ the predicate $\mathcal{F}^n = \begin{bmatrix} F_{n-1} & F_n \\ F_n & F_{n+1} \end{bmatrix}.$

B a s i s s t e p

The basis step is formed by the statement $P(2)$:

$$\begin{bmatrix} 0 & 1 \\ 1 & 1 \end{bmatrix} \begin{bmatrix} 0 & 1 \\ 1 & 1 \end{bmatrix} = \begin{bmatrix} 1 & 1 \\ 1 & 2 \end{bmatrix} = \begin{bmatrix} F_1 & F_2 \\ F_2 & F_3 \end{bmatrix},$$

which corresponds to the formula (2.85).

I n d u c t i v e s t e p

Assume that for $n = k$ the statement is true:

$$\mathcal{F}^k = \begin{bmatrix} F_{k-1} & F_k \\ F_k & F_{k+1} \end{bmatrix}.$$

Compute the matrix F^n for $n = k + 1$:

$$
\mathcal{F}^{k+1} = \begin{bmatrix} F_{k-1} & F_k \\ F_k & F_{k+1} \end{bmatrix} \begin{bmatrix} 0 & 1 \\ 1 & 1 \end{bmatrix} = \begin{bmatrix} F_k & F_{k-1} + F_k \\ F_{k+1} & F_k + F_{k+1} \end{bmatrix}.
$$

According to the definition of the Fibonacci sequence, each element of this sequence is equal to the sum of two previous ones, and for all $k > 1$ the identity $F_{k-1} + F_k = F_{k+1}$ is valid.

Thus, the predicate $P(n)$ is proved for all natural $n > 1$.

2.44 *Solution.*

Let us try to find regularity in the sequence $A^1, A^2, A^3, \dots$. To do this, raise the matrix to the second, third and fourth powers:

$$
A^2 = \begin{bmatrix} 1 & \alpha & \gamma \\ 0 & 1 & \beta \\ 0 & 0 & 1 \end{bmatrix} \begin{bmatrix} 1 & \alpha & \gamma \\ 0 & 1 & \beta \\ 0 & 0 & 1 \end{bmatrix} = \begin{bmatrix} 1 & 2\alpha & 2\gamma + \alpha\beta \\ 0 & 1 & 2\beta \\ 0 & 0 & 1 \end{bmatrix},
$$

$$
A^3 = A^2 \cdot A = \begin{bmatrix} 1 & 2\alpha & 2\gamma + \alpha\beta \\ 0 & 1 & 2\beta \\ 0 & 0 & 1 \end{bmatrix} \begin{bmatrix} 1 & \alpha & \gamma \\ 0 & 1 & \beta \\ 0 & 0 & 1 \end{bmatrix} = \begin{bmatrix} 1 & 3\alpha & 3\gamma + 3\alpha\beta \\ 0 & 1 & 3\beta \\ 0 & 0 & 1 \end{bmatrix},
$$

$$
A^4 = A^3 \cdot A = \begin{bmatrix} 1 & 3\alpha & 3\gamma + 3\alpha\beta \\ 0 & 1 & 3\beta \\ 0 & 0 & 1 \end{bmatrix} \begin{bmatrix} 1 & \alpha & \gamma \\ 0 & 1 & \beta \\ 0 & 0 & 1 \end{bmatrix} = \begin{bmatrix} 1 & 4\alpha & 4\gamma + 6\alpha\beta \\ 0 & 1 & 4\beta \\ 0 & 0 & 1 \end{bmatrix}.
$$

Analysis of the sequence of the powers $A^1, A^2, A^3, \dots$ leads to a hypothesis that

$$
A^n = \begin{bmatrix} 1 & n\alpha & n\gamma + n(n-1)\alpha\beta/2 \\ 0 & 1 & n\beta \\ 0 & 0 & 1 \end{bmatrix}.
$$

Let us prove the truth of this supposition with the help of the mathematical induction method.

Denote by $P(n)$ the statement "$A^n = \begin{bmatrix} 1 & n\alpha & n\gamma + n(n-1)\alpha\beta/2 \\ 0 & 1 & n\beta \\ 0 & 0 & 1 \end{bmatrix}$".

B a s i s   s t e p

The truth of the statement $P(1)$ is obvious.

Inductive step

Assume that $P(n)$ is valid for $n = k$ for some $k \geqslant 1$:

$$A^k = \begin{bmatrix} 1 & k\alpha & k\gamma + k(k-1)\alpha\beta/2 \\ 0 & 1 & k\beta \\ 0 & 0 & 1 \end{bmatrix}.$$

Prove that $P(k) \Rightarrow P(k+1)$.

$$A^{k+1} = A^k \cdot A = \begin{bmatrix} 1 & k\alpha & k\gamma + k(k-1)\alpha\beta/2 \\ 0 & 1 & k\beta \\ 0 & 0 & 1 \end{bmatrix} \begin{bmatrix} 1 & \alpha & \gamma \\ 0 & 1 & \beta \\ 0 & 0 & 1 \end{bmatrix}$$

$$= \begin{bmatrix} 1 & (k+1)\alpha & (k+1)\gamma + k(k+1)\alpha\beta/2 \\ 0 & 1 & (k+1)\beta \\ 0 & 0 & 1 \end{bmatrix}.$$

Therefore, $P(n)$ takes the true value for all $n \geqslant 1$. Thus it is proved that

$$A^n = \begin{bmatrix} 1 & n\alpha & n\gamma + n(n-1)\alpha\beta/2 \\ 0 & 1 & n\beta \\ 0 & 0 & 1 \end{bmatrix}$$

for all $n \in \mathbb{N}$.

2.45 *Solution.*

Assume that q-th power of the matrix $U(\varphi)$ is determined by the formula:

$$(U(\varphi))^q = \begin{bmatrix} \cos(q\varphi) & \sin(q\varphi) \\ -\sin(q\varphi) & \cos(q\varphi) \end{bmatrix}, \quad \text{where } q \in \mathbb{Z}.$$

Denote this statement by $P(q)$ and prove it first for $q \in \mathbb{N}$. Let us apply the mathematical induction method.

Basis step

For $n = 1$ we have

$$(U(\varphi))^1 = \begin{bmatrix} \cos(\varphi) & \sin(\varphi) \\ -\sin(\varphi) & \cos(\varphi) \end{bmatrix} = U(1 \cdot \varphi) \text{ is true.}$$

I n d u c t i v e   s t e p
Assume that $P(n)$ is true for $n = k$:

$$(U(\varphi))^k = \begin{bmatrix} \cos(k\varphi) & \sin(k\varphi) \\ -\sin(k\varphi) & \cos(k\varphi) \end{bmatrix}.$$

Prove the truth of the statement for $n = k + 1$.

$$(U(\varphi))^{k+1} = \begin{bmatrix} \cos(k\varphi) & \sin(k\varphi) \\ -\sin(k\varphi) & \cos(k\varphi) \end{bmatrix} \begin{bmatrix} \cos\varphi & \sin\varphi \\ -\sin\varphi & \cos\varphi \end{bmatrix}$$

$$= \begin{bmatrix} \cos(k\varphi)\cos\varphi - \sin(k\varphi)\sin\varphi & \cos(k\varphi)\sin\varphi + \sin(k\varphi)\cos\varphi \\ -\sin(k\varphi)\cos\varphi - \cos(k\varphi)\sin\varphi & -\sin(k\varphi)\sin\varphi + \cos(k\varphi)\cos\varphi \end{bmatrix}$$

$$= \begin{bmatrix} \cos(k+1)\varphi & \sin(k+1)\varphi \\ -\sin(k+1)\varphi & \cos(k+1)\varphi \end{bmatrix}.$$

Therefore, for $q \in \mathbb{N}$ there exists the equality:

$$(U(\varphi))^q = \begin{bmatrix} \cos(q\varphi) & \sin(q\varphi) \\ -\sin(q\varphi) & \cos(q\varphi) \end{bmatrix}.$$

Now it only remains for us to prove the truth of this equality for all integer q.
Indeed, $(U(\varphi))^0 = I = U(0)$ and for all $\varphi \in \mathbb{R}$ there exists the inverse matrix

$$U(\varphi)^{-1} = U(-\varphi).$$

Thus, for $q = 0, 1, 2, \ldots$ the equality $(U(\varphi))^{-q} = U(-q\varphi)$ is valid.
This means that

$$(U(\varphi))^q = \begin{bmatrix} \cos(q\varphi) & \sin(q\varphi) \\ -\sin(q\varphi) & \cos(q\varphi) \end{bmatrix} \text{ for } q \in \mathbb{Z}.$$

2.46 *Answer:*

$$[A, B^q] = q B^{q-1} \text{ for all } q \in \mathbb{Z}.$$

2.47 *Solution.*

Consecutively perform the algebraic operations:

(1)

$$f(A) = \begin{bmatrix} 1 & 0 \\ 0 & 1 \end{bmatrix} \begin{bmatrix} 1 & 0 \\ 0 & 1 \end{bmatrix} - 3 \begin{bmatrix} 1 & 0 \\ 0 & 1 \end{bmatrix} + 2 \begin{bmatrix} 1 & 0 \\ 0 & 1 \end{bmatrix}$$

$$= \begin{bmatrix} 1 & 0 \\ 0 & 1 \end{bmatrix} - \begin{bmatrix} 3 & 0 \\ 0 & 3 \end{bmatrix} + \begin{bmatrix} 2 & 0 \\ 0 & 2 \end{bmatrix} = \begin{bmatrix} 0 & 0 \\ 0 & 0 \end{bmatrix}.$$

(2)

$$f(A) = \begin{bmatrix} 1 & -2 \\ -3 & 1 \end{bmatrix} \begin{bmatrix} 1 & -2 \\ -3 & 1 \end{bmatrix} - 3 \begin{bmatrix} 1 & -2 \\ -3 & 1 \end{bmatrix} + 2 \begin{bmatrix} 1 & 0 \\ 0 & 1 \end{bmatrix} = \begin{bmatrix} 7 & -4 \\ -6 & 7 \end{bmatrix}$$

$$- \begin{bmatrix} 3 & -6 \\ -9 & 3 \end{bmatrix} + \begin{bmatrix} 2 & 0 \\ 0 & 2 \end{bmatrix} = \begin{bmatrix} 6 & 2 \\ 3 & 6 \end{bmatrix}.$$

2.48 *Solution.*

(1)

$$g(A) = \begin{bmatrix} 1 & 0 & 0 \\ 1 & 1 & 0 \\ 1 & 1 & 1 \end{bmatrix} \begin{bmatrix} 1 & 0 & 0 \\ 1 & 1 & 0 \\ 1 & 1 & 1 \end{bmatrix} \begin{bmatrix} 1 & 0 & 0 \\ 1 & 1 & 0 \\ 1 & 1 & 1 \end{bmatrix} + \begin{bmatrix} 1 & 0 & 0 \\ 1 & 1 & 0 \\ 1 & 1 & 1 \end{bmatrix} - 3 \begin{bmatrix} 1 & 0 & 0 \\ 0 & 1 & 0 \\ 0 & 0 & 1 \end{bmatrix}$$

$$= \begin{bmatrix} 1 & 0 & 0 \\ 3 & 1 & 0 \\ 6 & 3 & 1 \end{bmatrix} + \begin{bmatrix} 1 & 0 & 0 \\ 1 & 1 & 0 \\ 1 & 1 & 1 \end{bmatrix} - \begin{bmatrix} 3 & 0 & 0 \\ 0 & 3 & 0 \\ 0 & 0 & 3 \end{bmatrix} = \begin{bmatrix} -1 & 0 & 0 \\ 4 & -1 & 0 \\ 7 & 4 & -1 \end{bmatrix};$$

(2)

$$g(A) = \begin{bmatrix} -2 & 1 & 0 \\ 3 & -1 & 1 \\ 2 & 1 & 0 \end{bmatrix} \begin{bmatrix} -2 & 1 & 0 \\ 3 & -1 & 1 \\ 2 & 1 & 0 \end{bmatrix} \begin{bmatrix} -2 & 1 & 0 \\ 3 & -1 & 1 \\ 2 & 1 & 0 \end{bmatrix}$$

$$+ \begin{bmatrix} -2 & 1 & 0 \\ 3 & -1 & 1 \\ 2 & 1 & 0 \end{bmatrix} - 3 \begin{bmatrix} 1 & 0 & 0 \\ 0 & 1 & 0 \\ 0 & 0 & 1 \end{bmatrix}$$

$$= \begin{bmatrix} -21 & 11 & -3 \\ 27 & -13 & 5 \\ 7 & -1 & 1 \end{bmatrix} + \begin{bmatrix} -2 & 1 & 0 \\ 3 & -1 & 1 \\ 2 & 1 & 0 \end{bmatrix} - \begin{bmatrix} 3 & 0 & 0 \\ 0 & 3 & 0 \\ 0 & 0 & 3 \end{bmatrix} = \begin{bmatrix} -26 & 12 & -3 \\ 30 & -17 & 6 \\ 9 & 0 & -2 \end{bmatrix}.$$

2.49 *Solution.*

(1) The numerator of the fraction is equal to

$$\begin{bmatrix} 1 & 0 & 0 \\ 0 & -41 & 0 \\ 0 & 0 & -6 \end{bmatrix}.$$

In turn, the denominator forms the matrix

$$\begin{bmatrix} -4 & 0 & 0 \\ 0 & -46 & 0 \\ 0 & 0 & -6 \end{bmatrix},$$

and the matrix that is inverse of it is equal to

$$\begin{bmatrix} -1/4 & 0 & 0 \\ 0 & -1/46 & 0 \\ 0 & 0 & -1/6 \end{bmatrix}.$$

Having performed the multiplication operation, we obtain

$$g(A) = \begin{bmatrix} 1 & 0 & 0 \\ 0 & -41 & 0 \\ 0 & 0 & -6 \end{bmatrix} \begin{bmatrix} -1/4 & 0 & 0 \\ 0 & -1/46 & 0 \\ 0 & 0 & -1/6 \end{bmatrix} = 1/92 \begin{bmatrix} -23 & 0 & 0 \\ 0 & 82 & 0 \\ 0 & 0 & 92 \end{bmatrix}.$$

(2) The numerator of the fraction is equal to

$$\begin{bmatrix} -6 & 2 & 0 \\ 0 & -6 & 0 \\ 0 & 2 & -6 \end{bmatrix}.$$

The denominator of this fraction is equal to

$$\begin{bmatrix} -6 & -2 & 0 \\ 0 & -6 & 0 \\ 0 & -2 & -6 \end{bmatrix}.$$

Let us find the matrix inverse of the denominator:

$$\begin{bmatrix} -1/6 & 1/18 & 0 \\ 0 & -1/6 & 0 \\ 0 & 1/18 & -1/6 \end{bmatrix}.$$

As a result we obtain

$$g(A) = \begin{bmatrix} -6 & 2 & 0 \\ 0 & -6 & 0 \\ 0 & 2 & -6 \end{bmatrix} \begin{bmatrix} -1/6 & 1/18 & 0 \\ 0 & -1/6 & 0 \\ 0 & 1/18 & -1/6 \end{bmatrix} = \frac{1}{3} \begin{bmatrix} 3 & -2 & 0 \\ 0 & 3 & 0 \\ 0 & -2 & 3 \end{bmatrix}.$$

2.50 *Proof.*

Let us consider the sequence of integer non-negative powers of the matrix $A = \begin{bmatrix} 0 & -1 \\ 1 & 0 \end{bmatrix}$:

$$A^0 = \begin{bmatrix} 1 & 0 \\ 0 & 1 \end{bmatrix}, A^1 = \begin{bmatrix} 0 & -1 \\ 1 & 0 \end{bmatrix}, A^2 = \begin{bmatrix} -1 & 0 \\ 0 & -1 \end{bmatrix},$$

$$A^3 = \begin{bmatrix} 0 & 1 \\ -1 & 0 \end{bmatrix}, A^4 = \begin{bmatrix} 1 & 0 \\ 0 & 1 \end{bmatrix} \text{ and so on.}$$

Therefore, the elements of the matrix exp(A) are defined by the sums

$$(e^A)_{11} = 1 + 0/1! - 1/2! + 0/3! + 1/4! + \cdots = \sum_{k=0}^{\infty} \frac{(-1)^k}{(2k)!} = \cos 1,$$

$$(e^A)_{12} = 0 - 1/1! - 0/2! + 1/3! + 0/4! + \cdots = \sum_{k=0}^{\infty} \frac{(-1)^{k+1}}{(2k+1)!} = -\sin 1,$$

$$(e^A)_{21} = 0 + 1/1! + 0/2! - 1/3! + 0/4! + \cdots = \sum_{k=0}^{\infty} \frac{(-1)^k}{(2k+1)!} = \sin 1,$$

$$(e^A)_{22} = 1 + 0/1! - 1/2! + 0/3! - 1/4! + \cdots = \sum_{k=0}^{\infty} \frac{(-1)^k}{(2k)!} = \cos 1.$$

Therefore, there exists the equality

$$\exp(A) = \begin{bmatrix} \cos 1 & -\sin 1 \\ \sin 1 & \cos 1 \end{bmatrix}.$$

2.51 *Solution.*

(1) Compute the lower powers of the matrix A:

$$A^0 = \begin{bmatrix} 1 & 0 & 0 \\ 0 & 1 & 0 \\ 0 & 0 & 1 \end{bmatrix}, \quad A^1 = \begin{bmatrix} 1 & 0 & 0 \\ 0 & 1 & 0 \\ 0 & 0 & 0 \end{bmatrix}, \quad A^2 = \begin{bmatrix} 1 & 0 & 0 \\ 0 & 1 & 0 \\ 0 & 0 & 0 \end{bmatrix}.$$

It is clear that $\forall n \geqslant 1 \ (A^n = A)$. The elements of the matrix exp(A) are equal to

$$(e^A)_{11} = (e^A)_{22} = 1 + \frac{1}{1!} + \frac{1}{2!} + \frac{1}{3!} + \cdots = e, \quad (e^A)_{33} = 1,$$

and the remaining elements take zero values.
Therefore,

$$\exp(A) = \begin{bmatrix} e & 0 & 0 \\ 0 & e & 0 \\ 0 & 0 & 1 \end{bmatrix}.$$

(2) The lower powers of the matrix are equal to

$$A^0 = \begin{bmatrix} 1 & 0 & 0 \\ 0 & 1 & 0 \\ 0 & 0 & 1 \end{bmatrix}, \quad A^1 = \begin{bmatrix} 0 & 1 & 0 \\ 0 & 0 & 1 \\ 0 & 0 & 0 \end{bmatrix}, \quad A^2 = \begin{bmatrix} 0 & 0 & 1 \\ 0 & 0 & 0 \\ 0 & 0 & 0 \end{bmatrix}.$$

As is easy to see, $A^3 = O$, and all the higher natural powers of this matrix are equal to zero.

Finally we obtain

$$\exp(A) = \begin{bmatrix} 1 & 1 & 1/2 \\ 0 & 1 & 1 \\ 0 & 0 & 1 \end{bmatrix}.$$

2.52 *Solution.*

(1) By the mathematical induction method, it is easy to prove that $A^n = \begin{bmatrix} 1 & 0 & n \\ 0 & 1 & 0 \\ 0 & 0 & 1 \end{bmatrix}$

for all integer non-negative n.

According to the formula (2.49), we have $e^A = \begin{bmatrix} \sum_{k=0}^{\infty} \dfrac{1}{k!} & 0 & \sum_{k=1}^{\infty} \dfrac{k}{k!} \\ 0 & \sum_{k=0}^{\infty} \dfrac{1}{k!} & 0 \\ 0 & 0 & \sum_{k=0}^{\infty} \dfrac{1}{k!} \end{bmatrix}.$

The sums defining the diagonal elements converge to Euler's number e. The sum $(e^A)_{13} = \sum_{k=1}^{\infty} \dfrac{k}{k!} = \sum_{k=1}^{\infty} \dfrac{1}{(k-1)!} = \sum_{k=0}^{\infty} \dfrac{1}{k!}$ is also equal to e.

Thus, write the answer:

$$e^A = \begin{bmatrix} e & 0 & e \\ 0 & e & 0 \\ 0 & 0 & e \end{bmatrix}.$$

(2) Having computed the arbitrary natural power of the matrix A, we obtain $A^n = \begin{bmatrix} 1 & 0 & n \\ n & 1 & n(n+1)/2 \\ 0 & 0 & 1 \end{bmatrix}.$

Calculation of the diagonal elements of the exponential and the element $(e^A)_{13}$ is performed similarly to item (1) of this problem.

The element positioned at the intersection of the second row and the third column is defined by the sum $(e^A)_{23} = \sum_{n=1}^{\infty} \dfrac{n(n+1)}{2n!}$. Transform this sum to the form

$$\sum_{n=1}^{\infty} \frac{(n+1)}{2(n-1)!} = \sum_{n=1}^{\infty} \frac{(n-1)+2}{2(n-1)!} = \frac{1}{2}\sum_{n=2}^{\infty} \frac{1}{(n-2)!} + \sum_{n=1}^{\infty} \frac{2}{2(n-1)!}.$$

Therefore, $(e^A)_{23} = \dfrac{3}{2}e.$

As a result we obtain

$$e^A = \begin{bmatrix} e & 0 & e \\ e & e & \dfrac{3}{2}e \\ 0 & 0 & e \end{bmatrix}.$$

2.53 *Solution.*

(1) According to the formula (2.53), we have

$$\ln A = (A - I) - \frac{1}{2}(A - I)^2 + \cdots,$$

or

$$\ln \begin{bmatrix} 1 & 0 & 1 \\ 0 & 1 & 0 \\ 0 & 0 & 1 \end{bmatrix} = \begin{bmatrix} 0 & 0 & 1 \\ 0 & 0 & 0 \\ 0 & 0 & 0 \end{bmatrix} - \frac{1}{2}\begin{bmatrix} 0 & 0 & 1 \\ 0 & 0 & 0 \\ 0 & 0 & 0 \end{bmatrix}^2 + \cdots = \begin{bmatrix} 0 & 0 & 1 \\ 0 & 0 & 0 \\ 0 & 0 & 0 \end{bmatrix}.$$

(2) After computing the lower powers of the matrix $(A-I)$ we can write the general formula for $(A - I)^n$, where $n \geqslant 1$:

$$(A - I)^n = \begin{bmatrix} 0 & 0 & \delta_{n1} \\ \delta_{n1} & 0 & \delta_{n1} + \delta_{n2} \\ 0 & 0 & 0 \end{bmatrix},$$

where δ_{ij} is the Kronecker symbol.

This implies that

$$\ln \begin{bmatrix} 1 & 0 & 1 \\ 1 & 1 & 1 \\ 0 & 0 & 1 \end{bmatrix} = \begin{bmatrix} 0 & 0 & 1 \\ 1 & 0 & \dfrac{1}{2} \\ 0 & 0 & 0 \end{bmatrix}.$$

2.54 *Hint.*

Use the formula (2.49) and apply the mathematical induction method.

2.56 *Answer:*

A square matrix of size 3×3 after reducing it to the echelon form with the help of the elementary transformations can take one of the following forms:

$$\begin{bmatrix} 1 & a & b \\ 0 & 1 & c \\ 0 & 0 & 1 \end{bmatrix}, \begin{bmatrix} 1 & a & b \\ 0 & 1 & c \\ 0 & 0 & 0 \end{bmatrix}, \begin{bmatrix} 1 & a & b \\ 0 & 0 & 1 \\ 0 & 0 & 0 \end{bmatrix}, \begin{bmatrix} 1 & a & b \\ 0 & 0 & 0 \\ 0 & 0 & 0 \end{bmatrix}.$$

Here, by a, b and c are denoted the arbitrary real numbers.

2.57 *Answer:*

(1) addition the j-th row to the i-th row for $j > i$;
(2) addition the j-th column to the i-th column for $j < i$.

2.58 *Solution.*

(1) Perform the following elementary transformations: add to the second row the first one, subtract from the third row the doubled first one, then add to the third row the second row.

We will obtain the matrix in the echelon form: $A \to \begin{bmatrix} 1 & 2 & -4 & 3 & -2 \\ 0 & 5 & -10 & 1 & 2 \\ 0 & 0 & 0 & 0 & 12 \end{bmatrix}$, its

rank is equal to $\operatorname{rk} A = 3$.

(2) Subtract from the second row half of the first one, subtract from the fourth row two first rows, swap the second and the third rows, subtract from the fourth row the second one, swap the third and the fourth rows and, finally, subtract from the fourth row half of the third row.

We obtain $A \to \begin{bmatrix} 2 & -4 & 3 & 1 & 0 \\ 0 & 1 & -1 & 3 & 1 \\ 0 & 0 & -1 & -9 & 4 \\ 0 & 0 & 0 & 0 & 0 \end{bmatrix}$, therefore, $\operatorname{rk} A = 3$.

(3) Let us use the elementary transformation method. Subtract from the second row the quadruplicated first row, subtract from the third row the first one, subtract from the fourth row the doubled first row. Then, subtract from the third row the second one, multiplied by 5/9. Finally, subtract from the fourth row the second one, divided by 3.

Then we obtain

$$A = \begin{bmatrix} 1 & 2 & 1 & 3 \\ 0 & -9 & -9 & -18 \\ 0 & 0 & 0 & 0 \\ 0 & 0 & 0 & 0 \end{bmatrix},$$

therefore, rk $A = 2$.

(4) Swap the first and the second rows, add to the third row the triplicated first row, add to the fifth row the doubled first row. Then, add to the third row the second one, multiplied by 11/2. Subtract from the fourth row the second one, multiplied by 5/2. Add to the fifth row the second one, multiplied by 5/2. Finally, subtract from the fifth row the third one, multiplied by 2/3.

After the said transformations we obtain $A \rightarrow \begin{bmatrix} -1 & -4 & -5 \\ 0 & 2 & -4 \\ 0 & 0 & -30 \\ 0 & 0 & 0 \\ 0 & 0 & 0 \end{bmatrix}$, therefore, the rank of this matrix is equal to three.

(5) Perform the following elementary transformations: subtract from the second row the first row, multiplied by 1/3. Then, subtract from the third row the first one, multiplied by 1/3. And finally, subtract from the third row the quadruplicated second row.

After that we obtain $A \rightarrow \begin{bmatrix} 3 & 5 & 7 \\ 0 & 1/3 & 2/3 \\ 0 & 0 & 0 \end{bmatrix}$, the rank of such a matrix is equal to two.

(6) The matrix is presented in the echelon form, and, as is easy to see, rk $A = 3$.

(7) Add to the second row the doubled first row, subtract from the third row the doubled first row, add to the third row the second row.

Then we obtain

$$A \to \begin{bmatrix} 1 & -1 & 2 & 4 & 3 \\ 0 & -1 & 9 & 10 & 12 \\ 0 & 0 & 9 & 9 & 8 \end{bmatrix},$$

and the rank of the matrix A is equal to three.

(8) Subtract from the second row the doubled first row, subtract from the third row the first one, multiplied by five. Then, subtract from the fourth row the first one, multiplied by seven. Subtract from the third row the doubled second one and swap the third and the fourth rows. Finally, subtract from the third row the doubled second row.

After the said transformations we obtain

$$A \to \begin{bmatrix} 1 & 3 & 5 & -1 \\ 0 & -7 & 13 & 6 \\ 0 & 0 & 0 & -4 \\ 0 & 0 & 0 & 0 \end{bmatrix},$$

therefore, rk $A = 3$.

2.59 *Solution.*

Consider the minor of the first order $M_1^1 = \beta - \gamma$. If $\beta \neq \gamma$, then the rank of the matrix Λ is no less than one.

Then, consider the minor of the second order $M_{1,2}^{1,2} = \begin{vmatrix} \beta - \gamma & 0 \\ \gamma - \alpha & -\gamma \end{vmatrix}$. If the condition $\gamma \neq 0$ is fulfilled, then rk $\Lambda \geqslant 2$.

Finally, compute the bordering minors of the third order:

$$M_{1,2,3}^{1,2,3} = \begin{vmatrix} \beta - \gamma & 0 & \gamma \\ \gamma - \alpha & -\gamma & 0 \\ \alpha - \beta & \beta & -\alpha \end{vmatrix} = 0,$$

$$M_{1,2,4}^{1,2,3} = \begin{vmatrix} \beta - \gamma & 0 & -\beta \\ \gamma - \alpha & -\gamma & \alpha \\ \alpha - \beta & \beta & 0 \end{vmatrix} = 0.$$

Therefore, the maximum value that the rank of the matrix Λ can take is equal to two.

Note that with the help of the Kronecker symbol (see page 4) the formula for the rank of this matrix can be written in the form rk $\Lambda = 2(1 - \delta_{\alpha 0}\delta_{\beta 0}\delta_{\gamma 0})$.

2.60 *Solution.*

(a) Use the bordering minor method (see page 64):

$$M_1^1 = 1 \neq 0, \quad M_{1,2}^{1,2} = \begin{vmatrix} 1 & \lambda \\ 1 & 1 \end{vmatrix} = 1 - \lambda.$$

Then, consider two cases.

(1) If $\lambda = 1$, then the matrix is equal to $\begin{bmatrix} 1 & 1 & 0 \\ 1 & 1 & 0 \\ 0 & 0 & 1 \end{bmatrix} \xrightarrow[(2)\leftrightarrow(3)]{(2)-(1)} \begin{bmatrix} 1 & 1 & 0 \\ 0 & 0 & 1 \\ 0 & 0 & 0 \end{bmatrix}$. It is clear

that its rank is equal to two.

(2) If $\lambda \neq 1$, then we compute the bordering minor of the third order (it is the only one):

$$M_{1,2,3}^{1,2,3} = \begin{vmatrix} 1 & \lambda & 0 \\ 1 & 1 & 0 \\ 0 & 0 & 1 \end{vmatrix} = 1 - \lambda \neq 0.$$

Therefore, the rank of the matrix is equal to

$$\begin{cases} 2, & \text{if } \lambda = 1, \\ 3, & \text{if } \lambda \neq 1. \end{cases}$$

(b) The lower minor $M_1^1 = 1 - \lambda$.

Then, consider two cases.

(1) If $\lambda = 1$, then the matrix is equal to $\begin{bmatrix} 0 & 0 & 0 \\ 0 & 1 & 0 \\ 0 & 0 & 2 \end{bmatrix}$, and its rank, as is easy to see,

is equal to two.

(2) If $\lambda \neq 1$, then we continue to compute the bordering minors:

$$M_{1,2}^{1,2} = \begin{vmatrix} 1 - \lambda & 0 \\ 0 & 2 - \lambda \end{vmatrix} = (1 - \lambda)(2 - \lambda).$$

If $\lambda = 2$, we obtain the matrix $\begin{bmatrix} -1 & 0 & 0 \\ 0 & 0 & 0 \\ 0 & 0 & 1 \end{bmatrix}$, its rank is equal to two.

If $\lambda \neq 2$, then we need to compute the bordering minor $M^{1,2,3}_{1,2,3}$: $M^{1,2,3}_{1,2,3} = (1-\lambda)(2-\lambda)(3-\lambda)$.

In the case when $\lambda = 3$, the matrix is equal to $\begin{bmatrix} -2 & 0 & 0 \\ 0 & -1 & 0 \\ 0 & 0 & 0 \end{bmatrix}$, its rank is equal to two. Otherwise, the rank is equal to three.

As a result, we form the answer:

$$\begin{cases} 2, & \text{if } \lambda \in \{1, 2, 3\}, \\ 3, & \text{otherwise.} \end{cases}$$

(c) Since $M^1_1 = 1 \neq 0$, then the rank of the matrix is no less than one.
Then, consider the minors of the second order:

$$M^{1,2}_{1,2} = \begin{vmatrix} 1 & 2 \\ -1 & 2-\lambda \end{vmatrix} = 4 - \lambda.$$

This minor is not equal to zero at $\lambda \neq 4$. For this case, consider the bordering minor of the third order:

$$M^{1,2,3}_{1,2,3} = \begin{vmatrix} 1 & 2 & -3 \\ -1 & 2-\lambda & 10 \\ -1 & 0 & 3-\lambda \end{vmatrix} = \lambda^2 - 4\lambda - 14.$$

The determinant is equal to zero for $\lambda = 2 \pm 3\sqrt{2}$. For such values of λ the rank is equal to two; for other values the rank is equal to three.

Now consider the case $\lambda = 4$.
Then the non-zero minor of the second order:

$$M^{1,2}_{2,3} = \begin{vmatrix} 2 & -3 \\ 2-\lambda & 10 \end{vmatrix} = 14 \neq 0.$$

Then, the rank is no less than two. The third order minor is the only one, but it is only equal to zero for $\lambda = 2 \pm 3\sqrt{2}$, as is shown above. Then, in this case the rank is equal to three.

So, for $\lambda = 2 \pm 3\sqrt{2}$ the rank of the matrix is equal to two, for other λ the rank of the matrix is equal to three.

(d) There exists $M_2^1 = 1 \neq 0$, therefore, the rank of the matrix is no less than one. Then, consider the minors of the second order:

$$M_{1,2}^{1,2} = \begin{vmatrix} -\lambda & 1 \\ 0 & 1-\lambda \end{vmatrix} = \lambda(\lambda - 1).$$

This minor is not equal to zero at $\lambda \neq 0$ and for $\lambda \neq 1$. For this case, such for the minor of the third order:

$$M_{1,2,3}^{1,2,3} = \begin{vmatrix} -\lambda & 1 & 1 \\ 0 & 1-\lambda & 1 \\ 0 & 0 & 2-\lambda \end{vmatrix} = \lambda(\lambda - 1)(2 - \lambda).$$

$\lambda(\lambda - 1)(2 - \lambda) \neq 0$ for $\lambda \neq 2$ (since $\lambda \neq 1$ and $\lambda \neq 0$). In this case, the rank is equal to three.

For $\lambda = 2$ the rank is equal to two.

Assume that now $\lambda = 0$:

$$\begin{bmatrix} 0 & 1 & 1 \\ 0 & 1 & 1 \\ 0 & 0 & 2 \end{bmatrix},$$

$$M_{2,3}^{2,3} = \begin{vmatrix} 1 & 1 \\ 0 & 2 \end{vmatrix} = 2 \neq 0.$$

Then, the rank is more than two.

$$\begin{vmatrix} 0 & 1 & 1 \\ 0 & 1 & 1 \\ 0 & 0 & 2 \end{vmatrix} = 0.$$

Then, in this case the rank is equal to two.

Now consider $\lambda = 1$:

$$\begin{bmatrix} -1 & 1 & 1 \\ 0 & 0 & 1 \\ 0 & 0 & 1 \end{bmatrix}.$$

The rank is equal to two, because $M_{2,3}^{1,2} = \begin{vmatrix} 1 & 1 \\ 0 & 1 \end{vmatrix} = 1 \neq 0$, and the determinant of the matrix is equal to zero.

Therefore, the rank is equal to two for $\lambda \in \{0, 1, 2\}$, and is equal to three in other cases.

(e) The minor of the first order is $M_1^1 = 1 \neq 0$, therefore, the rank of the matrix takes the value no less than one.

Let us find the bordering minor of the second order:

$$M_{1,2}^{1,2} = \begin{vmatrix} 1 & -6 \\ -1 & 2 - \lambda \end{vmatrix} = -4 - \lambda.$$

This minor is not equal to zero if $\lambda \neq -4$. Let us find the minor of the third order for such values of λ:

$$M_{1,2,3}^{1,2,3} = \begin{vmatrix} 1 & -6 & -5 \\ -1 & 2 - \lambda & 5 \\ -1 & 6 & 1 - \lambda \end{vmatrix} = \lambda^2 + 8\lambda + 16.$$

This expression may only be equal to zero for $\lambda = -4$. Therefore, in this case the rank is equal to three.

Now consider the case when $\lambda = -4$.

$$\begin{bmatrix} 1 & -6 & -5 \\ -1 & 6 & 5 \\ -1 & 6 & 5 \end{bmatrix}.$$

The rank of the obtained matrix is equal to one.

As a result, the rank is equal to one for $\lambda = -4$, and is equal to three for $\lambda \neq -4$.

(f) There exists $M_1^2 = 1 \neq 0$, therefore, the rank is no less than one.
Let us find the minors of the second order.

$$M_{1,2}^{1,2} = \begin{vmatrix} 1 - \lambda & 2 \\ 1 & 2 - \lambda \end{vmatrix} = \lambda(\lambda - 3).$$

It is not equal to zero if $\lambda \neq 0$ and $\lambda \neq 3$. Find for such values of the parameter λ the bordering minors of the third order.

$$M_{1,2,3}^{1,2,3} = \begin{vmatrix} 1 - \lambda & 2 & 0 \\ 1 & 2 - \lambda & 0 \\ 0 & 0 & 3 - \lambda \end{vmatrix} = \lambda(\lambda - 3)(3 - \lambda).$$

The third order minor is non-zero for all considered in this case λ. Let us find the minor of the fourth order (it is the only one):

$$\begin{vmatrix} 1 - \lambda & 2 & 0 & 0 \\ 1 & 2 - \lambda & 0 & 0 \\ 0 & 0 & 3 - \lambda & 0 \\ 0 & 0 & 0 & 4 - \lambda \end{vmatrix} = \lambda(\lambda - 3)(3 - \lambda)(4 - \lambda).$$

It is equal to zero only for $\lambda = 4$. For this case, the rank is equal to three. For $\lambda \neq 4$ the rank is equal to four.
Then, consider the value $\lambda = 0$:

$$M_{3,4}^{3,4} = \begin{vmatrix} 3 & 0 \\ 0 & 4 \end{vmatrix} = 12 \neq 0,$$

$$M_{2,3,4}^{2,3,4} = \begin{vmatrix} 3 & 0 & 0 \\ 0 & 3 & 0 \\ 0 & 0 & 4 \end{vmatrix} = 24 \neq 0.$$

For $\lambda = 0$ the determinant of the initial matrix is equal to zero. Then, the rank is equal to three.

For $\lambda = 3$:

$$\begin{bmatrix} -2 & 2 & 0 & 0 \\ 1 & -1 & 0 & 0 \\ 0 & 0 & 0 & 0 \\ 0 & 0 & 0 & 1 \end{bmatrix}.$$

There is no minor of the third order that is not equal to zero. Therefore, the rank is equal to two.

Finally we obtain the rank is equal to two for $\lambda = 3$, is equal to three for $\lambda \in \{0, 4\}$, is equal to four in other cases.

(g) There exists $M_1^2 \neq 0$, therefore, the rank is no less than one.

Let us find the minors of the second order:

$$M_{1,2}^{1,2} = \begin{vmatrix} 1 - \lambda & 2 \\ 1 & 1 + \lambda \end{vmatrix} = -(\lambda^2 + 1) \neq 0.$$

This minor is always other than zero. Consider the bordering minors of the third order:

$$\begin{vmatrix} 1 - \lambda & 2 & 0 \\ 1 & 1 + \lambda & 0 \\ 0 & 0 & 2 - \lambda \end{vmatrix} = (2 - \lambda)(-\lambda^2 - 1) = 0 \text{ only for } \lambda = 2, \text{ but}$$

$$\begin{vmatrix} 1 - \lambda & 2 & 0 \\ 1 & 1 + \lambda & 0 \\ 0 & 0 & 2 + \lambda \end{vmatrix} = (2 + \lambda)(-\lambda^2 - 1) = 0 \text{ only for } \lambda = -2.$$

This is because the minors of the third order are not simultaneously equal to zero.

Calculate the determinant of the initial matrix (in other words, find the minor of the fourth order):

$$\begin{vmatrix} 1 - \lambda & 2 & 0 & 0 \\ 1 & 1 + \lambda & 0 & 0 \\ 0 & 0 & 2 - \lambda & 0 \\ 0 & 0 & 0 & 2 + \lambda \end{vmatrix} = (2 + \lambda)(2 - \lambda)(-\lambda^2 - 1).$$

For $\lambda = \pm 2$:

$(2 + \lambda)(2 - \lambda)(-\lambda^2 - 1) = 0$, therefore, the rank is equal to three.

For other λ the rank is equal to four.

So, the rank is equal to three for $\lambda = \pm 2$, and equal to four for $\lambda \neq \pm 2$.

(h) There exists $M_1^2 \neq 0$, therefore, the rank is no less than one.

Let us find the minors of the second order:

$$M_{1,2}^{1,2} = \begin{vmatrix} 1 + \lambda & 2 \\ 1 & 1 \end{vmatrix} = \lambda - 1 \neq 0 \text{ for } \lambda \neq 1.$$

For this case, find the minors of the third order:

$$M_{1,2,4}^{1,2,3} = \begin{vmatrix} 1 + \lambda & 2 & 4 \\ 1 & 1 & 0 \\ 0 & 0 & -1 \end{vmatrix} = -(\lambda - 1) \neq 0.$$

Then, the rank is greater than or equal to three. Calculate the determinant of the initial matrix:

$$\begin{vmatrix} 1 + \lambda & 2 & 0 & 4 \\ 1 & 1 & 0 & 0 \\ 0 & 0 & 2 + \lambda & -1 \\ 1 & 0 & 2 & 1 \end{vmatrix} = \lambda^2 - \lambda - 12.$$

The determinant is equal to zero for $\lambda = 4$ or -3. Hence, for such values of the parameter λ the rank is equal to three. In other cases the rank is equal to four.

Now consider the case when $\lambda = 1$. Let us find the minor of the second order:

$$M_{2,3}^{2,3} = \begin{vmatrix} 1 & 0 \\ 0 & 3 \end{vmatrix} = 3 \neq 0.$$

Calculate the minor of the third order:

$$M_{1,2,4}^{1,2,3} = \begin{vmatrix} 2 & 0 & 4 \\ 1 & 0 & 0 \\ 0 & 3 & -1 \end{vmatrix} = 12 \neq 0.$$

Then, the rank of the matrix is no less than three. Now find the determinant of the initial matrix.

$$\begin{vmatrix} 2 & 2 & 0 & 4 \\ 1 & 1 & 0 & 0 \\ 0 & 0 & 3 & -1 \\ 1 & 0 & 2 & 1 \end{vmatrix} = -12 \neq 0.$$

This implies that in this case the rank is equal to four.

So, the rank is equal to three for $\lambda \in \{-3, 4\}$, and is equal to four in other cases.

(i) The size of the matrix is equal to $(n + 1) \times (n + 1)$. As is easy to see, for $\lambda \in \{0, 1, \ldots, n\}$ the determinant of the matrix takes the value equal to zero. We should also note that in this case there exists a minor of the order n, other than zero. For example, for $\lambda = 0$:

$$\begin{bmatrix} 1 - \lambda & 1 & 1 \ldots & 1 & 1 \\ 0 & 2 - \lambda & 1 \ldots & 1 & 1 \\ \multicolumn{5}{c}{\cdots\cdots\cdots\cdots\cdots\cdots\cdots\cdots\cdots\cdots} \\ 0 & 0 & 0 \ldots & (n-1) - \lambda & 1 \\ 0 & 0 & 0 \ldots & 0 & n - \lambda \end{bmatrix}.$$

As is known, determinant of the upper triangular matrix is equal to the product of the diagonal terms of the matrix. Since the condition $\lambda = 0$ is fulfilled, then this product is not equal to zero. Therefore, the rank of the initial matrix in this case is equal to n. Yet, if the condition $\lambda \in \{0, 1, \ldots, n\}$ is not fulfilled, then the rank of the initial matrix takes the value equal to $n + 1$.

We obtain the final answer: the rank of the matrix is equal to n for $\lambda \in \{0, 1, \ldots, n\}$, and is equal to $n + 1$ for other values of λ.

Chapter 3
Systems of Linear Equations

The system of m linear equations with n unknowns is written as

$$\begin{cases} a_{11}x_1 + a_{12}x_2 + \cdots + a_{1n}x_n = b_1, \\ a_{21}x_1 + a_{22}x_2 + \cdots + a_{2n}x_n = b_2, \\ \cdots\cdots\cdots\cdots\cdots\cdots\cdots\cdots\cdots\cdots\cdots \\ a_{m1}x_1 + a_{m2}x_2 + \cdots + a_{mn}x_n = b_m. \end{cases} \tag{3.1}$$

Here, by $x_1, x_2, \ldots, x_n$ are denoted unknown numbers, a_{ij} and b_i are prescribed numbers, also referred to as **coefficients** of the system of equations (3.1). Variables b_i are called **known terms** or right-hand sides of equations.

Solution of the system of equations is such a collection of n numbers, which when $x_1, x_2, \ldots, x_n$ are substituted into the system in place of the unknown, turns all the equations into identities. The solution is written as a vector $[x_1, x_2, \ldots, x_n]^T$.

If $b_i = 0$ for all $i = 1, 2, \ldots, m$, then the system of equations is called **homogeneous**. If at least one of the known terms is $b_i \neq 0$, then the system of equations is called **non-homogeneous**.

In case when $m = n$, the system of equations is called **square**, and when $m \neq n$, the system is called **rectangular**.

A system of linear equations is called **consistent**, if it has at least one solution, and **inconsistent**, if there are no solutions [65].

If a consistent system has the only solution, it is called **determined**. If the consistent system has at least two different solutions, it is called **undetermined**.

© Springer Nature Switzerland AG 2021
S. Kurgalin, S. Borzunov, *Algebra and Geometry with Python*,
https://doi.org/10.1007/978-3-030-61541-3_3

A matrix

$$
A = \begin{bmatrix} a_{11} & a_{12} & \ldots & a_{1n} \\ a_{21} & a_{22} & \ldots & a_{2n} \\ \multicolumn{4}{c}{\ldots\ldots\ldots\ldots} \\ a_{m1} & a_{m2} & \ldots & a_{mn} \end{bmatrix}, \tag{3.2}
$$

consisting of coefficients of the unknown a_{ij} is called a **system matrix**.

A matrix

$$
B = \left[\begin{array}{cccc|c} a_{11} & a_{12} & \ldots & a_{1n} & b_1 \\ a_{21} & a_{22} & \ldots & a_{2n} & b_2 \\ \multicolumn{4}{c|}{\ldots\ldots\ldots\ldots} & \ldots \\ a_{m1} & a_{m2} & \ldots & a_{mn} & b_m \end{array} \right], \tag{3.3}
$$

into which a column of the constant terms b_j is added is called an **augmented system matrix**.

If the unknown and constant terms are written in the form of a column (of matrices of sizes $n \times 1$ and $m \times 1$, respectively):

$$
X = \begin{bmatrix} x_1 \\ x_2 \\ \vdots \\ x_n \end{bmatrix}, \quad B = \begin{bmatrix} b_1 \\ b_2 \\ \vdots \\ b_m \end{bmatrix}, \tag{3.4}
$$

then the system of equations (3.1) can be presented in an abridged **matrix form**:

$$
A \cdot X = B. \tag{3.5}
$$

3.1 Cramer's Rule

Consider a square system of n equations with n unknowns:

$$
\begin{cases} a_{11}x_1 + a_{12}x_2 + \cdots + a_{1n}x_n = b_1, \\ a_{21}x_1 + a_{22}x_2 + \cdots + a_{2n}x_n = b_2, \\ \multicolumn{1}{c}{\ldots\ldots\ldots\ldots\ldots\ldots\ldots\ldots\ldots\ldots} \\ a_{n1}x_1 + a_{n2}x_2 + \cdots + a_{nn}x_n = b_n. \end{cases} \tag{3.6}
$$

Theorem 3.1 (Cramer's[1] Rule) *If the determinant of the matrix A of the system (3.6) is other than zero, then the system (3.6) is **determined**, i.e. it has the **unique solution**. This solution can be computed by the formula*

$$x_i = \frac{\Delta_i}{\Delta}, \quad i = 1, 2, \ldots, n. \tag{3.7}$$

Here, Δ_i is the determinant of the matrix obtained from the initial matrix A by replacement of the i-th column with a column of constant terms:

$$\Delta_i = \begin{vmatrix} a_{11} & \ldots & a_{1\,i-1} & b_1 & a_{1\,i+1} & \ldots & a_{1n} \\ a_{21} & \ldots & a_{2\,i-1} & b_2 & a_{2\,i+1} & \ldots & a_{2n} \\ \vdots & \ddots & \vdots & \vdots & \vdots & \ddots & \vdots \\ a_{n1} & \ldots & a_{n\,i-1} & b_n & a_{n\,i+1} & \ldots & a_{nn} \end{vmatrix}. \tag{3.8}$$

If $\Delta = 0$ and at least one of the determinants Δ_i is other than zero, then the system (3.6) has no solutions (i.e. it is inconsistent).

If $\Delta = 0$, but also $\Delta_i = 0$, then the system (3.6) has infinitely many solutions (i.e. it is consistent but undetermined) or is inconsistent.

Example 3.1 Solve the system:

$$\begin{cases} 5x - 6y = -8, \\ 5x + 6y = 28. \end{cases} \tag{3.9}$$

For the given system we have

$$\Delta = \begin{vmatrix} 5 & -6 \\ 5 & 6 \end{vmatrix} = 60, \quad \Delta_x = \begin{vmatrix} -8 & -6 \\ 28 & 6 \end{vmatrix} = 120, \quad \Delta_y = \begin{vmatrix} 5 & -8 \\ 5 & 28 \end{vmatrix} = 180. \tag{3.10}$$

Thus, according to Cramer's rule

$$x = \frac{\Delta_x}{\Delta} = \frac{120}{60} = 2, \quad y = \frac{\Delta_y}{\Delta} = \frac{180}{60} = 3. \tag{3.11}$$

□

---

[1]Gabriel Cramer (1704–1752) was a Genevan mathematician.

Example 3.2 Solve the system of linear equations:

$$\begin{cases} 2x - y + 3z = 8, \\ x + y - 2z = 5, \\ 3x - 2y + z = 7. \end{cases} \qquad (3.12)$$

Solution Compute the determinants required to apply Cramer's rule:

$$\Delta = \begin{vmatrix} 2 & -1 & 3 \\ 1 & 1 & -2 \\ 3 & -2 & 1 \end{vmatrix} = -14, \quad \Delta_x = \begin{vmatrix} 8 & -1 & 3 \\ 5 & 1 & -2 \\ 7 & -2 & 1 \end{vmatrix} = -56, \qquad (3.13)$$

$$\Delta_y = \begin{vmatrix} 2 & 8 & 3 \\ 1 & 5 & -2 \\ 3 & 7 & 1 \end{vmatrix} = -42, \quad \Delta_z = \begin{vmatrix} 2 & -1 & 8 \\ 1 & 1 & 5 \\ 3 & -2 & 7 \end{vmatrix} = -14. \qquad (3.14)$$

Therefore

$$x = \frac{\Delta_x}{\Delta} = 4, \quad y = \frac{\Delta_y}{\Delta} = 3, \quad z = \frac{\Delta_z}{\Delta} = 1. \qquad (3.15)$$

$\square$

3.2 Inverse Matrix Method

Consider the system of equations (3.6). Write this system in the form $A \cdot X = B$ in accordance with (3.5).

Let the matrix A have the inverse one A^{-1}. Multiply both sides of the equality $A \cdot X = B$ by A^{-1} on the left:

$$A^{-1} \cdot A \cdot X = A^{-1} \cdot B. \qquad (3.16)$$

Transform the obtained matrix equation. Since the identities $A^{-1} \cdot A \equiv I$ and $I \cdot X \equiv X$ are valid, the solution of the system X can be written in the form

$$X = A^{-1} \cdot B. \qquad (3.17)$$

Therefore, if we find the inverse matrix A^{-1}, then the solution of the system can be obtained as the product of the matrices A^{-1} and B.

Example 3.3 Solve the system of equations

$$\begin{cases} -2x + 2y - 3z = -10, \\ 2x - y + 2z = 7, \\ 3x - y + 3z = 10 \end{cases} \tag{3.18}$$

by the inverse matrix method.

Solution The matrix of the analysed system of equations A has the form:

$$\begin{bmatrix} -2 & 2 & -3 \\ 2 & -1 & 2 \\ 3 & -1 & 3 \end{bmatrix}. \tag{3.19}$$

We find the inverse matrix using a cofactor matrix. The determinant of the matrix A is equal to

$$\det A = -2 \begin{vmatrix} -1 & 2 \\ -1 & 3 \end{vmatrix} - 2 \begin{vmatrix} 2 & 2 \\ 3 & 3 \end{vmatrix} - 3 \begin{vmatrix} 2 & -1 \\ 3 & -1 \end{vmatrix}$$

$$= (-2)(-3 + 2) - 2(6 - 6) - 3(-2 + 3) = -1. \tag{3.20}$$

Calculate the cofactors:

$$A_{11} = (-1)^{1+1} \begin{vmatrix} -1 & 2 \\ -1 & 3 \end{vmatrix} = -1, \quad A_{12} = (-1)^{1+2} \begin{vmatrix} 2 & 2 \\ 3 & 3 \end{vmatrix} = 0, \tag{3.21}$$

$$A_{13} = (-1)^{1+3} \begin{vmatrix} 2 & -1 \\ 3 & -1 \end{vmatrix} = 1, \quad A_{21} = (-1)^{2+1} \begin{vmatrix} 2 & -3 \\ -1 & 3 \end{vmatrix} = -3, \tag{3.22}$$

$$A_{22} = (-1)^{2+2} \begin{vmatrix} -2 & -3 \\ 3 & 3 \end{vmatrix} = 3, \quad A_{23} = (-1)^{2+3} \begin{vmatrix} -2 & 2 \\ 3 & -1 \end{vmatrix} = 4, \tag{3.23}$$

$$A_{31} = (-1)^{3+1} \begin{vmatrix} 2 & -3 \\ -1 & 2 \end{vmatrix} = 1, \quad A_{32} = (-1)^{3+2} \begin{vmatrix} -2 & -3 \\ 2 & 2 \end{vmatrix} = -2, \tag{3.24}$$

$$A_{33} = (-1)^{3+3} \begin{vmatrix} -2 & 2 \\ 2 & -1 \end{vmatrix} = -2. \tag{3.25}$$

Write the matrix formed by the cofactors:

$$
\begin{bmatrix}
-1 & 0 & 1 \\
-3 & 3 & 4 \\
1 & -2 & -2
\end{bmatrix}.
\tag{3.26}
$$

As a result, the desired inverse matrix will have the form:

$$
A^{-1} = \frac{1}{(-1)}
\begin{bmatrix}
-1 & -3 & 1 \\
0 & 3 & -2 \\
1 & 4 & -2
\end{bmatrix}
=
\begin{bmatrix}
1 & 3 & -1 \\
0 & -3 & 2 \\
-1 & -4 & 2
\end{bmatrix}.
\tag{3.27}
$$

Then, the solution of the system will be found using the matrix multiplication operation:

$$
\begin{bmatrix} x \\ y \\ z \end{bmatrix}
=
\begin{bmatrix}
1 & 3 & -1 \\
0 & -3 & 2 \\
-1 & -4 & 2
\end{bmatrix}
\begin{bmatrix} -10 \\ 7 \\ 10 \end{bmatrix}
=
\begin{bmatrix}
-10 + 21 - 10 \\
0 - 21 + 20 \\
10 - 28 + 20
\end{bmatrix}
=
\begin{bmatrix} 1 \\ -1 \\ 2 \end{bmatrix}.
\tag{3.28}
$$

□

3.3   Gaussian Elimination

The use of the notion of matrix rank allows obtaining the criterion of consistency of the system of linear equations.

Consider the system:

$$
\begin{cases}
a_{11}x_1 + a_{12}x_2 + \cdots + a_{1n}x_n = b_1, \\
a_{21}x_1 + a_{22}x_2 + \cdots + a_{2n}x_n = b_2, \\
\cdots\cdots\cdots\cdots\cdots\cdots\cdots\cdots\cdots\cdots\cdots \\
a_{m1}x_1 + a_{m2}x_2 + \cdots + a_{mn}x_n = b_m.
\end{cases}
\tag{3.29}
$$

The augmented matrix of the system of size $m \times (n + 1)$ has the form:

$$B = \begin{bmatrix} a_{11} & a_{12} & \dots & a_{1n} & | & b_1 \\ a_{21} & a_{22} & \dots & a_{2n} & | & b_2 \\ \dots\dots\dots\dots\dots & | & \dots \\ a_{m1} & a_{m2} & \dots & a_{mn} & | & b_m \end{bmatrix}. \tag{3.30}$$

The consistency criterion for the system of linear algebraic equations is the Kronecker–Capelli[2] theorem (also referred to as the theorem of Rouché[3]–Capelli) [64].

Theorem 3.2 (Kronecker–Capelli Theorem) *A system of linear algebraic equations* (3.29) *is consistent if and only if the rank of the basic matrix A equals to the rank of the augmented one, i.e.* rk $A =$ rk $B = r$.

If $r = n$, then we obtain a square matrix with a non-zero determinant. Its solution exists and it is unique.

If $r < n$ and the system is consistent, then there exists an infinite set of solutions.

If rk $B >$ rk A, then the system is inconsistent.

In what follows, by **zero equations** we will understand the equations of the form $0 \cdot x_1 + 0 \cdot x_2 + \cdots + 0 \cdot x_n = 0$.

Elementary transformations of the system of linear equations are:

1. interchange (swap) of any two equations;
2. multiplying the equation by any non-zero number;
3. adding to the equation another one multiplied by an arbitrary number;
4. dropping the zero equations.

Gaussian[4] method or **Gaussian elimination** consists in transformation of the augmented matrix B with the help of the elementary transformations to the echelon form. Such transformations are aimed at obtaining a system of the form:

$$\begin{cases} b_{11} \cdot x_1 + b_{12} \cdot x_2 + \cdots + b_{1r} \cdot x_r + \cdots + b_{1n} \cdot x_n = p_1, \\ \qquad b_{22} \cdot x_2 + \cdots + b_{2r} \cdot x_r + \cdots + b_{2n} \cdot x_n = p_2, \\ \dots\dots\dots\dots\dots\dots\dots\dots\dots\dots\dots\dots\dots\dots\dots\dots\dots \\ \qquad\qquad\qquad\qquad b_{rr} \cdot x_r + \cdots + b_{rn} \cdot x_n = p_r. \end{cases} \tag{3.31}$$

Note that for bringing the system of equations to the said form we may need to change the numbering of the variables.

---

[2] Alfredo Capelli (1855–1910), Italian mathematician.

[3] Eugène Rouché (1832–1910), French mathematician.

[4] Johann Carl Friedrich Gauß (1777–1855), prominent German mathematician and astronomer.

The system (3.31) may contain less equations than the initial one due to dropping of the zero equations of the form:

$$0 \cdot x_1 + 0 \cdot x_2 + \cdots + 0 \cdot x_n = 0. \tag{3.32}$$

If the transformations result in the equation

$$0 \cdot x_1 + 0 \cdot x_2 + \cdots + 0 \cdot x_n = d \neq 0, \tag{3.33}$$

then the system is inconsistent.

It is obvious that the minor $M_{1,2,\ldots,r}^{1,2,\ldots,r}$ is a basic minor, assuming that $x_{r+1}, \ldots, x_n$ are free unknowns, to which we can assign arbitrary values:

$$x_{r+1} = C_1, \quad \ldots, \quad x_n = C_{n-r}. \tag{3.34}$$

Rearrange these variables to the right side; then the obtained system necessarily has a solution relative to the unknowns $x_1, x_2, \ldots, x_r$.

From the last equation we find x_r, from the last but one we find x_{r-1}, etc.

Note Gaussian method is sometimes referred to as **method of successive elimination of unknowns**.

Example 3.4 Solve the system of equations by Gaussian method:

$$\begin{cases} x_1 - 2x_2 \quad\quad\quad + x_4 = -3, \\ 3x_1 - x_2 - 2x_3 \quad\quad\quad = 1, \\ 2x_1 + x_2 - 2x_3 - x_4 = 4, \\ x_1 + 3x_2 - 2x_3 - 2x_4 = 7. \end{cases} \tag{3.35}$$

Solution Write the augmented matrix:

$$\begin{bmatrix} 1 & -2 & 0 & 1 & | & -3 \\ 3 & -1 & -2 & 0 & | & 1 \\ 2 & 1 & -2 & -1 & | & 4 \\ 1 & 3 & -2 & -2 & | & 7 \end{bmatrix}. \tag{3.36}$$

Bring the obtained matrix to a triangular form. For this, subtract from the second row the triplicated first row, from the third row—the doubled first one, and from the fourth row—the first one:

$$
\begin{bmatrix}
1 & -2 & 0 & 1 & | & -3 \\
0 & 5 & -2 & -3 & | & 10 \\
0 & 5 & -2 & -3 & | & 10 \\
0 & 5 & -2 & -3 & | & 10
\end{bmatrix}. \tag{3.37}
$$

In the next step, subtract from the third and the fourth rows the second one. The first and the second rows remain unchanged:

$$
\begin{bmatrix}
1 & -2 & 0 & 1 & | & -3 \\
0 & 5 & -2 & -3 & | & 10 \\
0 & 0 & 0 & 0 & | & 0 \\
0 & 0 & 0 & 0 & | & 0
\end{bmatrix}. \tag{3.38}
$$

It is clear that the ranks of the basic and the augmented matrices are equal to two. The system is consistent and undetermined. As free variables, take x_1 and x_3. Let $x_1 = C_1$, and $x_3 = C_2$. Then we have

$$
\begin{cases}
-2 \cdot x_2 + x_4 = -3 - C_1, \\
5 \cdot x_2 - 3 \cdot x_4 = 10 + 2 \cdot C_2.
\end{cases} \tag{3.39}
$$

We are solving this system relative to the variables x_2 and x_4:

$$
\begin{cases}
x_2 = -1 + 3 \cdot C_1 - 2 \cdot C_2, \\
x_4 = -5 + 5 \cdot C_1 - 4 \cdot C_2.
\end{cases} \tag{3.40}
$$

We write the final answer in the form:

$$
X = \begin{bmatrix} x_1 \\ x_2 \\ x_3 \\ x_4 \end{bmatrix} = \begin{bmatrix} C_1 \\ -1 + 3C_1 - 2C_2 \\ C_2 \\ -5 + 5C_1 - 4C_2 \end{bmatrix} = \begin{bmatrix} 0 \\ -1 \\ 0 \\ -5 \end{bmatrix} + C_1 \begin{bmatrix} 1 \\ 3 \\ 0 \\ 5 \end{bmatrix} + C_2 \begin{bmatrix} 0 \\ -2 \\ 1 \\ -4 \end{bmatrix}. \tag{3.41}
$$

$\square$

Example 3.5 Solve the matrix equation relative to the unknown matrix Y:

$$\begin{bmatrix} 1 & -1 \\ -1 & 2 \end{bmatrix} Y \begin{bmatrix} 1 & 3 \\ 1 & 3 \end{bmatrix} = \begin{bmatrix} -2 & -6 \\ 3 & 9 \end{bmatrix}. \tag{3.42}$$

Note the degeneracy of one of the multipliers, namely the matrix $\begin{bmatrix} 1 & 3 \\ 1 & 3 \end{bmatrix}$.

Solution Note that the degeneracy of the matrix $\begin{bmatrix} 1 & 3 \\ 1 & 3 \end{bmatrix}$ does not allow us to use the equation solving method with the help of the inverse matrix. In this case, reduce the problem to the system of linear equations.

Denote the elements of the matrix Y by y_1, y_2, y_3 and y_4:

$$Y = \begin{bmatrix} y_1 & y_2 \\ y_3 & y_4 \end{bmatrix}. \tag{3.43}$$

Successively expand the product of the matrices:

$$\begin{bmatrix} 1 & -1 \\ -1 & 2 \end{bmatrix} \begin{bmatrix} y_1 & y_2 \\ y_3 & y_4 \end{bmatrix} \begin{bmatrix} 1 & 3 \\ 1 & 3 \end{bmatrix} = \begin{bmatrix} -2 & -6 \\ 3 & 9 \end{bmatrix} = \begin{bmatrix} y_1 - y_3 & y_2 - y_4 \\ -y_1 + 2y_3 & -y_2 + 2y_4 \end{bmatrix} \begin{bmatrix} 1 & 3 \\ 1 & 3 \end{bmatrix}$$

$$= \begin{bmatrix} y_1 + y_2 - y_3 - y_4 & 3y_1 + 3y_2 - 3y_3 - 3y_4 \\ -y_1 - y_2 + 2y_3 + 2y_4 & -3y_1 - 3y_2 + 6y_3 + 6y_4 \end{bmatrix}. \tag{3.44}$$

The obtained matrix is equal to $\begin{bmatrix} -2 & -6 \\ 3 & 9 \end{bmatrix}$.

Equating the respective elements of the matrices, we obtain a non-homogeneous system of linear equations relative to the unknowns y_1, y_2, y_3 and y_4:

$$\begin{cases} y_1 + y_2 - y_3 - y_4 = -2, \\ 3y_1 + 3y_2 - 3y_3 - 3y_4 = -6, \\ -y_1 - y_2 + 2y_3 + 2y_4 = 3, \\ -3y_1 - 3y_2 + 6y_3 + 6y_4 = 9. \end{cases} \tag{3.45}$$

Write the augmented matrix of this system:

$$\begin{bmatrix} 1 & 1 & -1 & -1 & | & -2 \\ 3 & 3 & -3 & -3 & | & -6 \\ -1 & -1 & 2 & 2 & | & 3 \\ -3 & -3 & 6 & 6 & | & 9 \end{bmatrix}. \tag{3.46}$$

Subtract from the second row the first row, multiplied by 3; then, subtract from the fourth row the third row, multiplied by 3:

$$\begin{bmatrix} 1 & 1 & -1 & -1 & | & -2 \\ 0 & 0 & 0 & 0 & | & 0 \\ -1 & -1 & 2 & 2 & | & 3 \\ 0 & 0 & 0 & 0 & | & 0 \end{bmatrix}. \tag{3.47}$$

Then, add to the third row the first row. The zero rows do not influence the solution, this is why we obtain the following equivalent matrix:

$$\begin{bmatrix} 1 & 1 & -1 & -1 & | & -2 \\ 0 & 0 & 1 & 1 & | & 1 \end{bmatrix}. \tag{3.48}$$

It corresponds to the system of equations with four unknowns

$$\begin{cases} y_1 + y_2 - y_3 - y_4 = -2, \\ \qquad\qquad y_3 + y_4 = 1. \end{cases} \tag{3.49}$$

The ranks of the basic and augmented matrices are equal to two. Therefore, we conclude that the system (3.49) is consistent and undetermined.

As the independent variables, select, for example y_2 and y_4:

$$y_2 = C_1, \quad y_4 = C_2, \quad \text{where } C_1, C_2 \in \mathbb{R}. \tag{3.50}$$

Having substituted (3.50) into the system (3.49), we obtain

$$\begin{cases} y_1 + C_1 - y_3 - C_2 = -2, \\ \qquad\qquad y_3 + C_2 = 1, \end{cases} \tag{3.51}$$

or

$$\begin{cases} y_1 = -1 - C_1, \\ y_3 = 1 - C_2. \end{cases} \quad (3.52)$$

Write the coefficients y_1–y_4 in the form of a column:

$$\begin{bmatrix} y_1 \\ y_2 \\ y_3 \\ y_4 \end{bmatrix} = \begin{bmatrix} -1 \\ 0 \\ 1 \\ 0 \end{bmatrix} + C_1 \begin{bmatrix} -1 \\ 1 \\ 0 \\ 0 \end{bmatrix} + C_2 \begin{bmatrix} 0 \\ 0 \\ -1 \\ 1 \end{bmatrix}. \quad (3.53)$$

As a result, the matrix Y is equal to

$$Y = \begin{bmatrix} -1 - C_1 & C_1 \\ 1 - C_2 & C_2 \end{bmatrix}, \quad (3.54)$$

where C_1, C_2 are arbitrary real numbers.

Thus, the matrix equation (3.42) has the infinite set of solutions, depending on two real parameters.

Check

By direct substitution of the obtain matrix into (3.42) it is easy to check that Y is determined correctly:

$$\begin{bmatrix} 1 & -1 \\ -1 & 2 \end{bmatrix} \begin{bmatrix} -1 - C_1 & C_1 \\ 1 - C_2 & C_2 \end{bmatrix} \begin{bmatrix} 1 & 3 \\ 1 & 3 \end{bmatrix}$$

$$= \begin{bmatrix} -2 - C_1 + C_2 & C_1 - C_2 \\ 3 + C_1 - 2C_2 & -C_1 + 2C_2 \end{bmatrix} \begin{bmatrix} 1 & 3 \\ 1 & 3 \end{bmatrix} = \begin{bmatrix} -2 & -6 \\ 3 & 9 \end{bmatrix}.$$

$\square$

Assume that A and B are arbitrary matrices with sizes $m \times n$ and $n \times p$, respectively. Estimate of the rank of the product of the matrices A and B is provided by the following theorem.

Theorem 3.3 *The rank of the matrix product satisfies the inequality*

$$\mathrm{rk}\, AB \leqslant \min(\mathrm{rk}\, A, \mathrm{rk}\, B). \quad (3.55)$$

In other words, the rank cannot increase when the matrices are multiplied [63, 64].

There exist many various methods of solving system of linear equations.

Gaussian method is one of the most frequently used.

Consider realization in Python of Gaussian method of solving system of linear equations (Listing 3.1).

For definiteness, we will form the solutions for the systems where the number of equations and unknowns coincides, but the provided algorithm can easily be transformed for the systems with an arbitrary relation between equations and unknowns.

Listing 3.1

```
1   import math
2
3
4   def gaussian_elimination(A, B):
5       m = len(A)
6       n = len(A[0])
7
8       if len(B) != m:
9           raise ValueError
10
11      C = [[A[i][j] if j != n else B[i] \
12              for j in range(n+1)] for i in range(m)]
13
14      # Forward elimination
15      for r in range(min(n, m)):
16          max_row_pos = r
17
18          # Pivoting strategy
19          for i in range(r + 1, m):
20              if abs(C[i][r]) > \
21                      abs(C[max_row_pos][r]):
22                  max_row_pos = i
23
24          C[r], C[max_row_pos] = \
25              C[max_row_pos], C[r]
26
27          if math.isclose(C[r][r], 0):
28              continue
29
30          for i in range(r + 1, m):
31              factor = C[i][r] / C[r][r]
32
```

```
33          for j in range(r, n + 1):
34              C[i][j] -= factor * C[r][j]
35
36      # Back substitution
37      answer = [0] * n
38
39      for i in range(min(n - 1, m - 1), -1, -1):
40          s = 0.0
41
42          for j in range(i + 1, n):
43              s += C[i][j] * answer[j]
44
45          if not math.isclose(C[i][i], 0):
46              answer[i] = (C[i][n] - s) / C[i][i]
47          elif not math.isclose(C[i][n] - s, 0):
48              return None
49
50      for i in range(n, m):
51          s = 0.0
52
53          for j in range(n):
54              s += C[i][j] * answer[j]
55
56          if not math.isclose(C[i][n] - s, 0):
57              return None
58
59      return answer
```

Two parameters arrive at the input of the given function: the coefficient matrix with the unknowns in the system of equations A and the right side matrix B.

The realization of the function consists of three main steps. In the first step, the matrix C is constructed, which is obtained by attributing the matrix B to the initial matrix A on the right.

Then, the so-called forward pass is performed with the purpose of bringing the matrix to the echelon form (that is to the form when each successive row, viewed from left to right, contains more zeroes than the previous one). This procedure is performed by applying to the matrix C of a series of elementary transformations by the following algorithm: successively, starting from the first one, all columns are scanned. Among the elements of the current column, the one with the greatest module is found, referred to as the **basic** or **pivot**. Then, from each row, a row is subtracted that contains the pivot and is multiplied by the coefficient equal to the relation of the row element in the considered column to the pivot. Thus, all the elements in the column, except the pivot, become equal to zero. The process is performed until the matrix has a row left that contains only two variables: one coefficient of the unknown and one value in the right side.

In the third step, the "backward pass" is performed. The values of all unknowns are consequently expressed in terms of the already found variables. So, the unique solution is obtained or it is determined that there are no solutions or the the set of solutions is infinite. The backward pass starts from the row containing the minimum number of non-zero coefficients, and continues until all the unknowns are expressed in terms of the already known ones, or until it is established that there is no unique solution.

The asymptotic complexity of Gaussian elimination, due to triple loop nesting by the variables r, i, j, is equal to $O(n^3)$, where n is the number of equations in the system.

Let us give an example (see Listing 3.2) of using the function gaussian_elimination(A, B) for solving the system of equations whose matrix of size 100×100 has unities on the secondary diagonal and other elements equal to zero. The column B is equal to $[1, 2, 3, 4, \ldots, 100]$.

Listing 3.2

```
1  size = 100
2
3  A = [[0 for j in range(size)] \
4        for i in range(size)]
5  B = [0 for i in range(size)]
6
7  for i in range(size):
8      for j in range(size):
9          A[i][j] = 1 if j == size - i - 1 else 0
10
11 for i in range(size):
12     B[i] = float(i)
13
14 print(gaussian_elimination(A, B))
```

3.4 Fundamental System of Solutions of Homogeneous Systems

Consider a homogeneous system of equations that has the form

$$\begin{cases} a_{11}x_1 + a_{12}x_2 + \cdots + a_{1n}x_n = 0, \\ a_{21}x_1 + a_{22}x_2 + \cdots + a_{2n}x_n = 0, \\ \ldots\ldots\ldots\ldots\ldots\ldots\ldots\ldots\ldots\ldots \\ a_{m1}x_1 + a_{m2}x_2 + \cdots + a_{mn}x_n = 0, \end{cases} \tag{3.56}$$

where m is the number of equations, n is the number of unknowns.

This system can be written in a matrix form:

$$A \cdot X = 0, \tag{3.57}$$

where A is the system matrix, and by X is denoted the column formed by the variables:

$$X = \begin{bmatrix} x_1 \\ x_2 \\ \vdots \\ x_n \end{bmatrix}. \tag{3.58}$$

This system necessarily has the solution $X = [0, 0, \ldots, 0]^T$, which is called **trivial**.

Our purpose is to find all non-trivial solutions, if any.

Theorem 3.4 *Let the matrix A of a homogeneous system of equations have the size* $m \times n$ *and the rank r. If* $r = n$, *then the system has only a trivial solution. If* $r < n$, *then there exist exactly* $n - r$ *linearly independent solutions, referred to as* **fundamental system of solutions**.

Suppose that we have found the basic minor and it is located in the upper left corner (otherwise, the order of variables and equations may be changed). Let us keep only those equations whose coefficients are included into the basic minor, that is from the first to the r-th. The unknowns from number $r + 1$ to number n are referred to as **free** and rearranged to the right side of the equations:

$$\begin{cases} a_{11}x_1 + \cdots + a_{1r}x_r = -a_{1r+1}x_{r+1} - \cdots - a_{1n}x_n, \\ \ldots\ldots\ldots\ldots\ldots\ldots\ldots\ldots\ldots\ldots\ldots\ldots\ldots\ldots\ldots\ldots\ldots\ldots\ldots \\ a_{r1}x_1 + \cdots + a_{rr}x_r = -a_{rr+1}x_{r+1} - \cdots - a_{rn}x_n. \end{cases} \tag{3.59}$$

Let us introduce for consideration a square matrix

$$C = \begin{bmatrix} a_{11} \ldots a_{1r} \\ \vdots \ddots \vdots \\ a_{r1} \ldots a_{rr} \end{bmatrix}, \tag{3.60}$$

and, according to the property of a basic minor, the following inequality $\det C \neq 0$ is valid.

Denote

$$Y = \begin{bmatrix} x_1 \\ x_2 \\ \vdots \\ x_r \end{bmatrix} \in \mathbb{R}^r, \quad Z = \begin{bmatrix} x_{r+1} \\ x_{r+2} \\ \vdots \\ x_n \end{bmatrix} \in \mathbb{R}^{n-r}. \tag{3.61}$$

Then, select $n - r$ linearly independent vectors $Z_1, Z_2, \ldots, Z_{n-r}$. Usually, collections are taken that form the so-called **canonical basis** in $\mathbb{R}^{n-r}$:

$$Z_1 = \begin{bmatrix} 1 \\ 0 \\ 0 \\ \vdots \\ 0 \end{bmatrix}, \quad Z_2 = \begin{bmatrix} 0 \\ 1 \\ 0 \\ \vdots \\ 0 \end{bmatrix}, \quad \ldots, \quad Z_{n-r} = \begin{bmatrix} 0 \\ 0 \\ 0 \\ \vdots \\ 1 \end{bmatrix}. \tag{3.62}$$

Let us call F_1 the vector in the right side of the equation for the given $Z_1, \ldots,$ F_{n-r}—for Z_{n-r}.

In this case, we have the systems of equations in a matrix form:

$$\begin{cases} C \cdot Y_1 = F_1, \\ \cdots\cdots\cdots\cdots \\ C \cdot Y_{n-r} = F_{n-r}. \end{cases} \tag{3.63}$$

According to Cramer's rule, the solutions $Y_1, Y_2, \ldots, Y_{n-r}$ are uniquely determined. Then the full solution of the system will consist of the vectors:

$$X_1 = \begin{pmatrix} Y_1 \\ Z_1 \end{pmatrix}, \ldots, X_{n-r} = \begin{pmatrix} Y_{n-r} \\ Z_{n-r} \end{pmatrix}, \tag{3.64}$$

that are linearly independent, since $Z_1, \ldots, Z_{n-r}$ are linearly independent.

The set of the solutions $X_1, X_2, \ldots, X_{n-r}$ represents the **fundamental system of solutions (FSS)**.

If the free unknown is the only one in the system, then we assign a value to it equal to one. If there are no free unknowns, i.e. $r = n$, then such a system has only a trivial solution and therefore there are no fundamental solutions.

We will denote the general solution of the homogeneous system by $X_{\text{gen.}}$:

$$X_{\text{gen.}} = C_1 X_1 + C_2 X_2 + \cdots + C_{n-r} X_{n-r}, \tag{3.65}$$

where $C_1, C_2, \ldots, C_{n-r}$ are arbitrary constants.

Example 3.6 Find the fundamental system of solutions for the homogeneous system of equations

$$\begin{cases} 2x_1 - 4x_2 + 5x_3 + 3x_4 = 0, \\ 3x_1 - 6x_2 + 4x_3 + 2x_4 = 0, \\ 4x_1 - 8x_2 + 17x_3 + 11x_4 = 0. \end{cases} \tag{3.66}$$

Solution Find the rank of the matrix of the given system

$$\begin{bmatrix} 2 & -4 & 5 & 3 \\ 3 & -6 & 4 & 2 \\ 4 & -8 & 17 & 11 \end{bmatrix}, \tag{3.67}$$

by bringing it to the upper triangular form.

Subtract from the third row the second one. As a result we obtain

$$\begin{bmatrix} 2 & -4 & 5 & 3 \\ 3 & -6 & 4 & 2 \\ 1 & -2 & 13 & 9 \end{bmatrix}, \tag{3.68}$$

permute the first and the third rows:

$$\begin{bmatrix} 1 & -2 & 13 & 9 \\ 3 & -6 & 4 & 2 \\ 2 & -4 & 5 & 3 \end{bmatrix}. \tag{3.69}$$

Subtract from the second row the first row multiplied by 3, and from the third row—the doubled first row. We have

$$\begin{bmatrix} 1 & -2 & 13 & 9 \\ 0 & 0 & -35 & -25 \\ 0 & 0 & -21 & -15 \end{bmatrix}. \tag{3.70}$$

Divide the second row by (-5), and the third one by (-3):

$$\begin{bmatrix} 1 & -2 & 13 & 9 \\ 0 & 0 & 7 & 5 \\ 0 & 0 & 7 & 5 \end{bmatrix}. \tag{3.71}$$

Subtract from the third row the second one:

$$\begin{bmatrix} 1 & -2 & 13 & 9 \\ 0 & 0 & 7 & 5 \\ 0 & 0 & 0 & 0 \end{bmatrix}. \tag{3.72}$$

From this we see that the rank of this matrix is equal to two.
As a basic non-zero minor, take, for example, the minor $M_{3,4}^{1,2}$ of the initial matrix:

$$\begin{vmatrix} 5 & 3 \\ 4 & 2 \end{vmatrix}. \tag{3.73}$$

Then, use the first two equations, whose coefficients are included into the basic minor. Rearrange to the right side of the equations the summands that are not included into the basic minor. We obtain

$$\begin{cases} 5x_3 + 3x_4 = 4x_2 - 2x_1, \\ 4x_3 + 2x_4 = 6x_2 - 3x_1. \end{cases} \tag{3.74}$$

Set two different values to the free unknowns x_1 and x_2. The first case:

$$\begin{bmatrix} x_1 \\ x_2 \end{bmatrix} = \begin{bmatrix} 1 \\ 0 \end{bmatrix}. \tag{3.75}$$

Substituting these values into the system, we obtain

$$\begin{cases} 5x_3 + 3x_4 = -2, \\ 4x_3 + 2x_4 = -3. \end{cases} \tag{3.76}$$

Solution of this system: $x_3 = -\dfrac{5}{2}$ and $x_4 = \dfrac{7}{2}$.

The second case:

$$\begin{bmatrix} x_1 \\ x_2 \end{bmatrix} = \begin{bmatrix} 0 \\ 1 \end{bmatrix}. \tag{3.77}$$

Similarly to the first case, we obtain

$$\begin{cases} 5x_3 + 3x_4 = 4, \\ 4x_3 + 2x_4 = 6. \end{cases} \tag{3.78}$$

While $x_3 = 5$ and $x_4 = -7$.
For the fundamental system of solutions we finally have

$$\begin{bmatrix} x_1 \\ x_2 \\ x_3 \\ x_4 \end{bmatrix} \in \left\{ \begin{bmatrix} 1 \\ 0 \\ -5/2 \\ 7/2 \end{bmatrix}, \begin{bmatrix} 0 \\ 1 \\ 5 \\ -7 \end{bmatrix} \right\}. \tag{3.79}$$

The general solution of this homogeneous system may be written as

$$X_{\text{gen.}} = C_1 \begin{bmatrix} 2 \\ 0 \\ -5 \\ 7 \end{bmatrix} + C_2 \begin{bmatrix} 0 \\ 1 \\ 5 \\ -7 \end{bmatrix}, \tag{3.80}$$

where C_1 and C_2 are arbitrary numbers.   □

3.5   General Solution of the Non-homogeneous System of Equations

Consider a non-homogeneous system of equations:

$$\begin{cases} a_{11}x_1 + a_{12}x_2 + \cdots + a_{1n}x_n = b_1, \\ a_{21}x_1 + a_{22}x_2 + \cdots + a_{2n}x_n = b_2, \\ \cdots\cdots\cdots\cdots\cdots\cdots\cdots\cdots\cdots\cdots\cdots \\ a_{m1}x_1 + a_{m2}x_2 + \cdots + a_{mn}x_n = b_m. \end{cases} \tag{3.81}$$

Theorem 3.5 *Let $X_{gen.}$ be the general solution of the homogeneous system, when all the values b_i are replaced with zeroes, and $X_{spec.}$ is the particular solution of the non-homogeneous system. Then, X, the general solution of the non-homogeneous system, is equal to*

$$X = X_{gen.} + X_{spec.}. \tag{3.82}$$

Example 3.7 Solve the system:

$$\begin{cases} 2x_1 + x_2 - x_3 - 3x_4 = 2, \\ 4x_1 + x_3 - 7x_4 = 3, \\ 2x_2 - 3x_3 + x_4 = 1, \\ 2x_1 + 3x_2 - 4x_3 - 2x_4 = 3. \end{cases} \tag{3.83}$$

Solution Write the augmented system matrix:

$$\begin{bmatrix} 2 & 1 & -1 & -3 & | & 2 \\ 4 & 0 & 1 & -7 & | & 3 \\ 0 & 2 & -3 & 1 & | & 1 \\ 2 & 3 & -4 & -2 & | & 3 \end{bmatrix}. \tag{3.84}$$

Find the rank of this matrix, for which purpose subtract from the second row the doubled first row, and from the fourth row—the first row:

$$\begin{bmatrix} 2 & 1 & -1 & -3 & | & 2 \\ 0 & -2 & 3 & -1 & | & -1 \\ 0 & 2 & -3 & 1 & | & 1 \\ 0 & 2 & -3 & 1 & | & 1 \end{bmatrix}. \tag{3.85}$$

One can see that the last three rows are proportional to each other, and it is enough to keep one of them:

$$\begin{bmatrix} 2 & 1 & -1 & -3 & | & 2 \\ 0 & -2 & 3 & -1 & | & -1 \end{bmatrix}. \tag{3.86}$$

The rank of the basic and the augmented matrices is equal to two, therefore, the system is consistent. Find the general solution of the homogeneous system, for which purpose rearrange x_3 and x_4 to the right side of the equations. We obtain

$$\begin{cases} 2x_1 + x_2 = x_3 + 3x_4, \\ -2x_2 = -3x_3 + x_4. \end{cases} \tag{3.87}$$

Select the following values for the independent variables:

$$\begin{bmatrix} x_3 \\ x_4 \end{bmatrix} = \begin{bmatrix} 1 \\ 0 \end{bmatrix}, \tag{3.88}$$

in such a case, we obtain the system

$$\begin{cases} 2x_1 + x_2 = 1, \\ -2x_2 = -3, \end{cases} \tag{3.89}$$

whence $x_2 = \dfrac{3}{2}, x_1 = -\dfrac{1}{4}$.

Now, selecting the values

$$\begin{bmatrix} x_3 \\ x_4 \end{bmatrix} = \begin{bmatrix} 0 \\ 1 \end{bmatrix}, \tag{3.90}$$

we obtain the system

$$\begin{cases} 2x_1 + x_2 = 3, \\ -2x_2 = 1, \end{cases} \tag{3.91}$$

whose solution is $x_2 = -\dfrac{1}{2}$ and $x_1 = \dfrac{7}{4}$.

Therefore, the general solution of the homogeneous system has the form:

$$X_{\text{gen.}} = C_1 \begin{bmatrix} -\dfrac{1}{4} \\ \dfrac{3}{2} \\ 1 \\ 0 \end{bmatrix} + C_2 \begin{bmatrix} \dfrac{7}{4} \\ -\dfrac{1}{2} \\ 0 \\ 1 \end{bmatrix}, \text{ where } C_1, C_2 \in \mathbb{R}. \tag{3.92}$$

If we introduce for consideration new constants $C'_1 = C_1/4$, $C'_2 = C_2/4$, then $X_{\text{gen.}}$ will be written in the form, free from fractions:

$$X_{\text{gen.}} = C'_1 \begin{bmatrix} -1 \\ 6 \\ 4 \\ 0 \end{bmatrix} + C'_2 \begin{bmatrix} 7 \\ -2 \\ 0 \\ 4 \end{bmatrix}, \text{ where } C'_1, C'_2 \in \mathbb{R}. \tag{3.93}$$

In order to find the particular solution, return to the augmented matrix (3.86). The equations for computation of $X_{\text{spec.}}$ have the form:

$$\begin{cases} 2x_1 + x_2 = x_3 + 3x_4 + 2, \\ -2x_2 = -3x_3 + x_4 - 1. \end{cases} \tag{3.94}$$

Assuming that the values of the independent variables are equal to zero, we find $x_1 = \dfrac{3}{4}$, $x_2 = \dfrac{1}{2}$, and

$$X_{\text{spec.}} = \begin{bmatrix} \frac{3}{4} \\ \frac{1}{2} \\ 0 \\ 0 \end{bmatrix}. \tag{3.95}$$

As a result, the general solution of the non-homogeneous system is equal to

$$X = X_{\text{gen.}} + X_{\text{spec.}} = C'_1 \begin{bmatrix} -1 \\ 6 \\ 4 \\ 0 \end{bmatrix} + C'_2 \begin{bmatrix} 7 \\ -2 \\ 0 \\ 4 \end{bmatrix} + \frac{1}{4} \begin{bmatrix} 3 \\ 2 \\ 0 \\ 0 \end{bmatrix}, \text{ where } C'_1, C'_2 \in \mathbb{R}. \tag{3.96}$$

$\square$

Review Questions

1. Define solution of a system of linear equations.
2. What system of equations is called consistent? inconsistent?

3. When is the system of equations definite? indefinite?
4. Explain how an augmented matrix of a system of linear equations constructed.
5. Describe the methods of solving the systems of linear equations: Gaussian method, inverse matrix method, Cramer's method.
6. What is the complexity of the Gaussian method?
7. What solution does any homogeneous system of linear equations have?
8. Define the fundamental system of solutions.
9. How is the general solution of a non-homogeneous system computed?

Problems

3.1. In order to expand a computer laboratory, its chief is planning to purchase 9 workstations and 7 notebooks. If 14 workstations and 9 notebooks are ordered, the cost of the purchase will grow by 1.5 times. Find how many times the workstation is more expensive than the notebook.

3.2. Solve the system of linear equations using Cramer's rule:

$$(a) \begin{cases} 3x - 5y = 13, \\ 2x + 7y = 81; \end{cases} \quad (b) \begin{cases} 2x_1 - x_2 + 3x_3 = 9, \\ 3x_1 - 5x_2 - x_3 = -10, \\ 4x_1 - 7x_2 + x_3 = -7; \end{cases}$$

$$(c) \begin{cases} x + 2y + z = 4, \\ 3x - 5y + 3z = 1, \\ 2x + 7y - z = 8; \end{cases} \quad (d) \begin{cases} 2x - 4y + 9z = 28, \\ 7x + 3y - 6z = -1, \\ 7x + 9y - 9z = 5; \end{cases}$$

$$(e) \begin{cases} 7x + 2y + 3z = 15, \\ 5x - 3y + 2z = 15, \\ 10x - 11y + 5z = 36; \end{cases} \quad (f) \begin{cases} x + y + z = 36, \\ 2x - 3z = -17, \\ 6x - 5z = 7; \end{cases}$$

$$(g) \begin{cases} 3x_1 + 2x_2 + x_3 = 5, \\ x_1 + x_2 - x_3 = 0, \\ 4x_1 - x_2 + 5x_3 = 3. \end{cases}$$

3.3. Solve the system of linear equations using Gaussian elimination:

$$(a) \begin{cases} 6x_1 + 2x_2 + 3x_3 = 74, \\ 7x_1 + 4x_2 = 91, \\ x_1 + x_2 + x_3 = 18; \end{cases} \quad (b) \begin{cases} 2x_1 + 5x_2 - 2x_3 = -6, \\ -3x_1 - 2x_2 + x_3 = 0, \\ 3x_2 + 2x_3 = -8; \end{cases}$$

$$(c) \begin{cases} 3x_1 - x_2 + 6x_3 = -4, \\ 3x_1 - 7x_2 = 2, \\ -4x_1 - 4x_2 - 3x_3 = -10; \end{cases} \quad (d) \begin{cases} 5x_1 + 3x_2 - 3x_3 = 8, \\ -4x_1 - 3x_2 - 2x_3 = 1, \\ -2x_1 + 3x_2 + 6x_3 = -29. \end{cases}$$

3.4. Solve the system of linear equations using Gaussian elimination:

$$(a) \begin{cases} -2x_1 + 7x_2 + 4x_3 = 32, \\ 2x_1 + 8x_2 - x_3 + 7x_4 = 63, \\ -6x_1 + 6x_2 + 8x_3 - 8x_4 = 2, \\ 6x_2 - 4x_3 + 5x_4 = 58; \end{cases} \quad (b) \begin{cases} -x_1 - 2x_3 - x_4 = -6, \\ -5x_1 - x_2 + 6x_3 + x_4 = 23, \\ 5x_1 - 8x_2 - 9x_3 + 4x_4 = 62, \\ 6x_1 - 9x_2 - 5x_3 + x_4 = 73; \end{cases}$$

$$(c) \begin{cases} 4x_1 - 9x_3 - x_4 = 37, \\ 7x_1 - x_2 - 5x_3 - 5x_4 = 36, \\ 8x_1 - 5x_2 + 4x_4 = -38, \\ x_1 - 4x_2 + 9x_3 - 4x_4 = -25; \end{cases} \quad (d) \begin{cases} -8x_1 + x_2 - 4x_3 - 8x_4 = 7, \\ -7x_2 - 6x_3 + 7x_4 = 56, \\ -8x_1 + 3x_2 + 2x_3 - 2x_4 = -63, \\ -8x_1 - 3x_2 - x_3 - 4x_4 = -6. \end{cases}$$

*3.5. Solve the system of linear equations relative to five unknowns:

$$(a) \begin{cases} 5x_1 - 8x_2 - 5x_3 + 8x_4 + 8x_5 = -5, \\ 2x_2 + 2x_3 + x_4 - x_5 = 8, \\ -2x_1 + 4x_2 + 3x_3 - 8x_4 + 4x_5 = -39, \\ 5x_1 + 6x_2 + 2x_3 - 2x_4 - 4x_5 = 32, \\ x_1 - 2x_2 - x_3 + 2x_4 - 2x_5 = 23; \end{cases}$$

$$(b) \begin{cases} 6x_1 - x_2 + 6x_3 + 3x_4 - 7x_5 = 6, \\ -4x_1 - 4x_2 + 3x_3 - x_4 - 8x_5 = -30, \\ -x_1 + x_2 + 5x_4 - x_5 = -22, \\ 4x_1 + x_2 - 3x_3 + 3x_4 - 5x_5 = -3, \\ 8x_2 + x_4 - x_5 = -61. \end{cases}$$

∗3.6. Solve the system of linear equations relative to six unknowns:

(a)
$$\begin{cases} -x_1 + 2x_3 + 5x_4 - 2x_5 - 4x_6 = \quad 4, \\ -x_1 - x_2 + x_3 - 3x_4 + x_5 - 4x_6 = -46, \\ -x_1 + 5x_2 + 4x_3 - 2x_5 - x_6 = -19, \\ x_1 - 2x_2 + 4x_3 - 2x_5 - x_6 = \quad -9, \\ -2x_1 - 5x_2 + 3x_3 - 2x_4 - 3x_5 = -20, \\ 3x_1 + x_2 + 3x_3 + x_6 = \quad 4; \end{cases}$$

(b)
$$\begin{cases} -2x_1 - 4x_2 - 4x_5 + 5x_6 = \quad 2, \\ -x_1 + 5x_2 + 3x_3 + 5x_5 = -12, \\ -2x_1 - 5x_2 + 5x_3 - 3x_4 - 5x_5 - 3x_6 = \quad -7, \\ -4x_1 - 5x_2 - 3x_3 + 5x_4 - 2x_5 + 3x_6 = \quad 10, \\ x_1 + 3x_2 - 5x_3 + 4x_4 + 3x_5 + 2x_6 = \quad 8, \\ -4x_1 + x_2 - 4x_3 + 3x_4 + x_5 - x_6 = \quad 3. \end{cases}$$

3.7. Find the fourth power polynomial $p(x)$ with real coefficients, for which the following properties are valid: $p(5) = 1$ and $p(1) = p(2) = p(3) = p(4) = 0$.

3.8. At what values of the parameter λ is the system of equations

$$\begin{cases} x_1 + x_2 = 1, \\ \lambda x_1 + x_2 = 2, \\ x_1 + \lambda x_2 = 4. \end{cases}$$

consistent?

3.9. At what values of the parameter λ does the system of linear equations

$$\begin{cases} x_1 + \lambda x_2 = \quad 0, \\ -x_1 + x_2 - x_3 = \quad 5, \\ -x_2 + x_3 = -4, \end{cases}$$

have the unique solution? With these values of λ, find the solution of the system using Cramer's rule.

*3.10. Solve the matrix equation $A \cdot X \cdot B = C$, where

$$A = \begin{bmatrix} -3 & 2 & 3 \\ -2 & 1 & 4 \\ 6 & 1 & 2 \end{bmatrix}, \quad B = \begin{bmatrix} -1 & 0 \\ -1 & 0 \end{bmatrix}, \quad C = \begin{bmatrix} -13 & 0 \\ -25 & 0 \\ -33 & 0 \end{bmatrix}. \tag{3.97}$$

Note the degeneracy of one of the multipliers, namely the matrix B.

*3.11. Solve the matrix equations:

(1) $\begin{bmatrix} 6 & 2 \\ 3 & 1 \end{bmatrix} \cdot X = \begin{bmatrix} -4 & 4 \\ -2 & 2 \end{bmatrix}$;

(2) $X \cdot \begin{bmatrix} 15 & -5 \\ -3 & 1 \end{bmatrix} = \begin{bmatrix} 1 & -1 \\ -2 & 1 \end{bmatrix}$;

(3) $X \cdot \begin{bmatrix} 6 & -1 & 1 \\ -10 & -5 & 6 \\ 0 & -20 & 23 \end{bmatrix} = \begin{bmatrix} 6 & 19 & -22 \\ 10 & 5 & -6 \\ -6 & -19 & 22 \end{bmatrix}$;

(4) $\begin{bmatrix} -1 & 2 & 5 \\ 5 & 3 & -1 \\ 7 & -1 & -11 \end{bmatrix} \cdot X = \begin{bmatrix} -7 & 8 & 2 \\ 2 & 7 & 1 \\ 1 & -1 & 1 \end{bmatrix}$.

3.12. A program code is given that processes the one-dimensional array a, consisting of five elements:

```
for i in range(len(a)):
    temp = a[0]

    for j in range(len(a) - 1):
        a[j] = -2 * a[j + 1]

    a[len(a) - 1] += temp
```

After executing this program code segment, the array a [] consists of the following elements: [-32, 32, 32, 32, 16]. Find what values the elements of the array a [i], where $i = 1, \ldots, 5$, before executing this segment.

3.13. A program code is given that processes the one-dimensional array a, consisting of seven elements:

```
for i in range(len(a)):
    temp = a[0]
```

```
for j in range(len(a) - 1):
    a[j] = 3 * a[j + 1] - 1

a[len(a) - 2] = -a[len(a) - 2] - temp
```

After executing this program code segment, the array `a []` consists of the following elements: `[365, 608, -769, -499, -409, 107, 5]`. Find what values the elements of the array `a[i]`, $i = 1, \ldots, 7$ took before executing this segment.

3.14. There exists a modification of Gaussian elimination referred to as **Gauss–Jordan**[5] **elimination**. In the Gauss–Jordan method, the coefficient matrix is brought not to a triangular, but to a diagonal form. Write the realization of this method in Python and compare its asymptotic complexity with the complexity of the standard Gaussian elimination.

3.15. Find the general solution and the fundamental system of solutions for the systems of equations:

$$(1) \begin{cases} x_1 - 4x_2 + x_3 = 0, \\ x_1 + x_2 - x_3 = 0, \\ 3x_1 - 2x_2 - x_3 = 0; \end{cases} \quad (2) \begin{cases} 2x_1 - x_2 + 3x_3 + x_4 = 0, \\ 2x_1 - 5x_2 - x_3 = 0, \\ 4x_1 - 7x_2 + x_3 + 3x_4 = 0; \end{cases}$$

$$(3) \begin{cases} x_1 + 2x_2 + 4x_3 - 3x_4 = 0, \\ 3x_1 + 5x_2 + 6x_3 - 4x_4 = 0, \\ 4x_1 + 5x_2 - 2x_3 + 3x_4 = 0, \\ 3x_1 + 8x_2 + 24x_3 - 19x_4 = 0; \end{cases} \quad (4) \begin{cases} 3x_1 + 5x_2 + 2x_3 = 0, \\ 4x_1 + 7x_2 + 5x_3 = 0, \\ x_1 + x_2 - 4x_3 = 0, \\ 2x_1 + 9x_2 + 6x_3 = 0; \end{cases}$$

$$(5) \begin{cases} 2x_1 + 4x_2 + 6x_3 + x_4 = 0, \\ x_1 + 2x_2 + 3x_3 + x_4 = 0, \\ 3x_1 + 6x_2 + 9x_3 - x_4 = 0, \\ x_1 + 2x_2 + 3x_3 + 5x_4 = 0; \end{cases} \quad (6) \begin{cases} x + 2y + 3z = 0, \\ 2x + 3y + 4z = 0, \\ x + y + z = 0; \end{cases}$$

$$(7) \begin{cases} x_1 - 2x_2 + 3x_3 - 4x_4 = 0, \\ 2x_1 - 4x_2 + 5x_3 + 7x_4 = 0, \\ 6x_1 - 12x_2 + 17x_3 - 9x_4 = 0, \\ 7x_1 - 14x_2 + 19x_3 + 17x_4 = 0. \end{cases}$$

---

[5]Wilhelm Jordan (1842–1899), German geodesist and mathematician.

Answers and Solutions

3.1 *Solution.*

Denote the price of one workstation by x, and of one notebook—by y. Assume that the cost of 9 workstations and 7 notebooks is a. Then, we obtain a system of linear equations relative to the unknowns x and y:

$$\begin{cases} 9x + 7y = a, \\ 14x + 9y = \dfrac{3}{2}a. \end{cases}$$

Its solution is $x = \dfrac{3}{34}a$, $y = \dfrac{1}{34}a$. Therefore, the workstation is three times more expensive than the notebook.

3.2 *Solution.*

(a) The augmented matrix of the system has the form

$$\begin{bmatrix} 3 & -5 & | & 13 \\ 2 & 7 & | & 81 \end{bmatrix}.$$

Compute the required determinants:

$$\Delta = \begin{vmatrix} 3 & -5 \\ 2 & 7 \end{vmatrix} = 21 + 10 = 31,$$

$$\Delta_x = \begin{vmatrix} 13 & -5 \\ 81 & 7 \end{vmatrix} = 91 + 405 = 496.$$

Therefore, $x = \dfrac{\Delta_x}{\Delta} = \dfrac{496}{31} = 16.$

$$\Delta_y = \begin{vmatrix} 3 & 13 \\ 2 & 81 \end{vmatrix} = 243 - 26 = 217,$$

$$y = \dfrac{\Delta_y}{\Delta} = \dfrac{217}{31} = 7.$$

(b) The augmented matrix of the system has the form

$$\begin{bmatrix} 2 & -1 & 3 & | & 9 \\ 3 & -5 & -1 & | & -10 \\ 4 & -7 & 1 & | & -7 \end{bmatrix}.$$

$$\Delta = \begin{vmatrix} 2 & -1 & 3 \\ 3 & -5 & -1 \\ 4 & -7 & 1 \end{vmatrix} \overset{(1)+3(2)}{=} \begin{vmatrix} 11 & -16 & 0 \\ 3 & -5 & -1 \\ 4 & -7 & 1 \end{vmatrix} = 11 \cdot (-12) + 16 \cdot 7 = -20,$$

$$\Delta_1 = \begin{vmatrix} 9 & -1 & 3 \\ -10 & -5 & -1 \\ -7 & -7 & 1 \end{vmatrix} =^{(1)+3(2)} \begin{vmatrix} -21 & -16 & 0 \\ -10 & -5 & -1 \\ -7 & -7 & 1 \end{vmatrix} =^{(3)+(2)}$$

$$= \begin{vmatrix} -21 & -16 & 0 \\ -10 & -5 & -1 \\ -17 & -12 & 0 \end{vmatrix} = 252 - 272 = -20,$$

$$x_1 = \frac{\Delta_1}{\Delta} = \frac{-20}{-20} = 1,$$

$$\Delta_2 = \begin{vmatrix} 2 & 9 & 3 \\ 3 & -10 & -1 \\ 4 & -7 & 1 \end{vmatrix} =^{(1)+3(2)} \begin{vmatrix} 11 & -21 & 0 \\ 3 & -10 & -1 \\ 4 & -7 & 1 \end{vmatrix} =^{(3)+(2)}$$

$$= \begin{vmatrix} 11 & -21 & 0 \\ 3 & -10 & -1 \\ 7 & -17 & 0 \end{vmatrix} = -(-1)(-187 + 147) = -40,$$

$$x_2 = \frac{\Delta_2}{\Delta} = \frac{-40}{-20} = 2,$$

$$\Delta_3 = \begin{vmatrix} 2 & -1 & 9 \\ 3 & -5 & -10 \\ 4 & -7 & -7 \end{vmatrix} =^{[1]+2[2]} \begin{vmatrix} 0 & -1 & 9 \\ -7 & -5 & -10 \\ -10 & -7 & -7 \end{vmatrix}$$
$$= (49 - 100) + 9(49 - 50) = -60,$$

$$x_3 = \frac{\Delta_3}{\Delta} = \frac{-60}{-20} = 3.$$

(c) The augmented matrix of the system has the form

$$\begin{bmatrix} 1 & 2 & 1 & | & 4 \\ 3 & -5 & 3 & | & 1 \\ 2 & 7 & -1 & | & 8 \end{bmatrix}.$$

$$\Delta = \begin{vmatrix} 1 & 2 & 1 \\ 3 & -5 & 3 \\ 2 & 7 & -1 \end{vmatrix} =^{[1]-[3]} \begin{vmatrix} 0 & 2 & 1 \\ 0 & -5 & 3 \\ 3 & 7 & -1 \end{vmatrix} = 3 \cdot (6 + 5) = 33,$$

$$\Delta_x = \begin{vmatrix} 4 & 2 & 1 \\ 1 & -5 & 3 \\ 8 & 7 & -1 \end{vmatrix} =^{(3)-2(1)} \begin{vmatrix} 4 & 2 & 1 \\ 1 & -5 & 3 \\ 0 & 3 & -3 \end{vmatrix} = 4 \cdot (15 - 9) - (-6 - 3) = 33,$$

$$x = \frac{\Delta_x}{\Delta} = \frac{33}{33} = 1,$$

$$\Delta_y = \begin{vmatrix} 1 & 4 & 1 \\ 3 & 1 & 3 \\ 2 & 8 & -1 \end{vmatrix} =^{[2]-4[1]} \begin{vmatrix} 1 & 0 & 1 \\ 3 & -11 & 3 \\ 2 & 0 & -1 \end{vmatrix} = -11 \cdot (-1 - 2) = 33,$$

$$y = \frac{\Delta_y}{\Delta} = \frac{33}{33} = 1,$$

$$\Delta_z = \begin{vmatrix} 1 & 2 & 4 \\ 3 & -5 & 1 \\ 2 & 7 & 8 \end{vmatrix} =^{(3)-2(1)} \begin{vmatrix} 1 & 2 & 4 \\ 3 & -5 & 1 \\ 0 & 3 & 0 \end{vmatrix} = -3 \cdot (1 - 12) = 33,$$

$$z = \frac{\Delta_z}{\Delta} = \frac{33}{33} = 1.$$

(d) The augmented matrix of the system has the form

$$\begin{bmatrix} 2 & -4 & 9 & | & 28 \\ 7 & 3 & -6 & | & -1 \\ 7 & 9 & -9 & | & 5 \end{bmatrix}.$$

$$\Delta = \begin{vmatrix} 2 & -4 & 9 \\ 7 & 3 & -6 \\ 7 & 9 & -9 \end{vmatrix} =^{(3)-(2)} \begin{vmatrix} 2 & 4 & 9 \\ 7 & 3 & -6 \\ 0 & 6 & -3 \end{vmatrix} = 2(-9 + 36) - 7(12 - 54) = 348,$$

$$\Delta_x = \begin{vmatrix} 28 & 4 & 9 \\ -1 & 3 & -6 \\ 5 & 9 & -9 \end{vmatrix} =^{(1)+(3)} \begin{vmatrix} 33 & 5 & 0 \\ -1 & 3 & -6 \\ 5 & 9 & -9 \end{vmatrix} = 6(297 - 25) - 9(99 + 5) = 696,$$

$$x = \frac{696}{348} = 2,$$

$$\Delta_y = \begin{vmatrix} 2 & 28 & 9 \\ 7 & -1 & -6 \\ 7 & 5 & -9 \end{vmatrix} =^{(3)-(2)} \begin{vmatrix} 2 & 28 & 9 \\ 7 & -1 & -6 \\ 0 & 6 & -3 \end{vmatrix} = 2(3 + 36) - 7(-84 - 54) = 1044,$$

$$y = \frac{1044}{348} = 3,$$

$$\Delta_z = \begin{vmatrix} 2 & -4 & 28 \\ 7 & 3 & -1 \\ 7 & 9 & 5 \end{vmatrix} =^{(3)-(2)} \begin{vmatrix} 2 & -4 & 28 \\ 7 & 3 & -1 \\ 0 & 6 & 6 \end{vmatrix} = 2(18 + 6) - 7(-24 - 168) = 1392,$$

$$z = \frac{1392}{348} = 4.$$

(e) The augmented matrix of the system has the form

$$\begin{bmatrix} 7 & 2 & 3 & | & 15 \\ 5 & -3 & 2 & | & 15 \\ 10 & -11 & 5 & | & 36 \end{bmatrix}.$$

$$\Delta = \begin{vmatrix} 7 & 2 & 3 \\ 5 & -3 & 2 \\ 10 & -11 & 5 \end{vmatrix} =^{(3)-2(2)} \begin{vmatrix} 7 & 2 & 3 \\ 5 & -3 & 2 \\ 0 & -5 & 1 \end{vmatrix} = 5(14-15) + (-21-10) = -36,$$

$$\Delta_x = \begin{vmatrix} 15 & 2 & 3 \\ 15 & -3 & 2 \\ 36 & -11 & 5 \end{vmatrix} =^{(2)-(1)} \begin{vmatrix} 15 & 2 & 3 \\ 0 & -5 & -1 \\ 36 & -11 & 5 \end{vmatrix} = -72,$$

$$x = \frac{-72}{-36} = 2,$$

$$\Delta_y = \begin{vmatrix} 7 & 15 & 3 \\ 5 & 15 & 2 \\ 10 & 36 & 5 \end{vmatrix} =^{(2)-(1)} \begin{vmatrix} 7 & 15 & 3 \\ -2 & 0 & -1 \\ 10 & 36 & 5 \end{vmatrix} = 2(75-108) + (252-150) = 36,$$

$$y = \frac{36}{-36} = -1,$$

$$\Delta_z = \begin{vmatrix} 7 & 2 & 15 \\ 5 & -3 & 15 \\ 10 & 11 & 36 \end{vmatrix} = -36,$$

$$z = \frac{-36}{-36} = 1.$$

(f) The augmented matrix of the system has the form

$$\begin{bmatrix} 1 & 1 & 1 & | & 36 \\ 2 & 0 & -3 & | & -17 \\ 6 & 0 & -5 & | & 7 \end{bmatrix}.$$

$$\Delta = \begin{vmatrix} 1 & 1 & 1 \\ 2 & 0 & -3 \\ 6 & 0 & -5 \end{vmatrix} = -(-10+18) = -8,$$

$$\Delta_x = \begin{vmatrix} 36 & 1 & 1 \\ -17 & 0 & -3 \\ 7 & 0 & -5 \end{vmatrix} = -(85+21) = -106,$$

$$x = \frac{-106}{-8} = \frac{53}{4}.$$

$$\Delta_y = \begin{vmatrix} 1 & 36 & 1 \\ 2 & -17 & -3 \\ 6 & 7 & -5 \end{vmatrix} \overset{(3)-3(2)}{\underset{(2)-2(1)}{=}} \begin{vmatrix} 1 & 36 & 1 \\ 0 & -89 & -5 \\ 0 & 58 & 4 \end{vmatrix} = -356 + 290 = -66,$$

$$y = \frac{-66}{-8} = \frac{33}{4},$$

$$\Delta_z = \begin{vmatrix} 1 & 1 & 36 \\ 2 & 0 & -17 \\ 6 & 0 & 7 \end{vmatrix} = -(14 + 17 \cdot 6) = -116,$$

$$z = \frac{-116}{-8} = \frac{29}{2}.$$

(g) The augmented matrix of the system has the form

$$\begin{bmatrix} 3 & 2 & 1 & | & 5 \\ 1 & 1 & -1 & | & 0 \\ 4 & -1 & 5 & | & 3 \end{bmatrix}.$$

$$\Delta = \begin{vmatrix} 3 & 2 & 1 \\ 1 & 1 & -1 \\ 4 & -1 & 5 \end{vmatrix} \overset{[1]-[2]}{=} \begin{vmatrix} 1 & 2 & 1 \\ 0 & 1 & -1 \\ 5 & -1 & 4 \end{vmatrix} \overset{[3]+[2]}{=} \begin{vmatrix} 1 & 2 & 3 \\ 0 & 1 & 0 \\ 5 & -1 & 4 \end{vmatrix} = 4 - 15 = -11,$$

$$\Delta_1 = \begin{vmatrix} 5 & 2 & 1 \\ 0 & 1 & -1 \\ 3 & -1 & 5 \end{vmatrix} \overset{[3]+[2]}{=} \begin{vmatrix} 5 & 2 & 3 \\ 0 & 1 & 0 \\ 3 & -1 & 4 \end{vmatrix} = 20 - 9 = 11,$$

$$x_1 = \frac{11}{-11} = -1.$$

$$\Delta_2 = \begin{vmatrix} 3 & 5 & 1 \\ 1 & 0 & -1 \\ 4 & 3 & 5 \end{vmatrix} \overset{[1]+[3]}{=} \begin{vmatrix} 4 & 5 & 1 \\ 0 & 0 & -1 \\ 9 & 3 & 5 \end{vmatrix} = 12 - 45 = -33,$$

$$x_2 = \frac{-33}{-11} = 3,$$

$$\Delta_3 = \begin{vmatrix} 3 & 2 & 5 \\ 1 & 1 & 0 \\ 4 & -1 & 3 \end{vmatrix} \overset{[1]-[2]}{=} \begin{vmatrix} 1 & 2 & 5 \\ 0 & 1 & 0 \\ 5 & -1 & 3 \end{vmatrix} = 3 - 25 = -22,$$

$$x_3 = \frac{-22}{-11} = 2.$$

3.3 *Solution.*

(a) Execute the following elementary transformations of the augmented system matrix:

$$\begin{bmatrix} 6 & 2 & 3 & | & 74 \\ 7 & 4 & 0 & | & 91 \\ 1 & 1 & 1 & | & 18 \end{bmatrix} \xrightarrow[\to (2)-7(3)]{(1)-6(3)} \begin{bmatrix} 0 & -4 & -3 & | & -34 \\ 0 & -3 & -7 & | & -35 \\ 1 & 1 & 1 & | & 18 \end{bmatrix} \to^{-3\cdot(1)}$$

$$\to \begin{bmatrix} 0 & 12 & 9 & | & 102 \\ 0 & -3 & -7 & | & -35 \\ 1 & 1 & 1 & | & 18 \end{bmatrix} \to^{(1)+4(2)} \begin{bmatrix} 0 & 0 & -19 & | & -38 \\ 0 & -3 & -7 & | & -35 \\ 1 & 1 & 1 & | & 18 \end{bmatrix}.$$

Hence it follows that $x_3 = \dfrac{38}{19} = 2$,

$-3x_2 = -35 + 14 = -21, x_2 = 7,$

$x_1 = 18 - 2 - 7 = 9.$

The final answer is $[x_1, x_2, x_3]^T = [9, 7, 2]^T$.

(b) Execute the following elementary transformations of the augmented system matrix:

$$\begin{bmatrix} 2 & 5 & -2 & | & -6 \\ -3 & -2 & 1 & | & 0 \\ 0 & 3 & 2 & | & -8 \end{bmatrix} \to^{2\cdot(2)} \begin{bmatrix} 2 & 5 & -2 & | & -6 \\ -6 & -4 & 2 & | & 0 \\ 0 & 3 & 2 & | & -8 \end{bmatrix} \to^{(2)+3(1)}$$

$$\to \begin{bmatrix} 2 & 5 & -2 & | & -6 \\ 0 & 11 & -4 & | & -18 \\ 0 & 3 & 2 & | & -8 \end{bmatrix} \to^{11\cdot(3)} \begin{bmatrix} 2 & 5 & -2 & | & -6 \\ 0 & 11 & -4 & | & -18 \\ 0 & 33 & 22 & | & -88 \end{bmatrix} \to^{(3)-3(2)}$$

$$\to \begin{bmatrix} 2 & 5 & -2 & | & -6 \\ 0 & 11 & -4 & | & -18 \\ 0 & 0 & 34 & | & -34 \end{bmatrix}.$$

Hence it follows that $x_3 = \dfrac{-34}{34} = -1$,

$11x_2 = -18 - 4 = -22, x_2 = -2,$

$2x_1 = -6 - 2 + 10 = 2, x_1 = 1.$

We obtain the answer: $[x_1, x_2, x_3]^T = [1, -2, 1]^T$.

(c) Execute the following elementary transformations of the augmented system matrix:

$$
\begin{bmatrix}
3 & -1 & 6 & | & -4 \\
3 & -7 & 0 & | & 2 \\
-4 & -4 & -3 & | & -10
\end{bmatrix}
\xrightarrow{(2)-(1)}
\begin{bmatrix}
3 & -1 & 6 & | & -4 \\
0 & -6 & -6 & | & 6 \\
-4 & -4 & -3 & | & -10
\end{bmatrix}
\xrightarrow{4\cdot(1)}
$$

$$
\rightarrow
\begin{bmatrix}
12 & -4 & 24 & | & -16 \\
0 & -6 & -6 & | & 6 \\
-4 & -4 & -3 & | & -10
\end{bmatrix}
\xrightarrow{(1)+3(3)}
\begin{bmatrix}
0 & -16 & 15 & | & -46 \\
0 & -6 & -6 & | & 6 \\
-4 & -4 & -3 & | & -10
\end{bmatrix}
\xrightarrow{(2)\cdot 8}
$$

$$
\rightarrow
\begin{bmatrix}
0 & -16 & 15 & | & -46 \\
0 & -48 & -48 & | & 48 \\
-4 & -4 & -3 & | & -10
\end{bmatrix}
\xrightarrow{(2)-3(1)}
\begin{bmatrix}
0 & -16 & 15 & | & -46 \\
0 & 0 & -93 & | & 186 \\
-4 & -4 & -3 & | & -10
\end{bmatrix}
\begin{matrix}
(1)\leftrightarrow(2) \\
\xrightarrow{(3)\leftrightarrow(1)}
\end{matrix}
$$

$$
\rightarrow
\begin{bmatrix}
-4 & -4 & -3 & | & -10 \\
0 & -16 & 15 & | & -46 \\
0 & 0 & -93 & | & 186
\end{bmatrix}.
$$

Hence it follows that $x_3 = -2$,
$-16x_2 = -46 + 30 = -16$, $x_2 = 1$,
$-4x_1 = -10 - 6 + 4 = -12$, $x_1 = 3$.
We obtain the answer: $[x_1, x_2, x_3]^T = [3, 1, -2]^T$.

(d) Execute the following elementary transformations of the augmented system matrix:

$$
\begin{bmatrix}
5 & 3 & -3 & | & 8 \\
-4 & -3 & -2 & | & 1 \\
-2 & 3 & 6 & | & -29
\end{bmatrix}
\xrightarrow{(2)-2(3)}
\begin{bmatrix}
5 & 3 & -3 & | & 8 \\
0 & -9 & -14 & | & 59 \\
-2 & 3 & 6 & | & -29
\end{bmatrix}
\xrightarrow{2\cdot(1)}
$$

$$
\rightarrow
\begin{bmatrix}
10 & 6 & -6 & | & 16 \\
0 & -9 & -14 & | & 59 \\
-2 & 3 & 6 & | & -29
\end{bmatrix}
\xrightarrow{(1)+5(3)}
\begin{bmatrix}
0 & 21 & 24 & | & 129 \\
0 & -9 & -14 & | & 59 \\
-2 & 3 & 6 & | & -29
\end{bmatrix}
\xrightarrow{\frac{1}{3}(1)}
$$

$$
\rightarrow
\begin{bmatrix}
0 & 7 & 8 & | & -43 \\
0 & -9 & -14 & | & 59 \\
-2 & 3 & 6 & | & -29
\end{bmatrix}
\xrightarrow{7\cdot(2)=}
\begin{bmatrix}
0 & 7 & 8 & | & -43 \\
0 & -63 & -98 & | & 413 \\
-2 & 3 & 6 & | & -29
\end{bmatrix}
\xrightarrow{(2)+9(1)}
$$

$$
\rightarrow
\begin{bmatrix}
0 & 7 & 8 & | & -43 \\
0 & 0 & -26 & | & 26 \\
-2 & 3 & 6 & | & -29
\end{bmatrix}
\begin{matrix}
(1)\leftrightarrow(3) \\
\xrightarrow{(3)\leftrightarrow(2)}
\end{matrix}
\begin{bmatrix}
-2 & 3 & 6 & | & -29 \\
0 & 7 & 8 & | & -43 \\
0 & 0 & -26 & | & 26
\end{bmatrix}.
$$

Hence it follows that $x_3 = -1$,
$7x_2 = -43 + 8 = -35$, $x_2 = -5$,
$-2x_1 = -29 + 6 + 15 = -8$, $x_1 = 4$.

We obtain the answer: $[x_1, x_2, x_3]^T = [4, -5, -1]^T$.

3.4 *Solution.*

(a) Execute the following elementary transformations of the augmented system matrix:

$$
\begin{bmatrix}
-2 & 7 & 4 & 0 & | & 32 \\
2 & 8 & -1 & 7 & | & 63 \\
-3 & 3 & 4 & -4 & | & 1 \\
0 & 6 & -4 & 5 & | & 58
\end{bmatrix}
\xrightarrow{(1)+(2)}
\begin{bmatrix}
0 & 15 & 3 & 7 & | & 95 \\
2 & 8 & -1 & 7 & | & 63 \\
-3 & 3 & 4 & -4 & | & 1 \\
0 & 6 & -4 & 5 & | & 58
\end{bmatrix}
$$

$$
=
\begin{bmatrix}
2 & 8 & -1 & 7 & | & 63 \\
0 & 15 & 3 & 7 & | & 95 \\
0 & 6 & -4 & 5 & | & 58 \\
-3 & 3 & 4 & -4 & | & 1
\end{bmatrix}
\xrightarrow{2(4)+3(1)}
\begin{bmatrix}
2 & 8 & -1 & 7 & | & 63 \\
0 & 15 & 3 & 7 & | & 95 \\
0 & 6 & -4 & 5 & | & 58 \\
0 & 30 & 5 & 13 & | & 191
\end{bmatrix}
\xrightarrow{2\cdot(2)}
$$

$$
\rightarrow
\begin{bmatrix}
2 & 8 & -1 & 7 & | & 63 \\
0 & 30 & 6 & 14 & | & 190 \\
0 & 6 & -4 & 5 & | & 58 \\
0 & 30 & 5 & 13 & | & 191
\end{bmatrix}
\xrightarrow[\;(4)-5(3)\;]{(2)-(4)}
\begin{bmatrix}
2 & 8 & -1 & 7 & | & 63 \\
0 & 0 & 1 & 1 & | & -1 \\
0 & 6 & -4 & 5 & | & 58 \\
0 & 0 & 25 & -12 & | & -99
\end{bmatrix}
\xrightarrow{(4)-25(2)}
$$

$$
\rightarrow
\begin{bmatrix}
2 & 8 & -1 & 7 & | & 63 \\
0 & 6 & -4 & 5 & | & 58 \\
0 & 0 & 1 & 1 & | & -1 \\
0 & 0 & 0 & -37 & | & -74
\end{bmatrix},
$$

$x_4 = \dfrac{-74}{-37} = 2$,
$x_3 = -1 - 2 = -3$,
$6x_2 = 58 - 12 - 10 = 36$, $x_2 = 6$,
$2x_1 = 63 - 48 - 3 - 14 = -2$, $x_1 = -1$.
Write the answer: $[x_1, x_2, x_3, x_4]^T = [-1, 6, -3, 2]^T$.

(b) Execute the following elementary transformations of the augmented system matrix:

$$\begin{bmatrix} -1 & 0 & -2 & -1 & | & -6 \\ -5 & -1 & 6 & 1 & | & 23 \\ 5 & -8 & -9 & 4 & | & 62 \\ 6 & -9 & -5 & 1 & | & 73 \end{bmatrix} \xrightarrow{(1)\cdot(-1)} \begin{bmatrix} 1 & 0 & 2 & 1 & | & 6 \\ -5 & -1 & 6 & 1 & | & 23 \\ 5 & -8 & -9 & 4 & | & 62 \\ 6 & -9 & -5 & 1 & | & 73 \end{bmatrix} \xrightarrow{(3)+(2)}$$

$$\begin{bmatrix} 1 & 0 & 2 & 1 & | & 6 \\ -5 & -1 & 6 & 1 & | & 23 \\ 0 & -9 & -3 & 5 & | & 85 \\ 6 & -9 & -5 & 1 & | & 73 \end{bmatrix} \xrightarrow[(4)-6(1)]{(2)+5(1)} \begin{bmatrix} 1 & 0 & 2 & 1 & | & 6 \\ 0 & -1 & 16 & 6 & | & 53 \\ 0 & -9 & -3 & 5 & | & 85 \\ 0 & -9 & -17 & -5 & | & 37 \end{bmatrix} \xrightarrow{(4)-(3)}$$

$$\rightarrow \begin{bmatrix} 1 & 0 & 2 & 1 & | & 6 \\ 0 & -1 & 16 & 6 & | & 53 \\ 0 & -9 & -3 & 5 & | & 85 \\ 0 & 0 & -14 & -10 & | & -48 \end{bmatrix} \xrightarrow{-\frac{1}{2}(4)} \begin{bmatrix} 1 & 0 & 2 & 1 & | & 6 \\ 0 & -1 & 16 & 6 & | & 53 \\ 0 & -9 & -3 & 5 & | & 85 \\ 0 & 0 & 7 & 5 & | & 24 \end{bmatrix} \xrightarrow{(3)-9(2)}$$

$$\rightarrow \begin{bmatrix} 1 & 0 & 2 & 1 & | & 6 \\ 0 & -1 & 16 & 6 & | & 53 \\ 0 & 0 & -147 & -49 & | & -392 \\ 0 & 0 & 7 & 5 & | & 24 \end{bmatrix} \xrightarrow{(3)+21(4)} \begin{bmatrix} 1 & 0 & 2 & 1 & | & 6 \\ 0 & -1 & 16 & 6 & | & 53 \\ 0 & 0 & 0 & 56 & | & 112 \\ 0 & 0 & 7 & 5 & | & 24 \end{bmatrix} \xrightarrow{(3)\leftrightarrow(4)}$$

$$\rightarrow \begin{bmatrix} 1 & 0 & 2 & 1 & | & 6 \\ 0 & -1 & 16 & 6 & | & 53 \\ 0 & 0 & 7 & 5 & | & 24 \\ 0 & 0 & 0 & 56 & | & 112 \end{bmatrix},$$

$$x_4 = \frac{112}{56} = 2,$$
$$7x_3 = 24 - 10 = 14, \ x_3 = 2,$$
$$-x_2 = 53 - 12 - 32, \ x_2 = -9,$$
$$x_1 = 6 - 2 - 4 = 0.$$

Write the answer: $[x_1, x_2, x_3, x_4]^T = [0, -9, 2, 2]^T$.

(c) Execute the following elementary transformations of the augmented system matrix:

$$\begin{bmatrix} 4 & 0 & -9 & -1 & | & 37 \\ 7 & -1 & -5 & -5 & | & 36 \\ 8 & -5 & 0 & 4 & | & -38 \\ 1 & -4 & 9 & -4 & | & -25 \end{bmatrix} \begin{smallmatrix} (1)-4(4) \\ (2)-7(4) \\ (3)-8(4) \\ (1)\leftrightarrow(2) \\ \to (1)\leftrightarrow(4) \\ (3)\leftrightarrow(4) \end{smallmatrix} = \begin{bmatrix} 1 & -4 & 9 & -4 & | & -25 \\ 0 & 16 & -45 & 15 & | & 137 \\ 0 & 27 & -68 & 23 & | & 211 \\ 0 & 27 & -72 & 36 & | & 162 \end{bmatrix} \to \tfrac{1}{9}\cdot(4)$$

$$\to \begin{bmatrix} 1 & -4 & 9 & -4 & | & -25 \\ 0 & 16 & -45 & 15 & | & 137 \\ 0 & 27 & -68 & 23 & | & 211 \\ 0 & 3 & -8 & 4 & | & 18 \end{bmatrix} \to (3)-9(4) \begin{bmatrix} 1 & -4 & 9 & -4 & | & -25 \\ 0 & 16 & -45 & 15 & | & 137 \\ 0 & 0 & 4 & -13 & | & 49 \\ 0 & 3 & -8 & 4 & | & 18 \end{bmatrix} \to 3\cdot(2)$$

$$\to \begin{bmatrix} 1 & -4 & 9 & -4 & | & -25 \\ 0 & 48 & -135 & 45 & | & 411 \\ 0 & 0 & 4 & -13 & | & 49 \\ 0 & 3 & -8 & 4 & | & 18 \end{bmatrix} \to (2)-16(4) \begin{bmatrix} 1 & -4 & 9 & -4 & | & -25 \\ 0 & 0 & -7 & -19 & | & 123 \\ 0 & 0 & 4 & -13 & | & 49 \\ 0 & 3 & -8 & 4 & | & 18 \end{bmatrix} \to 4\cdot(2)$$

$$\to \begin{bmatrix} 1 & -4 & 9 & -4 & | & -25 \\ 0 & 0 & -28 & -76 & | & 492 \\ 0 & 0 & 4 & -13 & | & 49 \\ 0 & 3 & -8 & 4 & | & 18 \end{bmatrix} \begin{smallmatrix}(2)+7(3) \\ \to (2)\leftrightarrow(4)\end{smallmatrix} \begin{bmatrix} 1 & -4 & 9 & -4 & | & -25 \\ 0 & 3 & -8 & 4 & | & 18 \\ 0 & 0 & 4 & -13 & | & 49 \\ 0 & 0 & 0 & -167 & | & 835 \end{bmatrix},$$

$x_4 = \dfrac{835}{-167} = -5,$

$4x_3 = 49 - 65 = -16, x_3 = -4,$

$3x_2 = 18 - 32 + 20 = 6, x_2 = 2,$

$x_1 = -25 + 8 + 36 - 20 = -1.$

Write the answer: $[x_1, x_2, x_3, x_4]^T = [-1, 2, -4, -5]^T$.

(d) Execute the following elementary transformations of the augmented system matrix:

$$\begin{bmatrix} -8 & 1 & -4 & -8 & | & 7 \\ 0 & -7 & -6 & 7 & | & 56 \\ -8 & 3 & 2 & -2 & | & -63 \\ -8 & -3 & -1 & -4 & | & -6 \end{bmatrix} \begin{smallmatrix}(3)-(1) \\ \to (4)-(1)\end{smallmatrix} \begin{bmatrix} -8 & 1 & -4 & -8 & | & 7 \\ 0 & -7 & -6 & 7 & | & 56 \\ 0 & 2 & 6 & 6 & | & -70 \\ 0 & -4 & 3 & 4 & | & -13 \end{bmatrix} \to \tfrac{1}{2}(3)$$

$$\to \begin{bmatrix} -8 & 1 & -4 & -8 & | & 7 \\ 0 & -7 & -6 & 7 & | & 56 \\ 0 & 1 & 3 & 3 & | & -35 \\ 0 & -4 & 3 & 4 & | & -13 \end{bmatrix} \begin{smallmatrix}(2)+7(3) \\ \to (4)+4(3)\end{smallmatrix} \begin{bmatrix} -8 & 1 & -4 & -8 & | & 7 \\ 0 & 0 & 15 & 28 & | & -189 \\ 0 & 1 & 3 & 3 & | & -35 \\ 0 & 0 & 15 & 16 & | & -153 \end{bmatrix} \to (4)-(2)$$

$$\rightarrow \begin{bmatrix} -8 & 1 & -4 & -8 & | & 7 \\ 0 & 0 & 15 & 28 & | & -189 \\ 0 & 1 & 3 & 3 & | & -35 \\ 0 & 0 & 0 & -12 & | & 36 \end{bmatrix},$$

$$x_4 = \frac{36}{-12} = -3,$$
$$15x_3 = 84 - 189, \ x_3 = -7,$$
$$x_2 = -35 + 9 + 21 = -5,$$
$$-8x_1 = 7 - 24 - 28 + 5 = -40, \ x_1 = 5.$$

Write the answer: $[x_1, x_2, x_3, x_4]^T = [5, -5, -7, -3]^T$.

3.5 *Answer:*

(a) $[x_1, x_2, x_3, x_4, x_5]^T = [6, -6, 7, 0, -6]^T$;
(b) $[x_1, x_2, x_3, x_4, x_5]^T = [6, -7, -1, -1, 4]^T$.

3.6 *Answer:*

(a) $[x_1, x_2, x_3, x_4, x_5, x_6]^T = [5, 0, -5, 5, -5, 4]^T$;
(b) $[x_1, x_2, x_3, x_4, x_5, x_6]^T = [1, -2, -2, 0, 1, 0]^T$.

3.7 *Solution.*

The fourth power polynomial with real coefficients can be presented in the form:

$$p(x) = a_4 x^4 + a_3 x^3 + a_2 x^2 + a_1 x + a_0,$$

where $a_0, a_1, \ldots, a_4 \in \mathbb{R}$ are the unknown coefficients. We obtain system of linear equations relative to the coefficients a_i:

$$\begin{cases} 5^4 a_4 + 5^3 a_3 + 5^2 a_2 + 5a_1 + a_0 = 1, \\ 1^4 a_4 + 1^3 a_3 + 1^2 a_2 + 1a_1 + a_0 = 0, \\ 2^4 a_4 + 2^3 a_3 + 2^2 a_2 + 2a_1 + a_0 = 0, \\ 3^4 a_4 + 3^3 a_3 + 3^2 a_2 + 3a_1 + a_0 = 0, \\ 4^4 a_4 + 4^3 a_3 + 4^2 a_2 + 4a_1 + a_0 = 0. \end{cases}$$

Write the matrix of this system and bring it to the echelon form:

$$\begin{bmatrix} 625 & 125 & 25 & 5 & 1 & | & 1 \\ 1 & 1 & 1 & 1 & 1 & | & 0 \\ 16 & 8 & 4 & 2 & 1 & | & 0 \\ 81 & 27 & 9 & 3 & 1 & | & 0 \\ 256 & 64 & 16 & 4 & 1 & | & 0 \end{bmatrix} \rightarrow \begin{bmatrix} 1 & 1 & 1 & 1 & 1 & | & 0 \\ 0 & 8 & 12 & 14 & 15 & | & 0 \\ 0 & 0 & 6 & 9 & 10 & | & 0 \\ 0 & 0 & 0 & 24 & 50 & | & 0 \\ 0 & 0 & 0 & 0 & 1 & | & 1 \end{bmatrix}.$$

we find the values of a_i:

$$\begin{cases} a_4 + a_3 + a_2 + a_1 + a_0 = 0, \\ 8a_3 + 12a_2 + 14a_1 + 15a_0 = 0, \\ 6a_2 + 9a_1 + 10a_0 = 0, \\ 24a_1 + 50a_0 = 0, \\ a_0 = 1; \end{cases} \Rightarrow \begin{cases} a_0 = 1, \\ a_1 = -\dfrac{50}{24}, \\ a_2 = \dfrac{35}{24}, \\ a_3 = -\dfrac{10}{24}, \\ a_4 = \dfrac{1}{24}. \end{cases}$$

As a result, the sought-for fourth power polynomial is equal to

$$p(x) = \frac{1}{24}(x^4 - 10x^3 + 35x^2 - 50x + 24).$$

3.8 *Solution.*

The system of linear equations is consistent if and only if the rank of the system matrix is equal to the rank of the augmented system matrix.

Find the rank of the basic matrix:

$$A = \begin{bmatrix} 1 & 1 \\ \lambda & 1 \\ 1 & \lambda \end{bmatrix}.$$

$$A \to \begin{matrix} \\ (2)-\lambda(1) \\ (3)-(1) \end{matrix} \begin{bmatrix} 1 & 1 \\ 0 & 1-\lambda \\ 0 & \lambda-1 \end{bmatrix} \to {}^{(3)+(2)} \begin{bmatrix} 1 & 1 \\ 0 & 1-\lambda \\ 0 & 0 \end{bmatrix}.$$

Therefore,

$$\operatorname{rk} A = \begin{cases} 2, & \text{if } \lambda \neq 1, \\ 1, & \text{if } \lambda = 1. \end{cases}$$

Compute the rank of the augmented matrix:

$$(A|B) = \begin{bmatrix} 1 & 1 & | & 1 \\ \lambda & 1 & | & 2 \\ 1 & \lambda & | & 4 \end{bmatrix} \xrightarrow[\substack{(2)-\lambda(1) \\ (3)-(1)}]{} \begin{bmatrix} 1 & 1 & | & 1 \\ 0 & 1-\lambda & | & 2-\lambda \\ 0 & \lambda-1 & | & 3 \end{bmatrix} \xrightarrow{(2)+(3)}$$

$$\rightarrow \begin{bmatrix} 1 & 1 & | & 1 \\ 0 & 1-\lambda & | & 2-\lambda \\ 0 & 0 & | & 5-\lambda \end{bmatrix}.$$

Thus,

$$\mathrm{rk}\,(A|B) = \begin{cases} 2, & \text{if } \lambda = 5, \\ 3, & \text{if } \lambda \neq 5. \end{cases}$$

The condition $\mathrm{rk}\,A = \mathrm{rk}\,(A|B)$ is satisfied at $\lambda = 5$. Therefore the system is consistent at $\lambda = 5$.

3.9 *Answer:*

The system has the unique solution at $\lambda \neq 0$. For such values of λ the solution of the system has the form $x_1 = -1$, $x_2 = \lambda^{-1}$, $x_3 = \lambda^{-1} - 4$.

3.10 *Solution.*

Since the matrix A has the size 3×3, and the matrix B has the size 2×2, then the unknown matrix X can be represented in the form:

$$X = \begin{bmatrix} a & b \\ c & d \\ e & f \end{bmatrix}, \text{ where } a, b, c, d, e, f \text{ are some real numbers.}$$

Having substituted this matrix into the equation $A \cdot X \cdot B = C$, we obtain a system of linear equations relative to the unknowns $a, b, \ldots, f$:

$$\begin{cases} 3a + 3b - 2c - 2d - 3e - 3f = -13, \\ 2a + 2b - c - d - 4e - 4f = -25, \\ -6a - 6b - c - d - 2e - 2f = -33. \end{cases}$$

Bring the augmented system matrix to the echelon form:

$$M = \begin{bmatrix} 3 & 3 & -2 & -2 & -3 & -3 & | & -13 \\ 2 & 2 & -1 & -1 & -4 & -4 & | & -25 \\ -6 & -6 & -1 & -1 & -2 & -2 & | & -33 \end{bmatrix} \rightarrow$$

$$\rightarrow \begin{bmatrix} 3 & 3 & -2 & -2 & -3 & -3 & | & -13 \\ 0 & 0 & 1 & 1 & -6 & -6 & | & -49 \\ 0 & 0 & 0 & 0 & 1 & 1 & | & 8 \end{bmatrix} .$$

As the basic minor we select, for example, $M_{1,3,5}^{1,2,3} \neq 0$. Hence, b, d, f are the independent variables by which the variables a, c, e are expressed:

$$X = \begin{bmatrix} 3 - b & b \\ -1 - d & d \\ 8 - f & f \end{bmatrix}, \text{ where } b, d, f \in \mathbb{R}.$$

3.11 *Answer:*

(1) $X = \begin{bmatrix} a & b \\ -2 - 3a & 2 - 3b \end{bmatrix}$, where $a, b \in \mathbb{R}$;

(2) $X \in \varnothing$, i.e. there are no solutions;

(3) $X = \dfrac{1}{5} \begin{bmatrix} 5a & 3(a-1) & -a - 4 \\ 5d & 3d - 5 & -d \\ 5g & 3(g+1) & 4 - g \end{bmatrix}$, where $a, d, g \in \mathbb{R}$;

(4) $X \in \varnothing$.

3.12 *Solution.*

Denote the initial values of the elements of the array a[] by a[1], a[2], ..., a[5]. After executing the program segment code, the array a[] will contain the following elements:

```
16a[1]+16a[5],
-8a[1]+16a[2]-8a[5],
 4a[1]-8a[2]+16a[3]+4a[5],
-2a[1]+4a[2]-8a[3]+16a[4]-2a[5],
 a[1]-2a[2]+4a[3]-8a[4]+17a[5].
```

We obtain system of linear equations with five unknowns. Its solution is $a[1] = -4$, $a[1] = 1$, $a[1] = 3$, $a[1] = 3$, $a[1] = 2$. Hence, the initial array has the form $[-4, 1, 3, 3, 2]$.

3.13 *Answer:* $[-14, -5, -2, 1, 1, 1, 5]$.

3.14 *Solution.*

```python
import math


def gauss_jordan_elimination(A, B):
    m = len(A)
    n = len(A[0])

    if len(B) != m:
        raise ValueError

    C = [[A[i][j] if j != n else B[i]
            for j in range(n+1)] for i in range(m)]

    for r in range(n):
        max_row_pos = r

        # Pivoting strategy
        for i in range(r + 1, n):
            if abs(C[i][r]) > abs(C[max_row_pos][r]):
                max_row_pos = i

        C[r], C[max_row_pos] = C[max_row_pos], C[r]

        if math.isclose(C[r][r], 0):
            continue

        for i in range(n):
            factor = C[i][r] / C[r][r]

            for j in range(n + 1):
                if i != r and j != r:
                    C[i][j] -= factor * C[r][j]

        for i in range(n):
            if i != r:
                C[i][r] = 0.0

        for j in range(n + 1):
```

```
        if j != r:
            C[r][j] /= C[r][r]

    C[r][r] = 1.0

answer = [0] * n

for i in range(n):
    if not math.isclose(C[i][i], 0):
        answer[i] = C[i][n] / C[i][i]
    elif not math.isclose(C[i][n], 0):
        return None

return answer
```

Let us give an example of a call of the function gauss_jordan_ elimination():

```
size = 100

A = [[0 for j in range(size)]
     for i in range(size)]

B = [0 for i in range(size)]

for i in range(size):
    for j in range(size):
        A[i][j] = 1 if j == i else 0

for i in range(size):
    B[i] = float(i)

print(A)
print(B)

print(gauss_jordan_elimination(A, B))
```

The asymptotic complexity of the Gauss–Jordan method coincides with the complexity of Gaussian method and is equal to $O(n^3)$, where n is the number of equations of the initial system of equations [58].

3.15 *Solution.*

(1) Denote the system matrix by A and bring it to the upper triangular form.

$$A = \begin{bmatrix} 1 & -4 & 1 \\ 1 & 1 & -1 \\ 3 & -2 & -1 \end{bmatrix} \xrightarrow[\substack{(2)-(1) \\ (3)-3(1)}]{} \begin{bmatrix} 1 & -4 & 1 \\ 0 & 5 & -2 \\ 0 & 10 & -4 \end{bmatrix} \xrightarrow[(3)-2(2)]{} \begin{bmatrix} 1 & -4 & 1 \\ 0 & 5 & -2 \\ 0 & 0 & 0 \end{bmatrix}.$$

The rank of the matrix is equal to two, and the number of variables is equal to three. This implies that the system will have $3 - 2 = 1$ free variables.

Write the resulting equations:

$$\begin{cases} x_1 - 4x_2 + x_3 = 0, \\ \qquad 5x_2 - 2x_3 = 0. \end{cases}$$

As the independent variable, select x_3:

$$\begin{cases} x_1 = \dfrac{3}{5}x_3, \\ x_2 = \dfrac{2}{5}x_3. \end{cases}$$

As a result we obtain the fundamental system of solutions: $\{[3, 2, 5]^T\}$.

(2) Write the system matrix A and bring it to the upper triangular form.

$$A = \begin{bmatrix} 2 & -1 & 3 & 1 \\ 2 & -5 & -1 & 0 \\ 4 & -7 & 1 & 3 \end{bmatrix} \xrightarrow[\substack{(2)-(1) \\ (3)-2(1)}]{} \begin{bmatrix} 2 & -1 & 3 & 1 \\ 0 & -4 & -4 & -1 \\ 0 & -5 & -5 & 1 \end{bmatrix} \xrightarrow[-1(2)]{}$$

$$\rightarrow \begin{bmatrix} 2 & -1 & 3 & 1 \\ 0 & 4 & 4 & 1 \\ 0 & -5 & -5 & 1 \end{bmatrix} \xrightarrow[4(3)+5(2)]{} \begin{bmatrix} 2 & -1 & 3 & 1 \\ 0 & 4 & 4 & 1 \\ 0 & 0 & 0 & 9 \end{bmatrix}.$$

It is clear that the rank of the matrix is $\text{rk } A = 3$, the number of variables is equal to 4. Therefore, the system will have $4 - 3 = 1$ free variables. Write the resulting equations:

$$\begin{cases} 2x_1 - x_2 + 3x_3 + x_4 = 0, \\ \qquad 4x_2 + 4x_3 + x_4 = 0, \\ \qquad\qquad\qquad\quad 9x_4 = 0. \end{cases}$$

As the independent variable, select x_3:

$$\begin{cases} x_1 = -2x_3, \\ x_2 = -x_3, \\ x_4 = 0. \end{cases}$$

The fundamental system of solutions is $\{[-2, -1, 1, 0]^T\}$.

(3) Write the system matrix A and bring it to the upper triangular form.

$$A = \begin{bmatrix} 1\ 2\ 4 & -3 \\ 3\ 5\ 6 & -4 \\ 4\ 5\ {-2} & 3 \\ 3\ 8\ 24 & -19 \end{bmatrix} \xrightarrow[\substack{(4)-(2) \\ (2)-3(1) \\ (3)-4(1)}]{} \begin{bmatrix} 1 & 2 & 4 & -3 \\ 0 & -1 & -6 & 5 \\ 0 & -3 & -18 & 15 \\ 0 & 3 & 18 & -15 \end{bmatrix}.$$

Note that the second, third and the fourth rows are proportional:

$$A \to \begin{bmatrix} 1 & 2 & 4 & -3 \\ 0 & -1 & -6 & 5 \end{bmatrix}.$$

The rank of the matrix rk $A = 2$, the number of variables is 4. Therefore, the system will have $4 - 2 = 2$ free variables. Write the resulting equations:

$$\begin{cases} x_1 + 2x_2 + 4x_3 - 3x_4 = 0, \\ \qquad -x_2 - 6x_3 + 5x_4 = 0. \end{cases}$$

As the independent variables, select x_3 and x_4:

$$\begin{cases} x_1 = 8x_3 - 7x_4, \\ x_2 = -6x_3 + 5x_4. \end{cases}$$

The fundamental system of solutions is $\{[8, -6, 1, 0]^T, [-7, 5, 0, 1]^T\}$.

(4) Write the system matrix A and bring it to the upper triangular form.

$$A = \begin{bmatrix} 3\ 5\ 2 \\ 4\ 7\ 5 \\ 1\ 1\ {-4} \\ 2\ 9\ 6 \end{bmatrix} \xrightarrow{(1)\leftrightarrow(3)} \begin{bmatrix} 1\ 1\ {-4} \\ 4\ 7\ 5 \\ 3\ 5\ 2 \\ 2\ 9\ 6 \end{bmatrix} \xrightarrow[\substack{(2)-4(1) \\ (3)-3(1) \\ (4)-2(1)}]{} \begin{bmatrix} 1\ 1\ {-4} \\ 0\ 3\ 21 \\ 0\ 2\ 14 \\ 0\ 7\ 14 \end{bmatrix} \xrightarrow{(4)-(3)}$$

$$\rightarrow \begin{bmatrix} 1 & 1 & -4 \\ 0 & 3 & 21 \\ 0 & 2 & 14 \\ 0 & 5 & 0 \end{bmatrix} \xrightarrow{\;3(3)-2(2)\;} \begin{bmatrix} 1 & 1 & -4 \\ 0 & 3 & 21 \\ 0 & 0 & 0 \\ 0 & 5 & 0 \end{bmatrix}.$$

The third row entirely consists of zero elements and can be eliminated from the system matrix.

$$\begin{bmatrix} 1 & 1 & -4 \\ 0 & 3 & 21 \\ 0 & 5 & 0 \end{bmatrix} \xrightarrow{(2)\leftrightarrow(3)} \begin{bmatrix} 1 & 1 & -4 \\ 0 & 5 & 0 \\ 0 & 3 & 21 \end{bmatrix} \xrightarrow{\;5(3)-3(2)\;} \begin{bmatrix} 1 & 1 & -4 \\ 0 & 5 & 0 \\ 0 & 0 & 105 \end{bmatrix}.$$

The rank of the matrix rk $A = 3$, the number of variables is 3. Therefore, the system has no free variables. Write the resulting equations:

$$\begin{cases} x_1 + x_2 - 4x_3 = 0, \\ 5x_2 = 0, \\ 105x_3 = 0. \end{cases}$$

Therefore

$$\begin{cases} x_1 = 0, \\ x_2 = 0, \\ x_3 = 0. \end{cases}$$

As a result, the system has only the trivial solution: $[0, 0, 0]^T$.

(5) Write the system matrix and bring it to the upper triangular form.

$$A = \begin{bmatrix} 2 & 4 & 6 & 1 \\ 1 & 2 & 3 & 1 \\ 3 & 6 & 9 & -1 \\ 1 & 2 & 3 & 5 \end{bmatrix} \xrightarrow{(1)-(2)} \begin{bmatrix} 1 & 2 & 3 & 0 \\ 1 & 2 & 3 & 1 \\ 3 & 6 & 9 & -1 \\ 1 & 2 & 3 & 5 \end{bmatrix} \xrightarrow[\substack{(2)-(1)\\(3)-3(1)\\(4)-(1)}]{} \begin{bmatrix} 1 & 2 & 3 & 0 \\ 0 & 0 & 0 & 1 \\ 0 & 0 & 0 & -1 \\ 0 & 0 & 0 & 5 \end{bmatrix}.$$

The second, third and fourth rows are proportional:

$$A \rightarrow \begin{bmatrix} 1 & 2 & 3 & 0 \\ 0 & 0 & 0 & 1 \end{bmatrix}.$$

The rank of the matrix $\operatorname{rk} A = 2$, the number of variables is equal to four, therefore, the system will have $4 - 2 = 2$ free variables. Write the resulting equations:

$$\begin{cases} x_1 + 2x_2 + 3x_3 = 0, \\ x_4 = 0. \end{cases}$$

Therefore

$$\begin{cases} x_1 = -2x_2 - 3x_3, \\ x_4 = 0. \end{cases}$$

The fundamental system of solutions is $\{[-2, 1, 0, 0]^T, [-3, 0, 1, 0]^T\}$.

(6) Write the matrix A and bring it to the upper triangular form.

$$A = \begin{bmatrix} 1 & 2 & 3 \\ 2 & 3 & 4 \\ 1 & 1 & 1 \end{bmatrix} \xrightarrow[\substack{(2)-2(1) \\ (3)-(1)}]{} \begin{bmatrix} 1 & 2 & 3 \\ 0 & -1 & -2 \\ 0 & -1 & -2 \end{bmatrix}.$$

The second and the third rows are proportional:

$$A \rightarrow \begin{bmatrix} 1 & 2 & 3 \\ 0 & -1 & -2 \end{bmatrix}.$$

The rank of the matrix $\operatorname{rk} A = 2$, the number of variables is equal to three. Therefore, the system will have $3 - 2 = 1$ free variables. Write the resulting equations:

$$\begin{cases} x + 2y + 3z = 0, \\ -y - 2z = 0. \end{cases}$$

Therefore

$$\begin{cases} x = z, \\ y = -2z. \end{cases}$$

The fundamental system of solutions is $\{[1, -2, 1]^T\}$.

(7) Write the system matrix A and bring it to the upper triangular form.

$$\begin{bmatrix} 1 & -2 & 3 & -4 \\ 2 & -4 & 5 & 7 \\ 6 & -12 & 17 & -9 \\ 7 & -14 & 19 & 17 \end{bmatrix} \xrightarrow[\begin{subarray}{l} (2)-2(1) \\ (3)-6(1) \\ (4)-7(1) \end{subarray}]{} \begin{bmatrix} 1 & -2 & 3 & -4 \\ 0 & 0 & -1 & 15 \\ 0 & 0 & -1 & 15 \\ 0 & 0 & -2 & 45 \end{bmatrix}.$$

The second and the third rows are proportional:

$$\begin{bmatrix} 1 & -2 & 3 & -4 \\ 0 & 0 & -1 & 15 \\ 0 & 0 & -2 & 45 \end{bmatrix} \xrightarrow{(3)-2(2)} \begin{bmatrix} 1 & -2 & 3 & -4 \\ 0 & 0 & -1 & 15 \\ 0 & 0 & 0 & 15 \end{bmatrix}.$$

The rank of the matrix $\operatorname{rk} A = 3$, the number of variables is equal to four. Therefore, the system will have $4 - 3 = 1$ free variables. Write the resulting equations:

$$\begin{cases} x_1 - 2x_2 + 3x_3 - 4x_4 = 0, \\ \qquad\qquad -x_3 + 15x_4 = 0, \\ \qquad\qquad\qquad\quad 15x_4 = 0. \end{cases}$$

Therefore

$$\begin{cases} x_1 = 2x_2, \\ x_3 = 0, \\ x_4 = 0. \end{cases}$$

The fundamental system of solutions has the form $\{[2, 1, 0, 0]^T\}$.

Chapter 4
Complex Numbers and Matrices

As was already mentioned in Chap. 1, complex numbers may appear as matrix elements. Moreover, the characteristics of real matrices (such as eigenvalues, see Chap. 5 "Vector Spaces" on page 226) in some cases appear to be complex. In this connection, let us discuss the methods of algebra of complex numbers.

Complex number z is an ordered pair of real numbers (a, b), where $a, b \in \mathbb{R}$. The first number a is called the **real part** of the complex number $z = (a, b)$ and is denoted by symbol $\operatorname{Re} z$, while the second number of the pair b is called the **imaginary part** z and is denoted by $\operatorname{Im} z$ [24].

A complex number of the form $(a, 0)$, where the imaginary part is zero, is identified with the real number a, i.e. $(a, 0) \equiv a$. This allows considering the set of all real numbers $\mathbb{R}$ as a subset of a set of complex numbers $\mathbb{C}$.

Two complex numbers $z_1 = (a_1, b_1)$ and $z_2 = (a_2, b_2)$ are considered equal if and only if their real and imaginary parts are pairwise equal: $z_1 = z_2 \Leftrightarrow a_1 = a_2$, $b_1 = b_2$.

4.1 Arithmetic Operations with Complex Numbers

On the set $\mathbb{C}$ are defined the operations of addition and multiplication of complex numbers. **Sum** of complex numbers $z_1 = (a_1, b_1)$ and $z_2 = (a_2, b_2)$ is the complex number z, equal to $z_1 + z_2 = (a_1 + a_2, b_1 + b_2)$. **Product** of numbers $z_1 = (a_1, b_1)$ and $z_2 = (a_2, b_2)$ is such a complex number $z = (a, b)$, that $a = a_1 a_2 - b_1 b_2$, $b = a_1 b_2 + a_2 b_1$.

The pair $(0, 1)$ is of the greatest importance in the operations with complex numbers; it is denoted by $(0, 1) \equiv i$ and is called **imaginary unit**. The basic property of the imaginary unit consists in that $i^2 = i \cdot i = (0, 1) \cdot (0, 1) = (-1, 0)$, or $i^2 = -1$.

© Springer Nature Switzerland AG 2021
S. Kurgalin, S. Borzunov, *Algebra and Geometry with Python*,
https://doi.org/10.1007/978-3-030-61541-3_4

A complex number of the form $z = (0, b)$ is called **purely imaginary**. Since $(0, b) = (b, 0) \cdot (0, 1)$, then the purely imaginary number z is presentable in the form of the product $z = bi$.

Any complex number can be presented in the form

$$z = (a, b) = (a, 0) + (0, b) = (a, 0) + (b, 0) \cdot (0, 1) = a + ib.$$

Such a notation is referred to as the **algebraic form** of a complex number. This allows considering i as a factor, whose square is equal to -1, and performing operations with complex numbers in the same manner as with algebraic polynomials, in intermediate calculations assuming $i^2 = -1$.

Example 4.1 Let $z_1 = 2 + 5i$, $z_2 = -3 + 2i$. Then, the addition of these numbers will result in a complex number

$$z_1 + z_2 = (2 + 5i) + (-3 + 2i) = (2 - 3) + (5 + 2)i = -1 + 7i. \qquad (4.1)$$

The product of the numbers z_1 and z_2 is computed by multiplying the expressions $(2 + 5i)$ and $(-3 + 2i)$ as polynomials with regard to the equality $i^2 = -1$:

$$z_1 z_2 = (2 + 5i)(-3 + 2i) = 2(-3) + 2(2i) - 3(5i) + (5i)(2i)$$

$$= -6 + 4i - 15i + 10i^2 = -6 - 10 + (4 - 15)i = -16 - 11i. \qquad (4.2)$$

$\square$

A complex number $z^* = (a, -b) = a - ib$ is called a **conjugate** of the complex number $z = (a, b) = a + ib$. There is one more frequent notation of a conjugate—$\overline{z}$. If the coefficients of the polynomial $p(z)$ are real, the equality $(p(z))^* = p(z^*)$ is valid.

It is convenient to present the number $z = a + ib$ as the point (x, y) of a plane with Cartesian coordinates $x = a$ and $y = b$. Correlate each complex number z with a point with coordinates (x, y) (and a position vector, connecting the origin of coordinates with this point). Such a plane is denoted by $\mathbb{Z}$ and is referred to as **complex plane** (see Fig. 4.1). Note that geometric interpretation of complex numbers is sometimes referred to as the **Argand**[1] **diagram**.

Many applications widely use a **trigonometric form** of the complex number z. Let us introduce the polar coordinate system so that the pole is at the origin of Cartesian system (x, y). The axis of the polar system will be directed along the positive direction of the axis Ox.

---

[1] Jean-Robert Argand (1768–1822), French mathematician.

Fig. 4.1 Representation of
the number z on a complex
plane

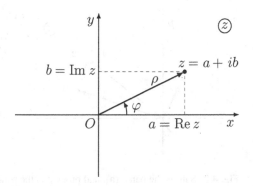

In this case, Cartesian and polar coordinates of an arbitrary point other than the
origin of coordinates are related by the formulae

$$x = \rho \cos \varphi, \quad y = \rho \sin \varphi,$$

$$\rho = \sqrt{x^2 + y^2}, \quad \varphi = \begin{cases} \arctan \dfrac{y}{x}, & \text{if } x > 0; \\[2mm] \arctan \dfrac{y}{x} + \pi, & \text{if } x < 0, \ y \geqslant 0; \\[2mm] \arctan \dfrac{y}{x} - \pi, & \text{if } x < 0, \ y < 0; \\[2mm] \dfrac{\pi}{2} \operatorname{sign} y, & \text{if } x = 0. \end{cases}$$

As a result, we obtain a trigonometric form of the number z

$$z = (x, y) = x + iy = \rho(\cos \varphi + i \sin \varphi).$$

The value ρ is called **modulus**, and φ—**argument** of the complex number z and
denoted $\rho = |z|$, $\varphi = \arg z$. It should be noted that the argument φ is ambiguously
determined: instead of the value φ we can take the value $\varphi + 2\pi k$, where $k \in \mathbb{Z}$. If
$\arg z$ is chosen in such a way that $-\pi < \arg z \leqslant \pi$, then such a value is called the
principal value of the argument.

For the numbers $z_1 = \rho_1(\cos \varphi_1 + i \sin \varphi_1)$ and $z_2 = \rho_2(\cos \varphi_2 + i \sin \varphi_2)$,
specified in the trigonometric form,

$$z_1 z_2 = \rho_1 \rho_2 (\cos(\varphi_1 + \varphi_2) + i \sin(\varphi_1 + \varphi_2)),$$

$$\frac{z_1}{z_2} = \frac{\rho_1}{\rho_2} (\cos(\varphi_1 - \varphi_2) + i \sin(\varphi_1 - \varphi_2)), \quad \rho_2 \neq 0.$$

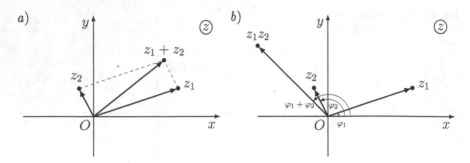

Fig. 4.2 Sum is the panel (**a**), and product is the panel (**b**) complex numbers z_1 and z_2

Geometric illustration of the sum and the product of complex numbers is shown in Fig. 4.2. For any $z_1, z_2 \in \mathbb{C}$, the position vector of the sum $z_1 + z_2$ is equal to the sum of the position vectors of the summands z_1 and z_2. The position vector of the product $z_1 z_2$ is obtained by rotating the position vector of the number z_1 by the angle of $\arg z_2$ counter-clockwise and extending by $|z_2|$ times.

Euler's[2] formula relates the exponential function of the imaginary argument with trigonometric functions of the imaginary part of the argument:

$$e^{i\varphi} = \cos\varphi + i\sin\varphi.$$

This is why we can introduce one more notation of the complex number, namely, **exponential**: $z = \rho e^{i\varphi}$. The exponential notation is convenient for operations of multiplication, division, raising to a power and extraction of root. For example, the n-th power of the number z can be presented in the form

$$z^n = (\rho e^{i\varphi})^n = \rho^n e^{in\varphi} = \rho^n(\cos n\varphi + i\sin n\varphi)$$

for all integer values n.

An important consequence of the obtained formula

$$(\cos\varphi + i\sin\varphi)^n = \cos n\varphi + i\sin n\varphi$$

is associated with the name of **de Moivre**.[3]

The n-th root of $z = \rho(\cos\varphi + i\sin\varphi)$ can be calculated as

$$\sqrt[n]{z} \equiv z^{1/n} = [\rho(\cos(\varphi + 2\pi k) + i\sin(\varphi + 2\pi k))]^{1/n}, \quad k \in \mathbb{Z},$$

---

[2]Leonhard Euler (1707–1783), prominent Swiss mathematician.

[3]Abraham de Moivre (1667–1754), English mathematician of French origin.

or, after applying Euler's formula,

$$\sqrt[n]{z} = \rho^{1/n}\left(\cos\left(\frac{\varphi + 2\pi k}{n}\right) + i\sin\left(\frac{\varphi + 2\pi k}{n}\right)\right), \quad k = 0, 1, \ldots, n - 1.$$

Here we obtain n possible values of the n-th root for $k = 0, 1, \ldots, n - 1$. Other acceptable k do not result in new values of $\sqrt[n]{z}$. For example, for $k = n$ the argument is $\arg z = \varphi/n + 2\pi$ and differs from the case $k = 0$ by 2π, which corresponds to the complex number equal to it.

Example 4.2 Denote roots of the equation $z^n = 1$, where n is a natural number, by ω_k, $k = 0, \ldots, n - 1$. Prove that on a complex plane the points corresponding to the values ω_k are located at the vertices of a regular n-gon, inscribed in a unit circle, whose centre is located at the origin of coordinates [24].

Proof According to the introduced definition,

$$\omega_k = (e^{2\pi i})^{k/n} = e^{2\pi ik/n}, \quad k = 0, 1, \ldots, n - 1. \tag{4.3}$$

In particular, for $n = 4$ we have the following values $\sqrt[4]{1}$:

$$\omega_k = (e^{2\pi i})^{k/4} = e^{\pi ik/2}, \quad k \in \{0, 1, 2, 3\}, \tag{4.4}$$

or, after computing the complex exponents:

$$\omega_k = 1, \, i, \, -1, \, -i. \tag{4.5}$$

Modulus of the complex number $\omega_k = e^{2\pi ik/n}$ is equal to one for all values of the variable k, and the argument is equal to $\arg\omega_k = 2\pi k/n$, $k = 0, 1, \ldots, n - 1$. Thereby, we can conclude that the n-th roots of one are located on the unit circle C, and the first root ω_0, associated with $k = 0$, lies in the real axis, and ω_k divide the circle by n arcs of the same length (see the example for the instance $n = 9$ in Fig. 4.3). $\quad\square$

Fig. 4.3 Location of the roots of the n-th power of one on the unit circle for $n = 9$

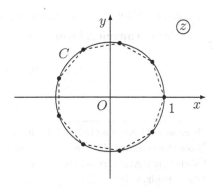

4.2   Fundamental Theorem of Algebra

Theorem 4.1 (Fundamental Theorem of Algebra)   States that *any polynomial of a non-zero degree with complex coefficients has a complex root* [43]. This is why an arbitrary polynomial with real (or complex) coefficients always has some root $z \in \mathbb{C}$.

Each polynomial of degree n

$$p(z) = c_n z^n + c_{n-1} z^{n-1} + \cdots + c_0, \quad c_i \in \mathbb{C} \text{ for } i = 0, 1, \ldots, n, \ c_n \neq 0,$$

can uniquely (accurate to the order of cofactors) be expanded into the product

$$p(z) = c_n (z - z_1)^{m_1} (z - z_2)^{m_2} \ldots (z - z_k)^{m_k},$$

where z_i is a root of the polynomial $p(z)$ with a multiplicity of m_i, $1 \leqslant i \leqslant k$.

For polynomials with degree lower than the fifth, we can always find the roots having expressed them by arithmetic operations or arithmetic roots of an arbitrary multiplicity, or **radicals**. The method for calculating the cubic polynomial's roots was suggested by **Cardanus**[4] (see Sect. 4.3 below), of the fourth degree polynomial—**Ferrari**.[5] However, there are no common methods for finding roots of polynomials of higher degrees, according to next theorem:

Theorem 4.2 (Abel[6]–Ruffini[7] Theorem)   *Any arbitrary equation of degree n for $n \geqslant 5$ is unsolvable in radicals.*

4.3   Cardano Formula

In order to determine the roots of the cubic equation

$$az^3 + bz^2 + cz + d = 0, \quad \text{where } a, b, c, d \in \mathbb{C}, \tag{4.6}$$

proceed as follows. Using the change of the variable $z = y - \dfrac{b}{3a}$ the equation is brought to a **canonical form**

$$y^3 + py + q = 0, \quad p, q \in \mathbb{C}. \tag{4.7}$$

---

[4]Hieronymus Cardanus (1501–1576), Italian mathematician and philosopher.
[5]Lodovico Ferrari (1522–1565), Italian mathematician.
[6]Niels Henrik Abel (1802–1829), Norwegian mathematician.
[7]Paolo Ruffini (1765–1822), Italian mathematician.

By **Cardano formula**, the roots of the cubic equation y_1, y_2, y_3 in the canonical form are equal [33]

$$y_1 = \alpha + \beta, \tag{4.8}$$

$$y_2 = -\frac{\alpha + \beta}{2} + i\sqrt{3} \cdot \frac{\alpha - \beta}{2}, \tag{4.9}$$

$$y_3 = -\frac{\alpha + \beta}{2} - i\sqrt{3} \cdot \frac{\alpha - \beta}{2}, \tag{4.10}$$

where

$$\alpha = \sqrt[3]{-\frac{q}{2} + \sqrt{Q}},$$

$$\beta = \sqrt[3]{-\frac{q}{2} - \sqrt{Q}},$$

$$Q = \left(\frac{p}{3}\right)^3 + \left(\frac{q}{2}\right)^2.$$

Using these relations, one should for each of the three values of the cube root of α take that value of the root β, for which the equality $\alpha\beta = -p/3$ is valid.

Example 4.3 Find the roots of the equation $z^3 - 5z^2 + 9z - 5 = 0$, using the Cardano formula.

Solution Replace the variable $z = y + \dfrac{5}{3}$. We obtain the cubic equation in the canonical form

$$y^3 + \frac{2}{3}y + \frac{20}{27} = 0, \tag{4.11}$$

here $p = \dfrac{2}{3}$, $q = \dfrac{20}{27}$. Further using Cardano formula (4.8)–(4.10):

$$Q = \left(\frac{p}{3}\right)^3 + \left(\frac{q}{2}\right)^2 = \frac{4}{27},$$

$$\alpha, \beta = \sqrt[3]{-\frac{10}{27} \pm \sqrt{\frac{4}{27}}} = \frac{1}{3}\sqrt[3]{-10 \pm 6\sqrt{3}}.$$

Let $\alpha = \dfrac{1}{3}\sqrt[3]{6\sqrt{3} - 10}$, then, in order for the condition $\alpha\beta = -p/3$ to be satisfied, we choose $\beta = -\dfrac{1}{3}\sqrt[3]{6\sqrt{3} + 10}$. The roots of the equation will have the form

$$y_1 = \frac{1}{3} \left(\sqrt[3]{6\sqrt{3} - 10} - \sqrt[3]{6\sqrt{3} + 10} \right);$$

$$y_2 = -\frac{1}{6} \left(\sqrt[3]{6\sqrt{3} - 10} - \sqrt[3]{6\sqrt{3} + 10} \right) + \frac{i\sqrt{3}}{6} \left(\sqrt[3]{6\sqrt{3} - 10} + \sqrt[3]{6\sqrt{3} + 10} \right);$$

$$y_3 = -\frac{1}{6} \left(\sqrt[3]{6\sqrt{3} - 10} - \sqrt[3]{6\sqrt{3} + 10} \right) - \frac{i\sqrt{3}}{6} \left(\sqrt[3]{6\sqrt{3} - 10} + \sqrt[3]{6\sqrt{3} + 10} \right).$$

The obtained expressions can be simplified, if we note that the equality $6\sqrt{3} \pm 10 = (\sqrt{3} \pm 1)^3$ is valid. Then $\sqrt[3]{6\sqrt{3} \pm 10} = \sqrt{3} \pm 1$, and

$$y_1 = \frac{1}{3} \left(\sqrt{3} - 1 - (\sqrt{3} + 1) \right) = -\frac{2}{3};$$

$$y_2 = -\frac{1}{6} \left(\sqrt{3} - 1 - (\sqrt{3} + 1) \right) + \frac{i\sqrt{3}}{6} \left(\sqrt{3} - 1 + \sqrt{3} + 1 \right) = \frac{1}{3} + i;$$

$$y_3 = -\frac{1}{6} \left(\sqrt{3} - 1 - (\sqrt{3} + 1) \right) - \frac{i\sqrt{3}}{6} \left(\sqrt{3} - 1 + \sqrt{3} + 1 \right) = \frac{1}{3} - i.$$

Returning to the original variable $z = y + \frac{5}{3}$, we obtain $z_1 = 1$, $z_2 = 2 + i$, $z_3 = 2 - i$. $\qquad\square$

4.4 Complex Coefficient Matrices

Among the complex coefficient matrices, classes of Hermitian and unitary matrices play a special role in algebra and its applications.

4.4.1 Hermitian Matrices

Consider the matrix Z, containing the complex elements $Z = (z_{ij})$, where $i = 1, 2, \ldots, m$, $j = 1, 2, \ldots, n$. **Hermitian conjugate** matrix relative to Z is the matrix Z^H, whose elements are equal to

$$z_{ij}^H = z_{ji}^*. \tag{4.12}$$

In order to obtain Hermitian conjugate matrix, the operations of transposition and complex conjugation are applied to the initial matrix. The mentioned operations are independent and can be executed in any sequence.

Example 4.4 A Hermitian conjugate matrix relative to $Z = \begin{bmatrix} 1+i & 2+3i \\ -1 & 5-4i \end{bmatrix}$ is the

matrix

$$
Z^H = \left(\begin{bmatrix} 1+i & 2+3i \\ -1 & 5-4i \end{bmatrix}^T \right)^* = \left(\begin{bmatrix} 1+i & -1 \\ 2+3i & 5-4i \end{bmatrix} \right)^* = \begin{bmatrix} 1-i & -1 \\ 2-3i & 5+4i \end{bmatrix}.
$$
(4.13)

$\square$

Note Often, for designation of a Hermitian conjugate matrix, the notations $Z^\dagger$ or Z^+ are used.

Let us enumerate the main properties of the Hermitian conjugation operation:

1. $I^H = I$;
2. $(Z_1 + Z_2)^H = Z_1^H + Z_2^H$;
3. $(\lambda Z)^H = \lambda^* Z^H \quad \forall \lambda \in \mathbb{C}$;
4. $(Z^H)^H = Z$;
5. if A^{-1} exists, then $(A^{-1})^H = (A^H)^{-1}$;
6. $\det A^H = \det A^* = (\det A)^*$.

Theorem 4.3 *If for the complex matrices Z_1 and Z_2 the product $Z_1 Z_2$ is determined, then*

$$
(Z_1 Z_2)^H = Z_2^H Z_1^H.
$$
(4.14)

Proof The validity of the theorem follows from the property of the matrix product transposition (see Problem **1.3**):

$$
(Z_1 Z_2)^T = Z_2^T Z_1^T.
$$
(4.15)

With the help of Eq. (4.15) we obtain a chain of equalities

$$
(Z_1 Z_2)^H = ((Z_1 Z_2)^T)^* = (Z_2^T Z_1^T)^* = (Z_2^T)^* (Z_1^T)^* = Z_2^H Z_1^H,
$$
(4.16)

which proves the identity (4.14).

Among complex matrices, Hermitian matrices are very widely used. **Hermitian matrix** is a square matrix, where $Z^H = Z$. A respective condition for the elements of such a matrix: $\forall i, j \ (z_{ij} = z_{ji}^*)$. In other words, the Hermitian matrix coincides with its Hermitian-conjugated.

Example 4.5 The matrix $Z = \begin{bmatrix} 1 & -3-i \\ -3+i & 1 \end{bmatrix}$ is Hermitian, as is easy to see.

Let us verify it.

$$Z^H = \left(\begin{bmatrix} 1 & -3-i \\ -3+i & 1 \end{bmatrix}^T \right)^* = \left(\begin{bmatrix} 1 & -3+i \\ -3-i & 1 \end{bmatrix} \right)^* = \begin{bmatrix} 1 & -3-i \\ -3+i & 1 \end{bmatrix}. \tag{4.17}$$

Recall that applying the complex conjugation operation to a real number does not change this number.   □

Note Hermitian matrices are also referred to as **self-adjoint** matrices. The theory of self-adjoint matrices is widely used in modern physics [44].

4.4.2 Unitary Matrices

A square matrix U with complex elements is called **unitary**, of the condition $U^H U = I$ is met. The condition of unitarity can be written in other equivalent forms as follows:

$$UU^H = I \quad \text{or} \quad U^H = U^{-1}. \tag{4.18}$$

Example 4.6 Prove that the matrix $Z = \dfrac{1}{\sqrt{2}} \begin{bmatrix} 1 & 1 \\ -i & i \end{bmatrix}$ is unitary. To do this, compute the product $Z^H Z$:

$$Z^H Z = \frac{1}{\sqrt{2}} \left(\begin{bmatrix} 1 & 1 \\ -i & i \end{bmatrix} \right)^H \frac{1}{\sqrt{2}} \begin{bmatrix} 1 & 1 \\ -i & i \end{bmatrix} = \frac{1}{\sqrt{2}} \begin{bmatrix} 1 & i \\ 1 & -i \end{bmatrix} \frac{1}{\sqrt{2}} \begin{bmatrix} 1 & 1 \\ -i & i \end{bmatrix} = \begin{bmatrix} 1 & 0 \\ 0 & 1 \end{bmatrix}. \tag{4.19}$$

Therefore, Z is a unitary matrix.   □

Theorem 4.4 *The determinant of a unitary matrix is a complex number whose modulus is equal to one.*

Proof See in Problem **4.51**.

There is a close connection between unitary and Hermitian matrices: each unitary matrix A is presented in the form $A = \exp(iB)$, where B is a Hermitian matrix [26].

4.5 Fundamentals of Quantum Computing

In quantum computers, for implementation of computing, processes of a quantum nature are used manifested in experiments with objects of the microcosm— elementary particles, atoms, molecules, molecular clusters, etc. The description of such processes is based on the application of complex numbers and complex matrices.

As is well known, the basic notion of classical information theory is **bit** [18]. A classical bit takes the values 0 or 1 (and no other).

Qubit (quantum bit) is the smallest element that executes the information storage function in a quantum computer [6, 54, 75].

Qubit is a quantum system $|\psi\rangle$ that allows two states: $|0\rangle$ and $|1\rangle$ [54]. In accordance with the so-called bra-ket (bracket) Dirac[8] notation, the symbols $|0\rangle$ or $|1\rangle$ are read as "ket 0" and "ket 1", respectively. The brackets $|\ldots\rangle$ show that ψ is some state of the quantum system.

The fundamental difference between the classical bit and the qubit consists in that the qubit can be in a state different from $|0\rangle$ or $|1\rangle$. The arbitrary state of the qubit is defined by the linear combination of basic states:

$$|\psi\rangle = u\,|0\rangle + v\,|1\rangle, \tag{4.20}$$

where the complex coefficients u and v satisfy the following condition:

$$|u|^2 + |v|^2 = 1. \tag{4.21}$$

The mathematical description of the basic states reduces to their representation in matrix form:

$$|0\rangle = \begin{bmatrix} 1 \\ 0 \end{bmatrix}, \quad |1\rangle = \begin{bmatrix} 0 \\ 1 \end{bmatrix}. \tag{4.22}$$

Based on the presentation (4.22) the arbitrary state of the qubit is written as

$$|\psi\rangle = \begin{bmatrix} u \\ v \end{bmatrix}. \tag{4.23}$$

---

[8]Paul Adrien Maurice Dirac (1902–1984), English physicist.

A system of two qubits is set by a linear combination of basic states

$$|00\rangle = \begin{bmatrix} 1 \\ 0 \\ 0 \\ 0 \end{bmatrix}, \quad |01\rangle = \begin{bmatrix} 0 \\ 1 \\ 0 \\ 0 \end{bmatrix}, \quad |10\rangle = \begin{bmatrix} 0 \\ 0 \\ 1 \\ 0 \end{bmatrix}, \quad |11\rangle = \begin{bmatrix} 0 \\ 0 \\ 0 \\ 1 \end{bmatrix}. \tag{4.24}$$

Similarly are introduced the states

$$|00\ldots00\rangle, \quad |00\ldots01\rangle, \quad \ldots, \quad |11\ldots11\rangle \tag{4.25}$$

of several interacting qubits. Such quantum states are called **computational basis states** or, for short, **basis states**.

For changing the state of a quantum system, quantum operations are used referred to as **gates** (quantum gate). Thus, the gates perform logical operations with qubits. Note that the change of the state $|\psi\rangle$ in time is also referred to as the **evolution** of the quantum system.

An important step of quantum algorithms is the procedure of **measurement of state**. When the qubit state is measured, it randomly passes to one of its states: $|0\rangle$ or $|1\rangle$. Therefore, the complex coefficients u and v from the qubit definition (4.20) are associated with probability to get the value 0 or 1 when its state is measured. According to the postulates of quantum theory, the probabilities of passing to the states $|0\rangle$ and $|1\rangle$ are equal to $|u|^2$ and $|v|^2$, respectively. In this connection, the equality (4.21) reflects the probability conservation law. After the measurement, the qubit passes to the basic state, complying with the classical result of measurement. Generally speaking, the probabilities of getting the result 0 and 1 are different for different states of the quantum system.

In other words, the quantum computing is a sequence of simple form operations with the collection of the interacting qubits. In the final step of the quantum computing procedure, the state of the quantum system is measured and a conclusion about the computing result is made. The measurement makes it possible to obtain, at a macroscopic level, the information about the quantum state. The peculiarity of the quantum measurements is their irreversibility, which radically differentiates the quantum computing from the classical one.

A quantum system, formed by N two-level elements, has $\Sigma(N) = 2^N$ independent states. The key point of functioning of such a system is the interaction of separate qubits with each other. The number of states $\Sigma(N)$ grows exponentially with the growth of the quantum system, which allows solving practical problems of a very high asymptotic complexity (see section "Estimation of Algorithm Efficiency" on page 13). For example, an efficient quantum algorithm of prime factorization is known, which is very important for cryptography. As a result, the quantum algorithms provide exponential or polynomial acceleration in comparison with the classical solution methods for many problems.

Unfortunately, no full-function quantum computer has been created yet, although many of its elements have already been built and studied at the leading world's laboratories [72]. The main obstacle to the development of quantum computing is instability of a system of many qubits. The more qubits are united into an entangled system, the more effort is required to ensure smallness of the number of measurement errors. Nevertheless, the history of quantum computer development demonstrates an enormous potential laid in the uniting of quantum theory and algorithm theory.

Prior to proceeding to describing the basic quantum operations with qubits, let us introduce the notion of the Pauli[9] matrices and the Dirac matrices.

4.5.1 Pauli Matrices and Dirac Matrices

The matrices σ_1, σ_2 and σ_3

$$\sigma_1 = \begin{bmatrix} 0 & 1 \\ 1 & 0 \end{bmatrix}, \quad \sigma_2 = \begin{bmatrix} 0 & -i \\ i & 0 \end{bmatrix}, \quad \sigma_3 = \begin{bmatrix} 1 & 0 \\ 0 & -1 \end{bmatrix} \tag{4.26}$$

are called the **Pauli matrices**. They are widely used in quantum theory for describing half-integer spin particles, for example, an electron. (Spin is a quantum property of an elementary particle, intrinsic angular momentum [44]. So, electrons, protons and neutrino have half-integer spin; spin of photons and gravitons is integer).

The following properties are valid for the Pauli matrices.

1. The Pauli matrices are Hermitian and unitary:

$$\forall k \in \{1, 2, 3\} \quad \sigma_k = \sigma_k^H = \sigma_k^{-1}. \tag{4.27}$$

2. $\forall k \in \{1, 2, 3\}$ the square of the Pauli matrix is equal to the identity matrix:

$$\sigma_i^2 = \begin{bmatrix} 1 & 0 \\ 0 & 1 \end{bmatrix}. \tag{4.28}$$

3. $\forall i, j \in \{1, 2, 3\}$ the equalities are valid

$$\sigma_i \sigma_j + \sigma_j \sigma_i = 2\delta_{ij} \begin{bmatrix} 1 & 0 \\ 0 & 1 \end{bmatrix}. \tag{4.29}$$

---

[9]Wolfgang Ernst Pauli (1900–1958), Swiss and American physicist.

Sometimes, in linear algebra and its applications, one has to use matrices split into rectangular parts or blocks [25, 47]. Consider the rectangular matrix $A = (a_{ij})$, where $1 \leqslant i \leqslant m, 1 \leqslant j \leqslant n$. Let $m = m_1 + m_2, n = n_1 + n_2$.

Let us draw horizontal and vertical lines and split the matrix A into four rectangular blocks:

$$
A = \overbrace{\left[\begin{array}{c|c} B_{11} & B_{12} \\ \hline B_{21} & B_{22} \end{array} \right]}^{n_1 \quad n_2} \begin{array}{l} \} \; m_1 \\ \\ \} \; m_2 \end{array}
\tag{4.30}
$$

Thus, the matrix A is presented in the form of a **block matrix**, consisting of the blocks $B_{11}, B_{12}, B_{21}, B_{22}$ of size $m_1 \times n_1, m_1 \times n_2, m_2 \times n_1, m_2 \times n_2$, respectively.

As an example of block matrix setting, we provide the definition of the Dirac matrices. Four **Dirac matrices** $\alpha_1, \alpha_2, \alpha_3, \beta$ are part of the equation named after him [4], for a half-integer spin relativistic particle, and are expressed in terms of the Pauli matrix $\sigma_k, k = 1, 2, 3$, as follows:

$$
\alpha_k = \begin{bmatrix} O & \sigma_k \\ \sigma_k & O \end{bmatrix}, \quad \beta = \begin{bmatrix} I & O \\ O & -I \end{bmatrix},
\tag{4.31}
$$

where O is a non-zero matrix of size 2×2, I is an identity matrix of the same size. (Relativistic are the particles whose velocity is close to the light velocity.)

Each of the Dirac matrices has a Hermitian property and a unitary property. Moreover, for all $l, m \in \{1, 2, 3\}$ the equalities are valid:

$$
\alpha_l \alpha_m + \alpha_m \alpha_l = 2\delta_{lm} I,
\tag{4.32}
$$

$$
\alpha_l \beta + \beta \alpha_l = O.
\tag{4.33}
$$

Note that the size of the matrices I and O in formulae (4.32) and (4.33) is equal to 4×4.

4.5.2 Basic Operations with Qubits

Consider the basic operations with qubits.

The influence of a quantum gate on the qubit $|\psi\rangle$ is exerted by applying a **quantum-mechanical operator**, for example, $U |\psi\rangle$ [67, 72]. The operators may be presented in the form of unitary matrices. In particular, the evolution of a single qubit is described by a unitary matrix of size 2×2.

Successive application of the series of operators $U_1, U_2, \ldots, U_n$ to one qubit is equivalent to the influence of the operator W, whose matrix is a product of the matrices $U_1 U_2 \ldots U_n$ [54]:

$$U_n(U_{n-1}(\ldots(U_2(U_1 \, |\psi\rangle)))\ldots)) = (U_1 U_2 \ldots U_n) \, |\psi\rangle = W \, |\psi\rangle . \qquad (4.34)$$

Such operator W is called a **composition** of operators $U_1, U_2, \ldots, U_n$. As a consequence of non-commutativity of the matrix multiplication operation in the general case the sequence of applying the quantum gates is of importance.

Example 4.7 Let us show that the application of the operator σ_3 (see definition (4.26)) to the qubit in the state $|\psi\rangle = u \, |0\rangle + v \, |1\rangle$ transfers it to the state $|\psi'\rangle = u \, |0\rangle - v \, |1\rangle$.

Proof Write the qubit $|\psi\rangle$ in matrix representation:

$$|\psi\rangle = \begin{bmatrix} u \\ v \end{bmatrix} . \qquad (4.35)$$

Define the action of σ_3 on this quantum state:

$$|\psi'\rangle = \sigma_3 \begin{bmatrix} u \\ v \end{bmatrix} = \begin{bmatrix} 1 & 0 \\ 0 & -1 \end{bmatrix} \begin{bmatrix} u \\ v \end{bmatrix} = \begin{bmatrix} u \\ -v \end{bmatrix} = u \begin{bmatrix} 1 \\ 0 \end{bmatrix} - v \begin{bmatrix} 0 \\ 1 \end{bmatrix} = u \, |0\rangle - v \, |1\rangle .$$
$$(4.36)$$
$$\square$$

Graphic representation of quantum operations in the form schemes or diagrams (quantum circuit) is widely used.

Some quantum-mechanical operator U that transforms a single qubit (one qubit gate) is represented as follows:

$$|\psi_{\text{in}}\rangle \quad \underline{\quad\boxed{U}\quad} \quad |\psi_{\text{out}}\rangle$$

The sequence of steps of quantum algorithms corresponds to the direction on the diagram from left to right.

In Table 4.1 the gates are enumerated that transform one qubit, and the matrix representations of these gates.

Let us show the method of computing a quantum operation matrix based on its action on the basic vectors.

Table 4.1 One qubit operations

Name	Designation	Matrix
Identity transformation	—$\boxed{I}$—	$I = \begin{bmatrix} 1 & 0 \\ 0 & 1 \end{bmatrix}$
Pauli element X	—$\boxed{X}$—	$\sigma_1 = \begin{bmatrix} 0 & 1 \\ 1 & 0 \end{bmatrix}$
Pauli element Y	—$\boxed{Y}$—	$\sigma_2 = \begin{bmatrix} 0 & -i \\ i & 0 \end{bmatrix}$
Pauli element Z	—$\boxed{Z}$—	$\sigma_3 = \begin{bmatrix} 1 & 0 \\ 0 & -1 \end{bmatrix}$
Hadamard element	—$\boxed{H}$—	$\frac{1}{\sqrt{2}} \begin{bmatrix} 1 & 1 \\ 1 & -1 \end{bmatrix}$
Phase element	—$\boxed{S}$—	$\begin{bmatrix} 1 & 0 \\ 0 & i \end{bmatrix}$
Element $\pi/8$	—$\boxed{T}$—	$\begin{bmatrix} 1 & 0 \\ 0 & e^{i\pi/4} \end{bmatrix}$
Measurement	—$\boxed{\measuredangle}$	Projection on $|0\rangle$ and $|1\rangle$

Hadamard[10] gate transforms the system's state by the rule:

$$|0\rangle \rightarrow \frac{1}{\sqrt{2}}(|0\rangle + |1\rangle), \tag{4.37}$$

$$|1\rangle \rightarrow \frac{1}{\sqrt{2}}(|0\rangle - |1\rangle). \tag{4.38}$$

Therefore, the arbitrary state $|\psi\rangle$ will in this case change as follows:

$$|\psi\rangle = \begin{bmatrix} u \\ v \end{bmatrix} = u \begin{bmatrix} 1 \\ 0 \end{bmatrix} + v \begin{bmatrix} 0 \\ 1 \end{bmatrix} \rightarrow$$

$$\rightarrow u\frac{1}{\sqrt{2}}(|0\rangle + |1\rangle) + v\frac{1}{\sqrt{2}}(|0\rangle - |1\rangle) = \frac{1}{\sqrt{2}} \begin{bmatrix} 1 & 1 \\ 1 & -1 \end{bmatrix} \begin{bmatrix} u \\ v \end{bmatrix}. \tag{4.39}$$

Thus, the Hadamard element is presented by the matrix $\frac{1}{\sqrt{2}} \begin{bmatrix} 1 & 1 \\ 1 & -1 \end{bmatrix}$.

---

[10]Jacques Salomon Hadamard (1865–1963), French mathematician.

Table 4.2 Two qubits operations

Name	Designation	Matrix
Exchange		$\begin{bmatrix} 1 & 0 & 0 & 0 \\ 0 & 0 & 1 & 0 \\ 0 & 1 & 0 & 0 \\ 0 & 0 & 0 & 1 \end{bmatrix}$
Controlled NOT		$\begin{bmatrix} 1 & 0 & 0 & 0 \\ 0 & 1 & 0 & 0 \\ 0 & 0 & 0 & 1 \\ 0 & 0 & 1 & 0 \end{bmatrix}$
Controlled phase element		$\begin{bmatrix} 1 & 0 & 0 & 0 \\ 0 & 1 & 0 & 0 \\ 0 & 0 & 1 & 0 \\ 0 & 0 & 0 & i \end{bmatrix}$

Obviously, in order to execute complex algorithms, the qubits must interact with each other and exchange information. In this connection, of particular importance are the logical operations that affect two or more qubits. In Table 4.2, the most important gates are enumerated, that transform the state of two qubits.

Example 4.8 Let us determine how the qubit $|\psi\rangle$ is transformed under the action of two applications of the Hadamard element:

$$|\psi\rangle \quad \boxed{H} \quad \boxed{H} \quad |\psi'\rangle$$

Solution As shown above, in matrix representation, the Hadamard element is described by the matrix

$$M_H = \frac{1}{\sqrt{2}} \begin{bmatrix} 1 & 1 \\ 1 & -1 \end{bmatrix}. \tag{4.40}$$

Compute the matrix that corresponds to two applications of the Hadamard element as a matrix product (see formula (4.34)):

$$M_H M_H = \frac{1}{\sqrt{2}} \begin{bmatrix} 1 & 1 \\ 1 & -1 \end{bmatrix} \cdot \frac{1}{\sqrt{2}} \begin{bmatrix} 1 & 1 \\ 1 & -1 \end{bmatrix} = \frac{1}{2} \begin{bmatrix} 2 & 0 \\ 0 & 2 \end{bmatrix} = \begin{bmatrix} 1 & 0 \\ 0 & 1 \end{bmatrix}. \tag{4.41}$$

An identity matrix is obtained, therefore, two applications of the Hadamard element bring the qubit back to its original state. □

Example 4.9 Find a matrix representation of the following quantum circuit:

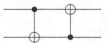

Solution The quantum circuit consists of two elements "controlled NOT", also referred to as "CNOT" (Controlled NOT).

The matrix of the element CNOT has the form (see Table 4.2):

$$\begin{bmatrix} 1 & 0 & 0 & 0 \\ 0 & 1 & 0 & 0 \\ 0 & 0 & 0 & 1 \\ 0 & 0 & 1 & 0 \end{bmatrix} \tag{4.42}$$

In view of the matrix representation of CNOT, compute how the arbitrary state $|\psi_1\psi_2\rangle = (u_1\,|0\rangle + v_1\,|1\rangle)(u_2\,|0\rangle + v_2\,|1\rangle)$ will change after the action of the first CNOT:

$$|\psi_1\psi_2\rangle = \begin{bmatrix} u_1 \\ v_1 \\ u_2 \\ v_2 \end{bmatrix} \rightarrow \begin{bmatrix} 1 & 0 & 0 & 0 \\ 0 & 1 & 0 & 0 \\ 0 & 0 & 0 & 1 \\ 0 & 0 & 1 & 0 \end{bmatrix} \begin{bmatrix} u_1 \\ v_1 \\ u_2 \\ v_2 \end{bmatrix} = \begin{bmatrix} u_1 \\ v_1 \\ v_2 \\ u_2 \end{bmatrix}. \tag{4.43}$$

This is equivalent to the fact that the states of the computational basis (4.24) are transformed in accordance with the rule:

$$\begin{cases} |00\rangle \rightarrow |00\rangle\,, \\ |01\rangle \rightarrow |01\rangle\,, \\ |10\rangle \rightarrow |11\rangle\,, \\ |11\rangle \rightarrow |10\rangle\,. \end{cases} \tag{4.44}$$

Note that the next element CNOT takes the input states in the reverse order relative to the first element. In this case, the basis state transformation rule has the form:

$$\begin{cases} |00\rangle \rightarrow |00\rangle\,, \\ |01\rangle \rightarrow |11\rangle\,, \\ |10\rangle \rightarrow |10\rangle\,, \\ |11\rangle \rightarrow |01\rangle\,. \end{cases} \tag{4.45}$$

Therefore, the next step of the quantum system evolution is described by the matrix

$$
\begin{bmatrix}
1 & 0 & 0 & 0 \\
0 & 0 & 0 & 1 \\
0 & 0 & 1 & 0 \\
0 & 1 & 0 & 0
\end{bmatrix}.
\tag{4.46}
$$

We perform the matrix computations:

$$
\begin{bmatrix} u_1 \\ v_1 \\ v_2 \\ u_2 \end{bmatrix}
\rightarrow
\begin{bmatrix}
1 & 0 & 0 & 0 \\
0 & 0 & 0 & 1 \\
0 & 0 & 1 & 0 \\
0 & 1 & 0 & 0
\end{bmatrix}
\begin{bmatrix} u_1 \\ v_1 \\ v_2 \\ u_2 \end{bmatrix}
=
\begin{bmatrix} u_1 \\ u_2 \\ v_2 \\ v_1 \end{bmatrix}.
\tag{4.47}
$$

As a result, the original state $|\psi_1\psi_2\rangle = [u_1, v_1, u_2, v_2]^T$ passes to $[u_1, u_2, v_2, v_1]^T$. The matrix representation of the circuit under analysis can be written in the form:

$$
\begin{bmatrix}
1 & 0 & 0 & 0 \\
0 & 1 & 0 & 0 \\
0 & 0 & 0 & 1 \\
0 & 0 & 1 & 0
\end{bmatrix}
\begin{bmatrix}
1 & 0 & 0 & 0 \\
0 & 0 & 0 & 1 \\
0 & 0 & 1 & 0 \\
0 & 1 & 0 & 0
\end{bmatrix}
=
\begin{bmatrix}
1 & 0 & 0 & 0 \\
0 & 0 & 0 & 1 \\
0 & 1 & 0 & 0 \\
0 & 0 & 1 & 0
\end{bmatrix}.
\tag{4.48}
$$

$\square$

Review Questions

1. What is a complex number?
2. Enumerate arithmetic operations on complex numbers.
3. How is the number conjugate of a given complex number found?
4. How can a complex number be presented geometrically?
5. Explain the differences between the following forms of complex numbers: algebraic, trigonometric and exponential.
6. Write Euler's formula.
7. How is the root of the n-th order of a complex number found? How many values does it take?
8. Formulate the fundamental theorem of algebra.

9. What is Cardano formula used for?
10. What matrices are called Hermitian? unitary?
11. Define the concepts of "bit" and "qubit".
12. What states are referred to as the computational basis states?
13. What is gate?
14. Explain how the quantum state is measured.
15. Write Pauli and Dirac matrices.
16. Enumerate the basic operations on one qubit.
17. What quantum operations are applied to a system of two qubits?

Problems

4.1. Perform algebraic manipulations and represent the specified complex number z in the algebraic form $z = \operatorname{Re} z + i \operatorname{Im} z$:

(1) $(3 + i)(2 + 5i)$;
(2) $(3 - i)(3 + i)$;
(3) $(1 - i)^4 + (1 + i)^4$;
(4) $2 - i^3$.

4.2. Perform algebraic manipulations and represent the specified complex number z in the algebraic form $z = \operatorname{Re} z + i \operatorname{Im} z$:

(1) $(2 - i)(-1 + i)$;
(2) $(6 + 5i)(4 - i)$;
(3) $(1 + 3i)^2 + (2 - i)^2$;
(4) $i^5 - i^3$.

4.3. Given are the complex numbers $z_1 = 5 + i$, $z_2 = 4 - i$, $z_3 = -1 + 3i$. Find $z_1 z_3 - z_2^2$.

4.4. Given are the complex numbers $z_1 = 1 - 2i$, $z_2 = -1 + i$, $z_3 = -i$. Find $z_1 z_2 (z_3^3 - z_1^3)$.

4.5. Perform the actions:

(1) $(1 + 4i)^3 + (1 - 4i)^3$;
(2) $(6 - i)^4 + (6 + i)^4$.

4.6. Having performed the division, represent the complex number $z = \dfrac{a + ib}{c + id}$, where $a, b, c, d \in \mathbb{R}$, $c \neq 0$, $d \neq 0$, in algebraic form.

4.7. Simplify the expressions:

(1) $\dfrac{3 + i}{3 - i}$;

(2) $\dfrac{1 + 3i}{1 - 3i}$;

(3) $\dfrac{2 + i^2}{3 + i^3}$;

(4) $\dfrac{i}{1 + i} + \dfrac{i}{1 - i}$.

4.8. Given are the complex numbers $z_1 = 2 + i$, $z_2 = z_1^*$, $z_3 = z_1 + z_2$. Find $(z_1 - z_3)(z_2 - z_3)/z_2$.

4.9. Given are the complex numbers $z_1 = 7 + 2i$, $z_2 = -z_1$, $z_3 = 2 - z_1^*$. Find $(z_1 z_2 + z_2 z_3 + z_1 z_3)/z_1$.

4.10. Find the number conjugate with the number z, if:

(1) $z = \dfrac{1 - 3i}{1 + 3i}$;

(2) $z = 2i + \dfrac{1}{2 + i}$.

4.11. Find z, if $z - 3z^* = 18 + 4i$.

4.12. Find z, if $3z^* - 7z = 10 - 10i$.

4.13. Prove that for arbitrary $z_1, z_2 \in \mathbb{C}$ the equalities are valid:

(1) $(z_1 + z_2)^* = z_1^* + z_2^*$;

(2) $(z_1 z_2)^* = z_1^* z_2^*$.

4.14. Prove that if $|z| = 1$, then $z^{-1} = z^*$.

∗4.15. Prove that for any $z_1, z_2 \in \mathbb{C}$ **triangle inequalities** are valid:

$$\text{abs}(|z_1| - |z_2|) \leqslant |z_1 + z_2| \leqslant |z_1| + |z_2|.$$

∗4.16. Calculate the sums:

(1) $\displaystyle\sum_{k=1}^{10} i^k$;

(2) $\displaystyle\sum_{k=-49}^{49} i^k$.

4.17. Simplify the expression i^m for arbitrary $m \in \mathbb{Z}$.

4.18. Represent the complex number in a trigonometric form:

(1) 2;

(2) $3i$;

(3) $4 + 3i$;

(4) $-i$;

(5) $-3 - 6i$;

(6) $\sqrt{2}(1 + i)$;

(7) $\sqrt{3}(-1 + 3i)$;

(8) $\dfrac{9 + i}{9 - i}$.

4.19. Represent the following complex numbers in algebraic form $z = \operatorname{Re} z + i \operatorname{Im} z$:

(1) $z = \cos \dfrac{\pi}{3} + i \sin \dfrac{\pi}{3}$;

(2) $z = 4 \left(\cos(-\dfrac{\pi}{2}) + i \sin(-\dfrac{\pi}{2}) \right)$;

(3) $z = \dfrac{\sqrt{3}}{\cos \dfrac{3\pi}{4} + i \sin \dfrac{3\pi}{4}}$;

(4) $z = \cos(-\dfrac{\pi}{8}) + i \sin(-\dfrac{\pi}{8})$.

4.20. Prove that a complex number of the type $u = \dfrac{a + ib}{a - ib}$, where $a, b \in \mathbb{R}$, can be presented in the form of an exponent with purely imaginary index, i. e. in the form

$$u = e^{i\delta}, \quad \delta \in \mathbb{R}.$$

4.21. Calculate i^i.

4.22. Calculate:

(1) $\sqrt{8i}$;

(2) $\sqrt[6]{4096}$.

∗4.23. Prove the validity of the identities for the roots of unity ω_k, where $0 \leqslant k \leqslant n - 1$ [24]:

(1) $\omega_{k+n/2} = -\omega_k$ for even n and $0 \leqslant k \leqslant n/2 - 1$;

(2) $\sum\limits_{k=0}^{n-1} \omega_k = 0$ for $n > 1$;

(3) $\prod\limits_{k=0}^{n-1} \omega_k = (-1)^{n-1}$ for all natural values n.

∗4.24. Prove the validity of the identities for the roots of unity ω_k, where $0 \leqslant k \leqslant n - 1$, for all natural values n:

(1) $\prod\limits_{k=0}^{n-1} (z - \omega_k) = z^n - 1$;

(2) $\sum\limits_{k=0}^{n-1} (\omega_k)^d = \begin{cases} 0, & 1 \leqslant d \leqslant n - 1; \\ n, & d = n. \end{cases}$

∗4.25. Prove the validity of the identities for the n-th roots of unity ω_k, where $k = 0, 1, \ldots, n - 1$, for all values $n > 2$:

(1) $\sum\limits_{k=0}^{n-2} \omega_k \omega_{k+1} = -\omega_{n-1}$;

(2) $\displaystyle\sum_{k=1}^{n-2} \frac{\omega_{k-1}\omega_{k+1}}{\omega_k} = -(1 + \omega_{n-1});$

(3) $\displaystyle\sum_{\substack{k,k'=0 \\ k<k'}}^{n-1} \omega_k\omega_{k'} = 0;$

(4) $\displaystyle\sum_{\substack{k,k'=0 \\ k<k'}}^{n-1} \frac{\omega_k\omega_{k'}}{\omega_{k'-k}} = \frac{n}{1-\omega_2}.$

*4.26. Let ω_k, where $0 \leqslant k \leqslant n-1$, are the n-th roots of unity, x is an arbitrary complex number, and $x \neq \omega_k$ for any k. Calculate the sum $\displaystyle\sum_{k=0}^{n-1} \frac{\omega_k}{x-\omega_k}$.

*4.27. Prove that for natural $n \in \mathbb{N}$ and $\alpha \neq 2\pi k, k \in \mathbb{Z}$, are valid the **Lagrange's**[11] **identities** [38]:

(a) $\cos\alpha + \cos 2\alpha + \cdots + \cos n\alpha = \dfrac{\sin(n\alpha/2)}{\sin(\alpha/2)} \cos[(n+1)\alpha/2];$

(b) $\sin\alpha + \sin 2\alpha + \cdots + \sin n\alpha = \dfrac{\sin(n\alpha/2)}{\sin(\alpha/2)} \sin[(n+1)\alpha/2].$

4.28. Prove de Moivre's formula for natural values of the exponent n, using the mathematical induction method.

*4.29. Using de Moivre's formula, express $\cos 3\varphi$ and $\sin 3\varphi$ in terms of $\cos\varphi$ and $\sin\varphi$.

*4.30. Express $\cos 4\varphi$ and $\sin 4\varphi$ in terms of $\cos\varphi$ and $\sin\varphi$.

*4.31. There exist relations that express polynomial's coefficients by its roots (**Viète**[12] **formulae**). If $\alpha_1, \alpha_2, \ldots, \alpha_n$ are roots of the polynomial $p(z) = x^n + a_1 x^{n-1} + \cdots + a_n$, and each root is taken the number of times corresponding to its multiplicity, then the following equalities are valid:

$$\alpha_1 + \alpha_2 + \cdots + \alpha_n = -a_1,$$

$$\alpha_1\alpha_2 + \alpha_2\alpha_3 + \cdots + \alpha_1\alpha_n + \alpha_2\alpha_3 + \cdots + \alpha_{n-1}\alpha_n = a_2,$$

$$\alpha_1\alpha_2\alpha_3 + \alpha_1\alpha_2\alpha_4 + \cdots + \alpha_{n-2}\alpha_{n-1}\alpha_n = -a_3,$$

$$\cdots$$

$$\alpha_1\alpha_2\ldots\alpha_{n-1} + \alpha_1\alpha_2\alpha_{n-2}\alpha_n + \cdots + \alpha_2\alpha_3\ldots\alpha_n = (-1)^{n-1}a_{n-1},$$

$$\alpha_1\alpha_2\ldots\alpha_n = (-1)^n a_n.$$

Prove validity of Viète formulae.

---

[11]Joseph-Louis Lagrange (1736–1813), French mathematician, mechanic and astronomer.

[12]François Viète, seigneur de la Bigotière (1540–1603), French mathematician.

4.32. Let z_1, z_2 be the roots of the quadratic trinomial $p(z) = z^2 + uz + v$. Express the following values by the coefficients u and v:

(1) $z_1^2 + z_2^2$;

(2) $z_1^{-2} + z_2^{-2}$.

4.33. Let z_1, z_2 be the roots of the quadratic trinomial $p(z) = z^2 + uz + v$. Express the following values by the coefficients u and v:

(1) $z_1^4 + z_2^4$;

(2) $z_1^{-4} + z_2^{-4}$.

4.34. Find the sum and the product of all roots for each equation:

(1) $z^3 + 3z^2 + 3z + 1 = 0$;

(2) $z^4 + 10z^2 + 20 = 0$.

4.35. Find the sum and the product of all roots for each equation:

(1) $z^{100} - 100z^{99} + z^{98} = 0$;

(2) $z^5 + z^4 + 1 = 0$.

∗4.36. Compute the determinant

$$\Delta = \begin{vmatrix} z_1 & z_2 & z_3 \\ z_3 & z_1 & z_2 \\ z_2 & z_3 & z_1 \end{vmatrix},$$

where z_1, z_2 and z_3 are the roots of the cubic equation $z^3 + \alpha z + \beta = 0$ with complex coefficients $\alpha, \beta \in \mathbb{C}$.

∗4.37. Solve the equation $z^3 - 3z + 2 = 0$, using the Cardano formula.

∗4.38. Solve the equation $2z^3 - 13z^2 - 17z + 70 = 0$, using Cardano formula.

4.39. Solve the systems of linear equations using Cramer's rule:

(a)
$$\begin{cases} (1 + i)x_1 + (-2 + 2i)x_2 + (1 + i)x_3 = 1 - i, \\ -x_1 + (3 - i)x_2 + (2 + i)x_3 = -7 + 2i, \\ (-2 - i)x_1 + (-1 - i)x_3 = 3 - 5i; \end{cases}$$

(b)
$$\begin{cases} (-2 + 2i)x_1 + (2 + i)x_2 + (2 + 2i)x_3 = 7i, \\ (-1 + 2i)x_1 + (-2 - i)x_2 + 2ix_3 = -8 + 2i, \\ (2 + 2i)x_1 + 2ix_2 + (1 - i)x_3 = 2 + 4i; \end{cases}$$

(c) $\begin{cases} \qquad\qquad (-1+i)x_3 = 1 - 3i, \\ (-1+i)x_1 - x_2 + (-1-i)x_3 = 3 - 2i, \\ x_1 + (4+i)x_2 + (3-i)x_3 = 4 + 8i; \end{cases}$

(d) $\begin{cases} (4+i)x_1 - ix_2 + 3x_3 = -i, \\ 3x_1 + 4x_2 + (1-i)x_3 = -4 + 9i, \\ ix_1 + 3x_2 + x_3 = -5 + 6i. \end{cases}$

4.40. Solve the systems of linear equations using Gaussian elimination:

(a) $\begin{cases} (2+i)x_1 + (1-i)x_2 + (1-i)x_3 = 8i, \\ (1-i)x_1 + (-1+i)x_2 + ix_3 = -2, \\ -ix_1 + (3+i)x_2 + (-1-i)x_3 = 8 + 4i; \end{cases}$

(b) $\begin{cases} (4-i)x_1 + (4-i)x_2 - ix_3 = 2 + 9i, \\ 5x_1 + 3x_2 - ix_3 = 5 + 10i, \\ 2x_1 + (2+i)x_2 + (1+i)x_3 = 3 + 5i; \end{cases}$

(c) $\begin{cases} (1+2i)x_1 + (3+2i)x_2 + (-2+2i)x_3 = 11 + i, \\ (4-i)x_1 + (-1+2i)x_2 - x_3 = 8 + 3i, \\ (1+i)x_1 + 2ix_2 = 4 + 8i; \end{cases}$

(d) $\begin{cases} 5x_1 + (-2-i)x_2 + (-3-i)x_3 = -12 - i, \\ ix_1 + (1+2i)x_2 + (2-i)x_3 = -3 + i, \\ 2ix_2 + x_3 = -4 + i. \end{cases}$

∗4.41. Solve the systems of linear equations relative to five variables using Gaussian elimination:

(a) $\begin{cases} (4-i)x_1 + ix_2 + (-1+i)x_3 + (4+2i)x_4 + (-3+i)x_5 = -5 - 9i, \\ 2x_1 - x_2 - 3x_3 + (2-i)x_4 + (-2-i)x_5 = 3 - 5i, \\ (1-i)x_1 + (1-i)x_2 + (2-i)x_3 + (-2-i)x_4 + 3x_5 = 10 + 11i, \\ (1+2i)x_1 - 2x_2 + (2+2i)x_3 - 3x_4 + (-2+i)x_5 = -2, \\ (-3+2i)x_1 + x_2 + (-2+i)x_3 - ix_4 + 2ix_5 = -11 + 3i; \end{cases}$

$$\begin{cases} -3x_1 + (-3-i)x_2 + (2-i)x_3 + (-2+i)x_4 + ix_5 = -2-i, \\ -x_1 + ix_2 + (-3+2i)x_3 + (1+i)x_4 + (4+2i)x_5 = 9+12i, \\ 2ix_1 + (2-i)x_2 + 3x_3 + 2x_4 + (-2-i)x_5 = -9-5i, \\ (-3-i)x_1 + (-3-i)x_2 + ix_3 + (1+i)x_4 + (-1'+i)x_5 = 4, \\ (2+i)x_1 + 2x_2 + (1+i)x_3 + (3+2i)x_5 = -3+10i. \end{cases}$$

(b)

4.42. Which of the following matrices are Hermitian?

(1) $\begin{bmatrix} 1 & 2+10i \\ 2-10i & 3 \end{bmatrix}$;

(2) $\dfrac{1}{\sqrt{2}} \begin{bmatrix} 1 & -1 \\ -2 & -2 \end{bmatrix}$;

(3) $\dfrac{1}{\sqrt{6}} \begin{bmatrix} 2+2i & 2i \\ -2i & -2+2i \end{bmatrix}$;

(4) $\begin{bmatrix} 3 & -i \\ -i & 3 \end{bmatrix}$;

(5) $\dfrac{1}{\sqrt{10}} \begin{bmatrix} 1 & 0 & 2 \\ 0 & 3 & 0 \\ 2 & 0 & -i \end{bmatrix}$;

(6) $\begin{bmatrix} 1 & 1 & 1 & 1 \\ 1 & i & -1 & -i \\ 1 & -1 & 1 & -1 \\ 1 & -i & -1 & i \end{bmatrix}$.

4.43. What condition is imposed on the diagonal elements of the Hermitian matrix?

4.44. Show that if Z_1 and Z_2 are complex matrices of the same order, then the matrix $\dfrac{1}{2}(Z_1 Z_2 + Z_2 Z_1)$ is Hermitian.

4.45. Anti-Hermitian or **skew-Hermitian** matrix is the matrix A, for which the relation $A^H = -A$ is fulfilled. In other words, Hermitian conjugate of such a matrix results in multiplying all its elements by (-1). Which of the following matrices are anti-Hermitian?

(1) $\dfrac{1}{\sqrt{2}}\begin{bmatrix} 0 & 2+i \\ -2+i & i \end{bmatrix}$;

(2) $\begin{bmatrix} i & -2 \\ 3 & -2i \end{bmatrix}$;

(3) $\dfrac{1}{3}\begin{bmatrix} 2+i & 2i \\ -2i & -2+i \end{bmatrix}$;

(4) $\dfrac{1}{5}\begin{bmatrix} 1 & 5 & 0 \\ -5 & 2i & -5 \\ 0 & 5 & 3i \end{bmatrix}$;

(5) $\dfrac{1}{5}\begin{bmatrix} -2i & 1 & 0 \\ -1 & 0 & -3 \\ 0 & 3 & 2i \end{bmatrix}$;

(6) $\begin{bmatrix} i & i & i & i \\ i & i & -i & -i \\ i & -i & i & -i \\ i & -i & -i & i \end{bmatrix}$.

4.46. What condition is imposed on the diagonal elements of an anti-Hermitian matrix?

4.47. Which of the following matrices are unitary?

(1) $\begin{bmatrix} 1 & 2 & 3 \\ 3 & 2 & 1 \end{bmatrix}$;

(2) $\dfrac{1}{\sqrt{2}}\begin{bmatrix} 1 & -i \\ -i & 1 \end{bmatrix}$;

(3) $\dfrac{1}{3}\begin{bmatrix} 2+i & 2i \\ -2i & -2+i \end{bmatrix}$;

(4) $\dfrac{1}{\sqrt{10}}\begin{bmatrix} 3 & -i \\ -i & 3 \end{bmatrix}$;

(5) $\dfrac{1}{\sqrt{10}} \begin{bmatrix} 1 & 0 & 0 \\ 0 & e^{2i} & 0 \\ 0 & 0 & e^{4i} \end{bmatrix}$;

(6) $\dfrac{1}{2} \begin{bmatrix} 1 & 1 & 1 & 1 \\ 1 & i & -1 & -i \\ 1 & -1 & 1 & -1 \\ 1 & -i & -1 & i \end{bmatrix}$.

4.48. What condition must satisfy the complex numbers u_1 and u_2 and the real number φ, in order for the matrix $\begin{bmatrix} u_1 & u_2 \\ -e^{i\varphi}u_2^* & e^{i\varphi}u_1^* \end{bmatrix}$ to be unitary?

4.49. At the examination in linear algebra the student says that for any square matrix A the equality $\ln(\exp A) = A$ is valid. Is the student right?

*4.50. Prove the identity $\det(e^A) = e^{\operatorname{tr} A}$, valid for the arbitrary square matrix A with complex coefficients.

4.51. Prove the Theorem 4.4 (see page 182): the determinant of a unitary matrix is a complex number, whose modulus is equal to one.

4.52. Compute the commutators of the Pauli matrices $[\sigma_1, \sigma_2]$, $[\sigma_2, \sigma_3]$ and $[\sigma_3, \sigma_1]$.

4.53. Compute the product of the Pauli matrices $\sigma_1\sigma_2\sigma_3$.

*4.54. Prove the generalization of **Euler's identity** for Pauli matrices σ_1, σ_2 and σ_3:

$$\exp(i\sigma_k\varphi) = I\cos\varphi + i\sigma_k\sin\varphi \quad \text{for } k = 1, 2, 3, \tag{4.49}$$

where $I = \begin{bmatrix} 1 & 0 \\ 0 & 1 \end{bmatrix}$ is the identity matrix of the second order.

4.55. What is the square of the Dirac matrices β?

4.56. Compute the product of the Dirac matrices $\alpha_1\alpha_2\alpha_3\beta$.

4.57. Compute the result of the actions of the quantum circuit

on a qubit that is originally in the state $|0\rangle$.

4.58. Compute the result of the actions of the quantum circuit

on a qubit that is originally in the state $|1\rangle$.

4.59. Compute the result of the actions of the quantum circuit

on a qubit that is originally in the state $|\psi\rangle = u\,|0\rangle + v\,|1\rangle$, where $u, v \in \mathbb{C}$.

4.60. Show that the quantum circuit

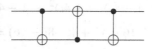

transfers the state $|\psi_1\psi_2\rangle$ to the state $|\psi_2\psi_1\rangle$.

Answers and Solutions

4.1 *Solution.*

(1) Operations with complex numbers should be performed similarly to the respective operations with algebraic polynomials, using the property $i^2 = -1$:

$$(3 + i)(2 + 5i) = 3 \cdot 2 + 3 \cdot (5i) + i \cdot 2 + i \cdot (5i)$$
$$= 6 + 15i + 2i + 5i^2 = 6 + 17i + 5(-1) = 1 + 17i;$$

(2) $(3 - i)(3 + i) = 3^2 - i^2 = 10;$

(3) $(1 - i)^4 + (1 + i)^4 = (1 - 4i + 6i^2 - 4i^3 + i^4) + (1 + 4i + 6i^2 + 4i^3 + i^4)$
$$= 2\left(1 + 6i^2 + (i^2)^2\right) = 2\left(1 + 6(-1) + (-1)^2\right) = -8;$$

(4) $2 - i^3 = 2 - i \cdot i^2 = 2 + i.$

4.2 *Answer:*

(1) $-1 + 3i;$
(2) $29 + 14i;$
(3) $-5 + 2i;$
(4) $2i.$

4.3 *Solution.*

We perform algebraic transformations, taking into account that $i^2 = -1$:
$z_1 z_3 - z_2^2 = (5 + i)(-1 + 3i) - (4 - i)^2 = 5 \cdot (-1) + 5 \cdot (3i) + i \cdot (-1) + i \cdot (3i) - (16 - 8i + i^2) = -8 + 14i - 15 + 8i = -23 + 22i.$

4.4 *Solution.*

We perform algebraic manipulations: $z_1 z_2 = 1 + 3i$, $z_3^3 = i$, $z_1^3 = -11 + 2i$, $z_3^3 - z_1^3 = 11 - i$. As a result $z_1 z_2 (z_3^3 - z_1^3) = 14 + 32i$.

4.5 *Solution.*

(1) We perform algebraic transformations:

$$(1 + 4i)^3 + (1 - 4i)^3$$

$$= (1 + 4i + 1 - 4i)((1 + 4i)^2 - (1 - 4i)(1 + 4i) + (1 - 4i)^2)$$

$$= 2(1 + 8i - 16 - 17 + 1 - 8i - 16) = -47 \cdot 2 = -94.$$

(2) Use the Newton[13] binomial formula:

$$(a + b)^n = \sum_{k=0}^{n} C(n, k) a^{n-k} b^k,$$

valid for all $a, b \in \mathbb{C}$ and natural n:

$$(6 - i)^4 + (6 + i)^4 = 6^4 - 4 \cdot 6^3 \cdot i + 6 \cdot 6^2 \cdot (-1) - 4 \cdot 6i \cdot (-1) + (-1) \cdot (-1)$$

$$+ 6^4 + 4 \cdot 6^3 \cdot i + 6 \cdot 6^2 \cdot (-1) + 4 \cdot 6i \cdot (-1) + (-1) \cdot (-1) = 2162.$$

4.6 *Solution.*

Denote $z_1 = a + ib$, $z_2 = c + id$. The fraction of the form $\dfrac{z_1}{z_2}$, where $z_1, z_2 \in \mathbb{C}$, can be conveniently transformed by multiplying it by $1 \equiv \dfrac{z_2^*}{z_2^*}$:

$$\frac{z_1}{z_2} = \frac{z_1}{z_2} \cdot 1 = \frac{z_1}{z_2} \cdot \frac{z_2^*}{z_2^*} = \frac{z_1 z_2^*}{z_2 z_2^*} = \frac{z_1 z_2^*}{|z_2|^2}.$$

This is why the result of the division of two complex numbers z_1/z_2, where $z_2 \neq 0$, will be the number

$$\frac{z_1}{z_2} = \frac{ac + bd}{c^2 + d^2} + i \frac{bc - ad}{c^2 + d^2}.$$

---

[13]Isaac Newton (1643–1727), English mathematician, physicist, mechanic and astronomer.

4.7 *Solution.*

Using the hint offered in the previous problem (i. e. $z_1/z_2 = (z_1 z_2^*)/(z_2 z_2^*) = (z_1 z_2^*)/|z_2|^2$), we obtain

(1) $\dfrac{3+i}{3-i} = \dfrac{(3+i)(3+i)}{9-i^2} = \dfrac{1}{5}(4+3i);$

(2) $\dfrac{1+3i}{1-3i} = \dfrac{(1+3i)^2}{1-9i^2} = \dfrac{1}{5}(-4+3i);$

(3) $\dfrac{2+i^2}{3+i^3} = \dfrac{2-1}{3+i^2 \cdot i} = \dfrac{1}{10}(3+i);$

(4) $\dfrac{i}{1+i} + \dfrac{i}{1-i} = \dfrac{i(1-i)+i(1+i)}{(1+i)(1-i)} = i.$

4.8 *Solution.*

Find z_2 and z_3: $z_2 = 2 - i$, $z_3 = (2+i) + (2-i) = 4$. After this, we obtain $(z_1-z_3)(z_2-z_3)/z_2 = (2+i-4)(2-i-4)/(2-i) = -(2-i)(-2-i)/(2-i) = 2+i.$

4.9 *Solution.*

Find z_1 and z_2: $z_2 = -7 - 2i$, $z_3 = -5 + 2i$.

Then, $z_1 z_2 = -45 - 28i$, $z_2 z_3 = 39 - 4i$, $z_1 z_3 = -39 + 4i$, $z_1 z_2 + z_2 z_3 + z_1 z_3 = -45 - 28i.$

The final answer is $(z_1 z_2 + z_2 z_3 + z_1 z_3)/z_1 = -7 - 2i.$

4.10 *Answer:*

(1) $z = \dfrac{1}{5}(-4 + 3i);$

(2) $z = \dfrac{1}{5}(2 - 9i).$

4.11 *Solution.*

Let $z = a+ib$, then $z - 3z^* = (a+ib) - 3(a-ib) = -2a + 4ib$. Since complex numbers are equal if and only if their real and imaginary parts are equal, we obtain

$$\begin{cases} -2a = 18, \\ 4b = 4; \end{cases} \quad \Leftrightarrow \quad \begin{cases} a = -9, \\ b = 1, \end{cases}$$

whence $z = -9 + i$.

4.12 *Answer:* $z = -\dfrac{5}{2} + i.$

4.13 *Proof.*

Let $z_1 = x_1 + iy_1$, $z_2 = x_2 + iy_2$, where $x_1, x_2, y_1, y_2 \in \mathbb{R}$.

(1) Express the left side of the equality by x_1, x_2, y_1 and y_2:

$$(z_1 + z_2)^* = [(x_1 + x_2) + i(y_1 + y_2)]^* = (x_1 + x_2) - i(y_1 + y_2).$$

Now transform its right side

$$z_1^* + z_2^* = (x_1 - iy_1) + (x_2 - iy_2) = (x_1 + x_2) - i(y_1 + y_2).$$

Therefore $\forall z_1, z_2 \in \mathbb{C}$ $(z_1 + z_2)^* = z_1^* + z_2^*$.

(2) The left side of the equality

$$(z_1 z_2)^* = [(x_1 + iy_1)(x_2 + iy_2)]^* = (x_1 x_2 - y_1 y_2) - i(x_1 y_2 + x_2 y_1).$$

The right side coincides with the left one:

$$z_1^* z_2^* = (x_1 - iy_1)(x_2 - iy_2) = (x_1 x_2 - y_1 y_2) - i(x_1 y_2 + x_2 y_1) = (z_1 z_2)^*.$$

4.14 *Proof.*

Take the number with the modulus equal to one in exponential form: $z = e^{i\varphi}$. After algebraic transformations

$$z^{-1} = (e^{i\varphi})^{-1} = e^{-i\varphi} = (e^{i\varphi})^* = z^*$$

we obtain the equality $z^{-1} = z^*$.

4.15 *Hint.*

Use the geometric interpretation of the numbers z_1 and z_2. The length of a side of an arbitrary triangle is no greater than the sum of the lengths of the two other sides, and is no less than the absolute value of their difference.

4.16 *Solution.*

Use the formula for the geometric progression sum:

$$\sum_{k=1}^{n} q^k = q + \cdots + q^n = \frac{q^{n+1} - q}{q - 1}.$$

(1) $\displaystyle\sum_{k=1}^{10} i^k = \frac{i^{11} - i}{i - 1} = \frac{-i - i}{i - 1} = \frac{-2i(-i - 1)}{(i - 1)(-i - 1)} = -1 + i.$

(2) $\displaystyle\sum_{k=-49}^{49} i^k = \frac{1}{i^{49}} \sum_{k=-99}^{99} i^{(k+49)}.$

In the last sum, replace the summation index $k' = k + 49$. Then the sought sum takes the following form:

$$\sum_{k=-49}^{49} i^k = i^{-49} \sum_{k'=0}^{98} i^{k'} = i^{-48-1} \frac{i^{99} - 1}{i - 1} = -i \cdot \frac{i^{99} - 1}{i - 1}$$

$$= -i \cdot \frac{i^{96} \cdot i^3 - 1}{i - 1} = -i \cdot \frac{i^3 - 1}{i - 1} = -\frac{i^4 - i}{i - 1} = \frac{1 - i}{i - 1} = 1.$$

4.17 *Solution.*

The imaginary unit has the property $i^4 = 1$. Consider four cases depending on the remainder of division m by 4:

(1) $m = 4k, k \in \mathbb{Z}$,

$$i^m = i^{4k} = (i^4)^k = 1^k = 1;$$

(2) $m = 4k + 1, k \in \mathbb{Z}$,

$$i^m = i^{4k+1} = i^{4k} \cdot i = 1 \cdot i = i;$$

(3) $m = 4k + 2, k \in \mathbb{Z}$,

$$i^m = i^{4k+2} = i^{4k} \cdot i^2 = i^2 = -1;$$

(4) $m = 4k + 3, k \in \mathbb{Z}$,

$$i^m = i^{4k+3} = i^{4k} \cdot i^3 = -i.$$

Thus, we finally obtain

$$i^m = \begin{cases} 1, & \text{if } m = 4k, \ k \in \mathbb{Z}; \\ i, & \text{if } m = 4k + 1, \ k \in \mathbb{Z}; \\ -1, & \text{if } m = 4k + 2, \ k \in \mathbb{Z}; \\ -i, & \text{if } m = 4k + 3, \ k \in \mathbb{Z}. \end{cases}$$

4.18 *Solution.*

(1) For transition to a trigonometric form of the complex number, we must determine the modulus $\rho = |z|$ and the argument $\varphi = \arg z$. Using the formulae for ρ and φ, we obtain

$$\rho = \sqrt{x^2 + y^2} = \sqrt{2^2 + 0^2} = 2,$$

$$\varphi = \arctan \frac{y}{x} + 2\pi k = 0 + 2\pi k = 2\pi k, \quad k \in \mathbb{Z},$$

this is why the trigonometric form of the number 2 has the form

$$2(\cos(2\pi k) + i \sin(2\pi k)), \text{ where } k \in \mathbb{Z}.$$

(2) $\rho = \sqrt{0 + 3^2} = 3$, $\varphi = \dfrac{\pi}{2}\text{sgn}(3) + 2\pi k = \dfrac{\pi}{2} + 2\pi k$, therefore $3i =$
$3\left(\cos\left(\dfrac{\pi}{2} + 2\pi k\right) + i\sin\left(\dfrac{\pi}{2} + 2\pi k\right)\right)$, $k \in \mathbb{Z}$.

(3) $\rho = \sqrt{16 + 9} = 5$, $\varphi = \arctan\left(\dfrac{3}{4} + 2\pi k\right)$,

$4 + 3i = 5\left(\cos\left(\arctan\dfrac{3}{4} + 2\pi k\right) + i\sin\left(\arctan\dfrac{3}{4} + 2\pi k\right)\right)$, $k \in \mathbb{Z}$.

(4) $\rho = 1$, $\varphi = \dfrac{\pi}{2}\text{sgn}(-1) + 2\pi k = -\dfrac{\pi}{2} + 2\pi k$;
$-i = \cos(\pi(2k - 1/2)) + i\sin(\pi(2k - 1/2))$, $k \in \mathbb{Z}$.

(5) $\rho = \sqrt{9 + 36} = 3\sqrt{5}$, $\varphi = \arctan 2 - \pi + 2\pi k$,
$-3 - 6i = 3\sqrt{5}(\cos(\arctan 2 + \pi(2k - 1)) + i\sin(\arctan 2 + \pi(2k - 1)))$,
$k \in \mathbb{Z}$.

(6) $\rho = 2$, $\varphi = \pi/4 + 2\pi k$,
$\sqrt{2} + \sqrt{2}i = 2(\cos(\pi/4 + 2\pi k) + i\sin(\pi/4 + 2\pi k))$, $k \in \mathbb{Z}$.

(7) $\rho = \sqrt{30}$, $\varphi = \arctan(-3) + \pi + 2\pi k$,
$-\sqrt{3} + 3\sqrt{3}i = \sqrt{30}(\cos(\pi - \arctan 3 + 2\pi k)) + i\sin(\pi - \arctan 3 + 2\pi k))$,
$k \in \mathbb{Z}$.

(8) Multiply the numerator and the denominator of the fraction by the value $(9 + i)$
and transform the obtained expression:

$$\frac{9 + i}{9 - i} = \frac{(9 + i)(9 + i)}{(9 - i)(9 - i)} = \frac{80 + 18i}{82} = \frac{40}{41} - \frac{9}{41}i,$$

$\rho = \sqrt{\left(\dfrac{40}{41}\right)^2 + \left(\dfrac{9}{41}\right)^2} = 1$, $\varphi = -\arctan\dfrac{9}{41} + 2\pi k$. We finally obtain

$$\frac{9 + i}{9 - i} = \cos\left(2\pi k - \arctan\frac{9}{40}\right) + i\sin\left(2\pi k - \arctan\frac{9}{40}\right), \quad k \in \mathbb{Z}.$$

4.19 *Solution.*

(1) $\cos\dfrac{\pi}{3} + i\sin\dfrac{\pi}{3} = \dfrac{1}{2} + \dfrac{\sqrt{3}}{2}i$;

(2) $4\left(\cos\left(-\dfrac{\pi}{2}\right) + i\sin\left(-\dfrac{\pi}{2}\right)\right) = -4i$;

(3) $\dfrac{\sqrt{3}}{\cos\dfrac{3\pi}{4} + i\sin\dfrac{3\pi}{4}} = -\dfrac{\sqrt{6}}{2}i - \dfrac{\sqrt{6}}{2}$;

(4) $\cos\left(-\dfrac{\pi}{8}\right) + i\sin\left(-\dfrac{\pi}{8}\right) = \dfrac{\sqrt{2 + \sqrt{2}}}{2} - \dfrac{\sqrt{2 - \sqrt{2}}}{2}i$.

4.20 *Proof.*

Turn to the exponential form of the number u. Let $a + ib = \rho e^{i\varphi}$, then $a - ib = \rho e^{-i\varphi}$ and

$$u = \frac{a + ib}{a - ib} = \frac{e^{i\varphi}}{e^{-i\varphi}} = e^{i\varphi} \cdot e^{i\varphi} = e^{2i\varphi}.$$

Therefore $u = e^{i\delta}$, where $\delta = 2\varphi$.

4.21 *Solution.*

The exponential notation of the imaginary unit has the form $i = e^{i\pi/2 + 2\pi ik}$, where $k \in \mathbb{Z}$. Having written the imaginary unit in the base in exponential form and using the identity $(e^a)^b = e^{ab}$, we obtain

$$i^i = (e^{i\pi/2 + 2\pi ik})^i = e^{i^2(\pi/2 + 2\pi k)} = e^{-\pi/2 + 2\pi k'}, \quad \text{where } k, k' \in \mathbb{Z}.$$

As is seen from the considered example, the exponential function is a multifunction on the set of complex numbers $\mathbb{C}$.

4.22 *Answer:*

(1) $\sqrt{8i} = \pm 2(1 + i)$;
(2) $\sqrt[6]{4096} \in \{\pm 4, \pm 2(1 \pm i\sqrt{3}), \pm 2(1 \mp i\sqrt{3})\}$.

4.23 *Proof.*

(1) Transform the exponential notation of the number $\omega_{k+n/2}$:

$$\omega_{k+n/2} = e^{\frac{2\pi i(k+n/2)}{n}} = e^{\frac{2\pi ik}{n}} e^{\pi i} = \omega_k e^{\pi i}.$$

Using the equality $e^{\pi i} = -1$, we obtain $\omega_{k+n/2} = -\omega_k$ for even n and $k = 0, 1, \ldots, n/2 - 1$.

(2) The values ω_k form a geometric progression, whose denominator is $\omega_1 = e^{2\pi i/n}$. Using the formula for the geometric progression sum $1 + q + q^2 + \cdots + q^n = \dfrac{1 - q^{n+1}}{1 - q}$ for $|q| < 1$, we obtain

$$\sum_{k=0}^{n-1} \omega_k = \sum_{k=0}^{n-1} e^{2\pi ik/n} = \frac{(e^{2\pi i/n})^n - 1}{e^{2\pi i/n} - 1} = 0.$$

(3)

$$\prod_{k=0}^{n-1} \omega_k = \prod_{k=0}^{n-1} e^{2\pi ik/n} = e^{\sum_{k=0}^{n-1} 2\pi ik/n} = e^{(2\pi i \sum_{k=0}^{n-1} k)/n}.$$

The sum in the exponent is $\sum\limits_{k=0}^{n-1} k = \dfrac{n(n-1)}{2}$, and therefore,

$$\prod_{k=0}^{n-1} \omega_k = e^{\pi i (n-1)} = \cos \pi (n-1) + i \sin \pi (n-1) = (-1)^{n-1}.$$

4.26 *Answer:* $\dfrac{n}{x^n - 1}$.

4.27 *Proof.*

Consider the sum $Z = \sum\limits_{k=1}^{n} e^{i\alpha k}$. It is easy to see that the following relations are valid:

$$\cos \alpha + \cos 2\alpha + \cdots + \cos n\alpha = \operatorname{Re} Z,$$

$$\sin \alpha + \sin 2\alpha + \cdots + \sin n\alpha = \operatorname{Im} Z.$$

Calculate Z, using the formula for the geometric progression sum:

$$Z = \sum_{k=1}^{n} e^{i\alpha k} = \frac{e^{i\alpha(n+1)} - e^{i\alpha}}{e^{i\alpha} - 1}.$$

Simplify the obtained expression, multiplying the fraction by $1 = \dfrac{e^{-i\alpha/2}}{e^{-i\alpha/2}}$ and performing simple transformations:

$$Z = \frac{e^{i\alpha(n+1)} - e^{i\alpha}}{e^{i\alpha} - 1} \cdot \frac{e^{-i\alpha/2}}{e^{-i\alpha/2}} = \frac{e^{i\alpha(n+1/2)} - e^{i\alpha/2}}{e^{i\alpha/2} - e^{-i\alpha/2}}.$$

Denominator of the obtained fraction is $e^{i\alpha/2} - e^{-i\alpha/2} = 2i \sin \alpha/2$. Rewrite the exponents in the numerator, using Euler's formula:

$$Z = \frac{1}{2i \sin \alpha/2} \Big[\cos[(n + 1/2)\alpha] + i \sin[(n + 1/2)\alpha] - (\cos \alpha/2 + i \sin \alpha/2) \Big]$$

$$= \frac{\sin[(n + 1/2)\alpha] - \sin \alpha/2}{2 \sin \alpha/2} + \frac{\cos[(n + 1/2)\alpha] - \cos \alpha/2}{2i \sin \alpha/2}.$$

Further, use the known trigonometric formulae (see Appendix B "Trigonometric Formulae", formulae (B.16) and (B.18))

$$\sin a - \sin b = 2 \sin \frac{a - b}{2} \cos \frac{a + b}{2},$$

$$\cos a - \cos b = -2 \sin \frac{a - b}{2} \sin \frac{a + b}{2}.$$

We obtain

$$Z = \frac{\sin(n\alpha/2)}{\sin(\alpha/2)} \cos[(n+1)\alpha/2] + i \frac{\sin(n\alpha/2)}{\sin(\alpha/2)} \sin[(n+1)\alpha/2],$$

whence directly follow the Lagrange's identities.

4.28 *Proof.*

Denote the predicate "$(\cos\varphi + i\sin\varphi)^n = \cos n\varphi + i\sin n\varphi$" by $P(n)$ and prove the statement $\forall n\, P(n)$ by the mathematical induction method.

B a s i s   s t e p

For $n = 1$ we obtain the valid identity $(\cos\varphi + i\sin\varphi)^1 = \cos\varphi + i\sin\varphi$, therefore $P(1)$ is true.

I n d u c t i v e   s t e p

Suppose that $P(k)$, $k \in \mathbb{N}$ is true. Prove the truth of the proposition $P(k+1)$. We need to prove that

$$(\cos\varphi + i\sin\varphi)^{k+1} = \cos(k+1)\varphi + i\sin(k+1)\varphi.$$

Consider the expression $(\cos\varphi + i\sin\varphi)^{k+1}$ and represent it in the form

$$(\cos\varphi + i\sin\varphi)^{k+1} = (\cos\varphi + i\sin\varphi)^k \cdot (\cos\varphi + i\sin\varphi).$$

According to the inductive supposition, the first factor is

$$(\cos\varphi + i\sin\varphi)^k = \cos k\varphi + i\sin k\varphi.$$

Then

$$(\cos\varphi + i\sin\varphi)^{k+1} = (\cos k\varphi + i\sin k\varphi) \cdot (\cos\varphi + i\sin\varphi).$$

Open the brackets in the obtained expression, using the known identities for trigonometric functions, provided in Appendix B, formulae (B.11) and (B.9):

$$\cos(a+b) = \cos a \cos b - \sin a \sin b,$$
$$\sin(a+b) = \sin a \cos b + \cos a \sin b,$$

assuming $a = k\varphi$, $b = \varphi$. We obtain

$$(\cos\varphi + i\sin\varphi)^{k+1} = \underbrace{(\cos k\varphi \cos\varphi - \sin k\varphi \sin\varphi)}_{\cos(k+1)\varphi}$$

$$+ i \underbrace{(\sin k\varphi \cos\varphi + \cos k\varphi \sin\varphi)}_{\sin(k+1)\varphi}.$$

Hence, according to the mathematical induction principle, de Moivre's formula

$$(\cos \varphi + i \sin \varphi)^n = \cos n\varphi + i \sin n\varphi$$

is valid for all natural values $n \in \mathbb{N}$.

4.29 *Solution.*

Consider a more general case of the problem statement and express $\cos n\varphi$ and $\sin n\varphi$ in terms and cosine and sine of the angle φ.

For this, note that in the left side of de Moivre's formula (4.1) stands an expression that can be expanded by the Newton binomial formula (see page 192). Thus, represent the left side in the form:

$$(\cos \varphi + i \sin \varphi)^n = \sum_{j=0}^{n} C(n, j)(\cos^{n-j} \varphi)(i \sin \varphi)^j$$

$$= \sum_{j=0}^{n} i^j \, C(n, j) \cos^{n-j} \varphi \, \sin^j \varphi.$$

It is convenient to partition the sum into two sums—by even ($j = 2k$) and odd ($j = 2k + 1$) values of j, and introduce a new summation variable $k \in \mathbb{N}$:

$$(\cos \varphi + i \sin \varphi)^n$$

$$= \sum_{k=0}^{\lfloor n/2 \rfloor} i^{2k} C(n, 2k) \cos^{n-2k} \varphi \, \sin^{2k} \varphi$$

$$+ \sum_{k=0}^{\lfloor (n-1)/2 \rfloor} i^{2k+1} C(n, 2k + 1) \cos^{n-2k-1} \varphi \, \sin^{2k+1} \varphi$$

$$= \sum_{k=0}^{\lfloor n/2 \rfloor} (-1)^k C(n, 2k) \cos^{n-2k} \varphi \, \sin^{2k} \varphi$$

$$+ i \sum_{k=0}^{\lfloor (n-1)/2 \rfloor} (-1)^k C(n, 2k + 1) \cos^{n-2k-1} \varphi \, \sin^{2k+1} \varphi.$$

Now we only have to take advantage of the fact that $\cos n\varphi = \text{Re} \, (\cos \varphi + i \sin \varphi)^n$,
$\sin n\varphi = \text{Im} \, (\cos \varphi + i \sin \varphi)^n$. We obtain formulae for cosine and sine of a multiple

argument:

$$\cos n\varphi = \sum_{k=0}^{\lfloor n/2 \rfloor} (-1)^k C(n, 2k) \cos^{n-2k} \varphi \, \sin^{2k} \varphi,$$

$$\sin n\varphi = \sum_{k=0}^{\lfloor (n-1)/2 \rfloor} (-1)^k C(n, 2k+1) \cos^{n-2k-1} \varphi \, \sin^{2k+1} \varphi.$$

For $n = 3$ the obtained formulae take the form

$$\cos 3\varphi = \sum_{k=0}^{1} (-1)^k C(3, 2k) \cos^{3-2k} \varphi \, \sin^{2k} \varphi = \cos^3 \varphi - 3 \cos \varphi \sin^2 \varphi,$$

$$\sin 3\varphi = \sum_{k=0}^{1} (-1)^k C(3, 2k+1) \cos^{3-2k-1} \varphi \, \sin^{2k+1} \varphi = 3 \cos^2 \varphi \sin \varphi - \sin^3 \varphi.$$

4.30 *Answer:*

$$\cos 4\varphi = \sum_{k=0}^{2} (-1)^k C(4, 2k) \cos^{4-2k} \varphi \, \sin^{2k} \varphi$$

$$= \cos^4 \varphi - 6 \cos^2 \varphi \sin^2 \varphi + \sin^4 \varphi,$$

$$\sin 4\varphi = \sum_{k=0}^{1} (-1)^k C(4, 2k+1) \cos^{4-2k-1} \varphi \, \sin^{2k+1} \varphi$$

$$= 4 \cos^3 \varphi \sin \varphi - 4 \cos \varphi \sin^3 \varphi.$$

4.31 *Hint.*
Multiply the brackets in the right side of factorization of the polynomial $p(z)$ and compare the obtained coefficients at the same powers with the coefficients $p(z)$.

4.32 *Solution.*

(1) Represent $z_1^2 + z_2^2$ in the form

$$z_1^2 + z_2^2 = (z_1 + z_2)^2 - 2z_1 z_2$$

and express the sum and the product of the roots $p(z)$ by Viète formulae, proved in the previous problem:

$$z_1^2 + z_2^2 = (-u)^2 - 2v = u^2 - 2v.$$

(2) $\dfrac{1}{z_1^2} + \dfrac{1}{z_2^2} = \dfrac{z_1^2 + z_2^2}{z_1^2 z_2^2} = \dfrac{u^2 - 2v}{v^2}.$

4.33 *Answer:*

(1) $z_1^4 + z_2^4 = (u^2 - 2v)^2 - 2v^2;$

(2) $\dfrac{1}{z_1^4} + \dfrac{1}{z_2^4} = \dfrac{(u^2 - 2v)^2 - 2v^2}{v^4}.$

4.34 *Answer:*

(1) $\displaystyle\sum_{k=1}^{3} z_k = -3, \quad \prod_{k=1}^{3} z_k = -1;$

(2) $\displaystyle\sum_{k=1}^{4} z_k = 0, \quad \prod_{k=1}^{4} z_k = 20.$

4.35 *Answer:*

(1) $\displaystyle\sum_{k=1}^{4} z_k = 100, \quad \prod_{k=1}^{4} z_k = 0;$

(2) $\displaystyle\sum_{k=1}^{5} z_k = -1, \quad \prod_{k=1}^{5} z_k = -1.$

4.36 *Solution.*

In the determinant, add to the first row the second and the third rows:

$$\Delta = \begin{vmatrix} z_1 + z_2 + z_3 & z_1 + z_2 + z_3 & z_1 + z_2 + z_3 \\ z_3 & z_1 & z_2 \\ z_2 & z_3 & z_1 \end{vmatrix}.$$

According to Viète formulae (see Problem **4.31**), the sum of all roots of $z_1 + z_2 + z_3$ is equal to the coefficient of the quadratic term z^2, taken with reversed sign. For the equation $z^3 + \alpha z + \beta = 0$ we have $z_1 + z_2 + z_3 = 0$, and, therefore, in the determinant Δ the first row entirely consists of zero elements. It is clear that such a determinant is equal to zero.

4.37 *Answer:* $z_1 = -2, z_2 = z_3 = 1.$

4.38 *Answer:* $z_1 = 7, z_2 = -\dfrac{5}{2}, z_3 = 2.$

4.39 *Answer:*

(a) $[x_1, x_2, x_3]^T = [2i, -2, i]^T;$
(b) $[x_1, x_2, x_3]^T = [i, 1, 2 + 2i]^T;$
(c) $[x_1, x_2, x_3]^T = [-3, 3, -2 + i]^T;$
(d) $[x_1, x_2, x_3]^T = [1, -1 + 2i, -2 - i]^T.$

4.40 *Answer:*

(a) $[x_1, x_2, x_3]^T = [2i, 1+i, -2+2i]^T$;
(b) $[x_1, x_2, x_3]^T = [3+2i, -3+i, 3-i]^T$;
(c) $[x_1, x_2, x_3]^T = [2, 3-i, -1+2i]^T$;
(d) $[x_1, x_2, x_3]^T = [-2+2i, 3i, 2+i]^T$.

4.41 *Answer:*

(a) $[x_1, x_2, x_3, x_4, x_5]^T = [2+i, i, -1+i, -2, 1+2i]^T$;
(b) $[x_1, x_2, x_3, x_4, x_5]^T = [i, -2, -1-i, 1+i, 2+2i]^T$.

4.42 *Answer:* Hermitian are matrices 1) and 2).

4.43 *Answer:* diagonal elements of Hermitian matrix are valid.

4.44 *Proof.*

Introduce the notation $W = \dfrac{1}{2}(Z_1 Z_2 + Z_2 Z_1)$ and find Hermitian conjugate matrix relative to W:

$$W^H = \left(\frac{1}{2}(Z_1 Z_2 + Z_2 Z_1)\right)^H = \frac{1}{2}((Z_1 Z_2)^H + (Z_2 Z_1)^H) = \frac{1}{2}(Z_2 Z_1 + Z_1 Z_2) = W.$$

It is proved that W is a Hermitian matrix.

4.45 *Answer:* anti-Hermitian are matrices (1), (5), (6).

4.46 *Answer:* diagonal elements of anti-Hermitian matrix are purely imaginary values.

4.47 *Answer:* unitary are matrices (2), (3), (4), (6).

4.48 *Answer:* $|u_1|^2 + |u_2|^2 = 1$, $\varphi \in \mathbb{R}$ is any real number.

4.49 *Solution.*

In this case, the student is wrong, since for $A = 2\pi i I$, where I is the identity matrix, we have $\exp(A) = e^{2\pi i} I = I$, therefore, $\ln(\exp(A)) = O \neq A$.

4.50 *Hint.*

For the diagonal matrix $\det(e^A) = e^{\lambda_1} e^{\lambda_2} \dots e^{\lambda_n} = e^{\operatorname{tr} A}$. As for the non-diagonal matrix, we either diagonalize it, if possible, or, with any predefined accuracy, approximate it by a sequence of matrices, each being diagonalizable

4.51 *Proof.*

Let U be an arbitrary unitary matrix. Using property (6) of Hermitian conjugate on page 181, represent the modulus of the determinant U in the following form:

$$|\det U| = \sqrt{(\det U)(\det U)^*} = \sqrt{(\det U)(\det U^H)}.$$

Since the product of the determinants of matrices is equal to the determinant of their product, $\forall A, B$ ($\det A \cdot \det B = \det(AB)$), then

$$|\det U| = \sqrt{\det(UU^H)} = \sqrt{\det I} = 1.$$

Thus, the modulus of the determinant of the unitary matrix is equal to one.

4.52 *Answer:* $[\sigma_1, \sigma_2] = 2i\sigma_3$, $[\sigma_2, \sigma_3] = 2i\sigma_1$, $[\sigma_3, \sigma_1] = 2i\sigma_2$.

4.53 *Answer:* $\sigma_1\sigma_2\sigma_3 = iI$, where I is the identity matrix of size 2×2.

4.55 *Answer:* $\beta^2 = I$, where I is an identity matrix of size 4×4.

4.56 *Answer:*

$$\alpha_1\alpha_2\alpha_3\beta = \begin{bmatrix} 0 & 0 & -i & 0 \\ 0 & 0 & 0 & -i \\ i & 0 & 0 & 0 \\ 0 & i & 0 & 0 \end{bmatrix}.$$

4.57 *Solution.*

The state of the qubit $|0\rangle$ is described by the matrix $\begin{bmatrix} 1 \\ 0 \end{bmatrix}$. Taking into account the matrix representation of the quantum elements H, S and X from Table 4.1, we obtain

$$|\psi\rangle \rightarrow |\psi'\rangle = \frac{1}{\sqrt{2}}\begin{bmatrix} 1 & 1 \\ 1 & -1 \end{bmatrix}\begin{bmatrix} 1 & 0 \\ 0 & i \end{bmatrix}\frac{1}{\sqrt{2}}\begin{bmatrix} 1 & 1 \\ 1 & -1 \end{bmatrix}\begin{bmatrix} 0 & 1 \\ 1 & 0 \end{bmatrix}\begin{bmatrix} 1 \\ 0 \end{bmatrix} = \begin{bmatrix} \dfrac{1-i}{2} \\ \dfrac{1+i}{2} \end{bmatrix}.$$

Thus, as a result of the quantum circuit's action on a qubit in the state $|\psi\rangle = |0\rangle$, it passes to the state

$$|\psi'\rangle = \frac{1-i}{2}|0\rangle + \frac{1+i}{2}|1\rangle.$$

4.58 *Answer:* as a result of the quantum circuit's action on a qubit in the state $|\psi\rangle = |1\rangle$, it passes to the state

$$|\psi'\rangle = \left(-\frac{i}{\sqrt{2}}\right)|0\rangle - \left(\frac{1-i}{2}\right)|1\rangle.$$

4.59 *Answer:* a qubit in the state $|\psi\rangle = u\,|0\rangle + v\,|1\rangle$, passes to the state $|\psi'\rangle = \left(\frac{1}{2}(1+i)u + \frac{1}{\sqrt{2}}v\right)|0\rangle + \left(\frac{1}{2}(1+i)u - \frac{1}{\sqrt{2}}v\right)|1\rangle$.

Chapter 5
Vector Spaces

By n-**dimensional arithmetic vector**, we will mean an ordered set of n real numbers.

The numbers x_i $(i = 1, 2, \ldots, n)$ are called **coordinates** or **components** of the vector x. They are written either in the row: $x = (x_1, x_2, \ldots, x_n)$, or in the column:

$$x = \begin{bmatrix} x_1 \\ x_2 \\ \vdots \\ x_n \end{bmatrix}. \tag{5.1}$$

For designation of vectors and distinguishing them from scalar values, bold font is used, for example, a, b, c, etc.[1]

The vectors $x = (x_1, x_2, \ldots, x_n)$ and $y = (y_1, y_2, \ldots, y_n)$ are called **equal vectors**, if the equalities $x_1 = y_1$, $x_2 = y_2$, $\ldots$, $x_n = y_n$ are valid.

Sum of the vectors x and y is the vector

$$x + y = (x_1 + y_1, x_2 + y_2, \ldots, x_n + y_n). \tag{5.2}$$

Product of the number α and a vector x is the vector

$$\alpha x = (\alpha x_1, \alpha x_2, \ldots, \alpha x_n). \tag{5.3}$$

---

[1] Another designation of vector is also used, when an arrow is placed above its symbol. For example, using this method, the vectors a and b will be designated as $\vec{a}$ and $\vec{b}$.

© Springer Nature Switzerland AG 2021
S. Kurgalin, S. Borzunov, *Algebra and Geometry with Python*,
https://doi.org/10.1007/978-3-030-61541-3_5

Difference of the vectors x and y is the vector

$$x - y = x + (-1)y = (x_1 - y_1, x_2 - y_2, \ldots, x_n - y_n). \qquad (5.4)$$

Zero vector, or **null vector**, is the vector $\mathbf{0} = (0, 0, \ldots, 0)$ that has zero coordinates. It is obvious that $x + \mathbf{0} = x - \mathbf{0} = x$.

For the vector $x \in \mathbb{R}$, by $-x$ denote the vector with the coordinates

$$(-x_1, -x_2, \ldots, -x_n), \qquad (5.5)$$

such a vector being referred to as **opposite** relative to x.

From the introduced definitions, it follows that $-x = (-1) \cdot x$ and $x + (-x) = \mathbf{0}$.

A set of all arithmetic vectors with the operations of addition and multiplication introduced on them is called a n-**dimensional arithmetic space** and denoted by $\mathbb{R}^n$ [30].

Example 5.1 $\mathbb{R}^1$ is a one-dimensional space (line), $\mathbb{R}^2$ is a two-dimensional space (plane) and $\mathbb{R}^3$ is a three-dimensional space. □

5.1   Linear Dependence of Vectors in the Space $\mathbb{R}^n$

Consider a set of vectors $x_1, x_2, \ldots, x_k \in \mathbb{R}^n$ and the real numbers $\alpha_1, \alpha_2, \ldots, \alpha_k \in \mathbb{R}$.

The vector $x = \alpha_1 \cdot x_1 + \alpha_2 \cdot x_2 + \cdots + \alpha_k \cdot x_k$ is called a **linear combination of vectors** $x_1, x_2, \ldots, x_k$.

Example 5.2 Let there be given the vectors

$$x_1 = \begin{bmatrix} 3 \\ -2 \\ -1 \end{bmatrix}, \quad x_2 = \begin{bmatrix} 4 \\ 3 \\ 0 \end{bmatrix}, \quad x_3 = \begin{bmatrix} 5 \\ 3 \\ 7 \end{bmatrix}. \qquad (5.6)$$

Then, the vector $x = 2x_1 - 3x_2 + x_3 = \begin{bmatrix} -1 \\ -10 \\ 5 \end{bmatrix}$ is the linear combination of vectors x_1, x_2, x_3. □

A system of vectors $x_1, x_2, \ldots, x_k$ is referred to as **linearly independent** one, if from the equality $\alpha_1 \cdot x_1 + \alpha_2 \cdot x_2 + \cdots + \alpha_k \cdot x_k = \mathbf{0}$, it follows that $\alpha_1 = \alpha_2 = \cdots = \alpha_k = 0$.

If there exists a set of real numbers $\alpha_1, \alpha_2, \ldots, \alpha_k$, among which at least one is not equal to zero, then the system of vectors is referred to as **linearly dependent** one.

Example 5.3 Given are the vectors $x_1 = (2, -3)$ and $x_2 = (4, 5)$. Show that they are linearly independent.

Find the solution of the system of equations

$$\alpha_1 \begin{bmatrix} 2 \\ -3 \end{bmatrix} + \alpha_2 \begin{bmatrix} 4 \\ 5 \end{bmatrix} = \begin{bmatrix} 0 \\ 0 \end{bmatrix}, \tag{5.7}$$

$$\begin{cases} \alpha_1 \cdot 2 + \alpha_2 \cdot 4 = 0, \\ \alpha_1 \cdot (-3) + \alpha_2 \cdot 5 = 0. \end{cases} \tag{5.8}$$

Since this system has the unique solution $\alpha_1 = \alpha_2 = 0$, the vectors x_1 and x_2 are linearly independent. □

Note Assume that the vectors $x_1, x_2, \ldots, x_k$ are linearly dependent. Then, at least one of the coefficients α_i is other than zero (for example, $\alpha_1 \neq 0$). In this case, we can write

$$x_1 = -\frac{\alpha_2}{\alpha_1} x_2 - \frac{\alpha_3}{\alpha_1} x_3 - \cdots - \frac{\alpha_k}{\alpha_1} x_k. \tag{5.9}$$

Thus, if the vectors are linearly dependent, then one of them is linearly expressed in terms of all others [30]. The converse is also true: if one of the vectors of the set is linearly expressed in terms of the others, then these vectors are linearly dependent. The last property can be considered as a definition of linear dependence of vectors.

5.2 Basis in the Space $\mathbb{R}^n$

Prior to introducing the concept of basis in the vector space $\mathbb{R}^n$, let us prove the following theorem.

Theorem 5.1 *Any system of $n + 1$ vectors in the space $\mathbb{R}^n$ is linearly dependent.*

Proof Take arbitrary $n + 1$ vectors

$$x_i = (x_{1i}, x_{2i}, \ldots, x_{ni}), \tag{5.10}$$

where $i = 1, 2, \ldots, n + 1$.

Construct their linear combination and equate it to zero vector:

$$\alpha_1 \cdot x_1 + \alpha_2 \cdot x_2 + \cdots + \alpha_{n+1} \cdot x_{n+1} = \mathbf{0}. \tag{5.11}$$

Writing this equality in a coordinate-wise manner, we arrive at a system of n equations with $n + 1$ unknowns $\alpha_1, \alpha_2, \ldots, \alpha_{n+1}$:

$$\begin{cases} \alpha_1 \cdot x_{11} + \alpha_2 \cdot x_{12} + \cdots + \alpha_{n+1} \cdot x_{1\,n+1} = 0, \\ \cdots\cdots\cdots\cdots\cdots\cdots\cdots\cdots\cdots\cdots\cdots\cdots\cdots\cdots \\ \alpha_1 \cdot x_{n1} + \alpha_2 \cdot x_{n2} + \cdots + \alpha_{n+1} \cdot x_{n\,n+1} = 0. \end{cases} \tag{5.12}$$

The matrix of the system (5.12) differs from the respective augmented matrix only in the zero column, this is why their ranks coincide. Therefore, according to the Kronecker–Capelli theorem, the system has infinitely many solutions. They necessarily include a non-zero solution. Thus, there exists a non-zero set of coefficients $\alpha_1, \alpha_2, \ldots, \alpha_{n+1}$, for which the linear combination of vectors $x_1, x_2, \ldots, x_{n+1}$ is equal to a non-zero vector. Therefore, the vectors x_i, where $1 \leqslant i \leqslant n + 1$, are linearly dependent.

Any system of n linearly independent vectors $b_1, b_2, \ldots, b_n$ is called a **basis** of a vector space.

Consider in the space $\mathbb{R}^n$ the system of vectors:

$$e_1 = (1, 0, \ldots, 0),$$
$$e_2 = (0, 1, \ldots, 0),$$
$$\cdots\cdots\cdots\cdots\cdots \tag{5.13}$$
$$e_n = (0, 0, \ldots, 1).$$

These vectors are linearly independent, since from the condition

$$\alpha_1 \cdot e_1 + \cdots + \alpha_n \cdot e_n = \mathbf{0} \tag{5.14}$$

directly follows the system of equalities $\alpha_1 = \alpha_2 = \cdots = \alpha_n = 0$.

The vectors $e_1, e_2, \ldots, e_n$ are called **normalized vectors** of the space $\mathbb{R}^n$; they form the basis in this space.

Conclusion: a linearly independent system of vectors in $\mathbb{R}^n$ can have a maximum of n vectors.

Consider a system of n vectors

$$x_i = (x_{1i}, x_{2i}, \ldots, x_{ni}), \quad \text{where } i = 1, 2, \ldots, n. \tag{5.15}$$

Construct a matrix of coordinates of the vectors:

$$\begin{bmatrix} x_{11} & x_{12} & \ldots & x_{1n} \\ x_{21} & x_{22} & \ldots & x_{2n} \\ \vdots & \vdots & \ddots & \vdots \\ x_{n1} & x_{n2} & \ldots & x_{nn} \end{bmatrix}. \tag{5.16}$$

Such a matrix is called a **matrix of a system of vectors**, and its determinant is called a **determinant of a system of vectors**.

Theorem 5.2 *In order for a system of n vectors to be the basis, it is necessary and sufficient that the determinant of the system is other than zero.*

Proof Consider an arbitrary system of n vectors

$$x_i = (x_{1i}, x_{2i}, \ldots, x_{ni}), \quad i = 1, 2, \ldots, n. \tag{5.17}$$

Construct their linear combination and equate it to zero vector:

$$\begin{bmatrix} \alpha_1 \cdot x_{11} + \alpha_2 \cdot x_{12} + \cdots + \alpha_n \cdot x_{1n} \\ \alpha_1 \cdot x_{21} + \alpha_2 \cdot x_{22} + \cdots + \alpha_n \cdot x_{2n} \\ \ldots\ldots\ldots\ldots\ldots\ldots\ldots\ldots\ldots\ldots \\ \alpha_1 \cdot x_{n1} + \alpha_2 \cdot x_{n2} + \cdots + \alpha_n \cdot x_{nn} \end{bmatrix} = \begin{bmatrix} 0 \\ 0 \\ \vdots \\ 0 \end{bmatrix}. \tag{5.18}$$

We will obtain a homogeneous system of n equations with n unknowns and a determinant other than zero. In this case, such a system has only a zero solution, i.e. $\alpha_i = 0$, where $i = 1, 2, \ldots, n$, and the system of vectors is a basis.

Theorem 5.3 *Assume that the vectors $x_1, x_2, \ldots, x_n$ form a basis. Then, any vector y of $\mathbb{R}^n$ can be represented, uniquely, in the form of a linear combination of the vectors x_i $(i = 1, 2, \ldots, n)$:*

$$y = \alpha_1 \cdot x_1 + \alpha_2 \cdot x_2 + \cdots + \alpha_n \cdot x_n. \tag{5.19}$$

Proof Write the expansion (5.19) in projections:

$$\alpha_1 x_1 + \alpha_2 x_2 + \cdots + \alpha_n x_n = \begin{bmatrix} \alpha_1 \cdot x_{11} + \alpha_2 \cdot x_{12} + \cdots + \alpha_n \cdot x_{1n} \\ \alpha_1 \cdot x_{21} + \alpha_2 \cdot x_{22} + \cdots + \alpha_n \cdot x_{2n} \\ \ldots\ldots\ldots\ldots\ldots\ldots\ldots\ldots\ldots\ldots \\ \alpha_1 \cdot x_{n1} + \alpha_2 \cdot x_{n2} + \cdots + \alpha_n \cdot x_{nn} \end{bmatrix} = \begin{bmatrix} y_1 \\ y_2 \\ \vdots \\ y_n \end{bmatrix}. \tag{5.20}$$

We have obtained a non-homogeneous system of n equations with n unknowns. Since the determinant of this system is other than zero, by virtue of Cramer's rule, this system has the unique solution.

Note The formula (5.19) is called an **expansion of the vector y in the basis x_i** ($i = 1, 2, \ldots, n$).

Example 5.4 Show that the vectors $a = (1, 1, 4)$, $b = (0, -3, 2)$ and $c = (2, 1, -1)$ form a basis.

Construct the determinant of the system of vectors and compute it:

$$\begin{vmatrix} 1 & 0 & 2 \\ 1 & -3 & 1 \\ 4 & 2 & -1 \end{vmatrix} = 1 \cdot (3 - 2) + 2 \cdot (2 + 12) = 29. \tag{5.21}$$

Since the determinant of the system is other than zero, the vectors a, b and c form a basis.

Example 5.5 Expand the vector $d = (6, 5, -14)$ in the basis (a, b, c) (see previous example).

Represent the vector d in the form of the expansion:

$$d = \alpha_1 \cdot a + \alpha_2 \cdot b + \alpha_3 \cdot c. \tag{5.22}$$

We have

$$\begin{bmatrix} 6 \\ 5 \\ -14 \end{bmatrix} = \alpha_1 \begin{bmatrix} 1 \\ 1 \\ 4 \end{bmatrix} + \alpha_2 \begin{bmatrix} 0 \\ -3 \\ 2 \end{bmatrix} + \alpha_3 \begin{bmatrix} 2 \\ 1 \\ -1 \end{bmatrix}. \tag{5.23}$$

Write this equality in the form of the system of linear equations:

$$\begin{cases} \alpha_1 & +2\alpha_3 = & 6, \\ \alpha_1 & -3\alpha_2 +\alpha_3 = & 5, \\ 4\alpha_1 & +2\alpha_2 -\alpha_3 = & -14. \end{cases} \tag{5.24}$$

The obtained system has the unique solution $\alpha_1 = -2$, $\alpha_2 = -1$, $\alpha_3 = 4$. Thus, $d = -2a - b + 4c$. $\qquad\qquad\qquad\Box$

5.3 Euclidean Vector Space

In an arbitrary n-dimensional vector space, it is possible to introduce a **scalar product** (known also as **inner product**, or **dot product**)—the rule according to which the two vectors $a, b \in \mathbb{R}^n$ are associated with the number $(a \cdot b)$. The scalar product suggests such analogues of a spatial arrangement of the multidimensional vectors $\mathbb{R}^n$ as orthogonality and collinearity.

By definition, the scalar product of the vectors $a = (a_1, a_2, \ldots, a_n)$ and $b = (b_1, b_2, \ldots, b_n)$ is computed by the formula:

$$(a \cdot b) = a_1 b_1 + a_2 b_2 + \cdots + a_n b_n. \tag{5.25}$$

Note that with the help of the summation sign, the variable $(a \cdot b)$ is compactly written as $(a \cdot b) = \sum_{i=1}^{n} a_i b_i$.

Let us enumerate the properties of a scalar product.

For arbitrary $a, a_1, a_2, b \in \mathbb{R}^n$ and $\alpha \in \mathbb{R}$, the following equalities are valid:

(1) $(a \cdot b) = (b \cdot a)$ (symmetry);
(2) $((a_1 + a_2) \cdot b) = (a_1 \cdot b) + (a_2 \cdot b)$ (linearity);
(3) $(\alpha a \cdot b) = \alpha(a \cdot b)$ (linearity);
(4) $(a \cdot a) \geqslant 0$, and $(a \cdot a) = 0 \Leftrightarrow a = 0$ (non-negativity).

Note For the scalar product of the vectors a and b, the designations (a, b), $a \cdot b$ or ab are also used.

Example 5.6 Let $n = 4$, and in the coordinate notation, the vectors have the form, $a = (10, -2, 1, 9)$, $b = (0, 3, 4, -2)$ and $c = (-12, 2, -4, -5)$. Then,

$$a \cdot b = 10 \cdot 0 + (-2) \cdot 3 + 1 \cdot 4 + 9 \cdot (-2) = -20,$$

$$a \cdot c = 10 \cdot (-12) + (-2) \cdot 2 + 1 \cdot (-4) + 9 \cdot (-5) = -173,$$

$$b \cdot c = 0 \cdot (-12) + 3 \cdot 2 + 4 \cdot (-4) + (-2) \cdot (-5) = 0.$$

$\square$

Example 5.7 Show that if the condition $(a \cdot t) = (b \cdot t)$ is valid for all $t \in \mathbb{R}^n$, then the vectors a and b are equal to each other.

Proof Based on the property of linearity, represent the equality $(a \cdot t) = (b \cdot t)$ in the equivalent form

$$(a \cdot t) = (b \cdot t) \Leftrightarrow ((a - b) \cdot t) = 0. \tag{5.26}$$

Into the obtained equality, substitute $t = a - b$. Then, according to the property of non-negativity, we have $(a - b) = 0$ or $a = b$, which is what we set out to prove. $\square$

Length, or **norm**, of a vector is the value $\|a\| = \sqrt{(a \cdot a)}$.

Note the following easily provable properties of the norm, which are valid for arbitrary vectors a and b of Euclidean space:

(1) $\|a\| \geqslant 0$, and $\|a\| = 0 \Leftrightarrow a = 0$;
(2) $\|\alpha a\| = \mathrm{abs}(\alpha)\|a\|$ for all $\alpha \in \mathbb{R}$;
(3) $\|a + b\| \leqslant \|a\| + \|b\|$.

The last inequality is referred to as a **triangle inequality** or **Minkowski**[2] **inequality**.

Example 5.8 For the vectors $a = (0, -1, -1, 3, 1)$ and $b = (5, -3, 0, -2, -1)$ in the space $\mathbb{R}^5$, we have

$$\|a\| = \sqrt{(a \cdot a)} = \sqrt{0 + (-1)^2 + (-1)^2 + 3^2 + 1^2} = \sqrt{12};$$

$$\|b\| = \sqrt{(b \cdot b)} = \sqrt{(-5)^2 + (-3)^2 + 0^2 + (-2)^2 + (-1)^2} = \sqrt{39}.$$

$\square$

Orthogonal are the vectors whose scalar product is equal to zero. Usually, the orthogonal vectors are designated as $a \perp b$.

A set of vectors, where all vectors are pairwise orthogonal, is naturally called **orthogonal**. If in such a set all vectors have a unit norm, then such a set is **orthonormal**.

Of course, in an arbitrary basis, the vectors might not possess the property of pairwise orthogonality, a fortiori orthonormality. Show that there exists a possibility to construct a new basis from the original one, and in the new basis, all the vectors will be pairwise orthogonal. Such a procedure is called the **Gram**[3]**–Schmidt**[4] **process** (orthogonalization).

Assume that in a vector Euclidean space with a norm, a basis $(p_1, p_2, \ldots, p_n)$ is set. The procedure of constructing a new orthonormal basis consists in performing the following steps.

Successively compute the vectors $q_1, q_2, \ldots, q_n$ by the formulae:

$$t_1 = p_1, \quad q_1 = \frac{t_1}{\|t_1\|},$$

$$t_2 = p_2 - (p_2, q_1)q_1, \quad q_2 = \frac{t_2}{\|t_2\|},$$

---

[2] Hermann Minkowski (1864–1909), German mathematician.
[3] Jørgen Pedersen Gram (1850–1916), Danish mathematician.
[4] Erhardt Schmidt (1876–1959), German mathematician.

$$t_3 = p_3 - (p_3, q_1)q_1 - (p_3, q_2)q_2, \quad q_3 = \frac{t_3}{\|t_3\|},$$

$$\ldots$$

$$t_n = p_n - (p_n, q_1)q_1 - \cdots - (p_n, q_{n-1})q_{n-1}, \quad q_n = \frac{t_n}{\|t_n\|}.$$

The obtained basis $(q_1, q_2, \ldots, q_n)$ is orthonormal.

5.4 Eigenvalues and Eigenvectors of a Matrix

Two matrices A and A', bound by the relation

$$A' = P^{-1}AP, \tag{5.27}$$

where P is some invertible matrix, are referred to as **similar**. In this case, the designation $A' \sim A$ is used.

Note Transition from the matrix A to A' is called **similarity transformation**.

Example 5.9 Matrices $\begin{bmatrix} 1 & -3 \\ -1 & 2 \end{bmatrix}$ and $\begin{bmatrix} -2 & 1 \\ -9 & 5 \end{bmatrix}$ are similar, since the following equality is valid:

$$\begin{bmatrix} -2 & 1 \\ -9 & 5 \end{bmatrix} = \begin{bmatrix} 1 & -1 \\ -2 & 1 \end{bmatrix}^{-1} \begin{bmatrix} 1 & -3 \\ -1 & 2 \end{bmatrix} \begin{bmatrix} 1 & -1 \\ -2 & 1 \end{bmatrix}. \tag{5.28}$$

Indeed, $\begin{bmatrix} 1 & -1 \\ -2 & 1 \end{bmatrix}^{-1} = \begin{bmatrix} -1 & -1 \\ -2 & -1 \end{bmatrix}$, and the equality (5.28) is easily verified by a direct multiplication of the matrices. $\qquad \square$

Theorem 5.4 (On the Matrix Similarity Properties) *The following properties of similarity are valid:*

(1) $A \sim A$—*reflexivity;*
(2) $A \sim B \Rightarrow B \sim A$—*symmetry;*
(3) $((A \sim B) \text{ and } (B \sim C)) \Rightarrow (A \sim C)$—*transitivity.*

It follows from the theorem on the matrix similarity properties that the similarity of matrices is an equivalence relation [1, 41, 53].

Two similar matrices have equal determinants. Indeed, from the definition of (5.27), it follows that

$$|A'| = |P^{-1}AP| = |P^{-1}|\,|A|\,|P| = |A|. \tag{5.29}$$

Note that the equality of the determinants does not at all imply the similarity of the matrices.

Example 5.10 Find out whether the following matrices are similar: $\begin{bmatrix} 0 & 1 \\ 1 & 0 \end{bmatrix}$ and $\begin{bmatrix} 1 & 0 \\ 1 & 1 \end{bmatrix}$.

The determinants of these matrices are equal to -1 and 1, respectively. Therefore, these matrices do not possess the property of similarity. □

The number λ and the non-zero vector b are referred to as **eigenvalue** and **eigenvector** of the matrix A, respectively, if the following equality is valid:

$$Ab = \lambda b. \tag{5.30}$$

The vector b is considered as a column vector. In order to find b and λ, represent Eq. (5.30) in the following form:

$$(A - \lambda I)b = 0, \tag{5.31}$$

where I is an identity matrix.

We have obtained a homogeneous system of linear equations. In order for it to have a non-zero solution, it is necessary and sufficient that the determinant of the matrix $A - \lambda I$ is equal to zero. Thus, in order to find λ, we should solve the equation

$$|A - \lambda I| = 0 \tag{5.32}$$

or, in an expanded notation

$$\begin{vmatrix} a_{11} - \lambda & a_{12} & \dots & a_{1n} \\ a_{21} & a_{22} - \lambda & \dots & a_{2n} \\ \dots\dots\dots\dots\dots\dots\dots\dots \\ a_{n1} & a_{n2} & \dots & a_{nn} - \lambda \end{vmatrix} = 0. \tag{5.33}$$

This equation is referred to as **characteristic equation**.

If we expand the determinant, we obtain a polynomial of power n relative to the variable λ:

$$p(\lambda) = (-\lambda)^n + h_1(-\lambda)^{n-1} + \cdots + h_{n-1}(-\lambda) + h_n, \qquad (5.34)$$

and the following properties are valid:

- the coefficient h_1 is equal to the trace of the matrix A: $h_1 = \operatorname{tr} A = a_{11} + a_{22} + \cdots + a_{nn}$;
- the constant term h_n coincides with the determinant: $h_n = \det A$.

The polynomial (5.34) is also referred to as **characteristic**.

It is known that characteristic polynomials of similar matrices coincide.

According to the fundamental theorem of algebra, the characteristic equation (5.30) has no more than n solutions. For each solution λ, it is associated with the eigenvector b.

Note Although the eigenvalue can be equal to zero, the eigenvector, by definition, is always different from the zero vector.

Example 5.11 Find the eigenvalues and eigenvectors of the matrix

$$A = \begin{bmatrix} 1 & 1 & 3 \\ 1 & 5 & 1 \\ 3 & 1 & 1 \end{bmatrix}. \qquad (5.35)$$

Compute the determinant of the matrix $A - \lambda I$ and equate it to zero:

$$|A - \lambda I| = \begin{vmatrix} 1 - \lambda & 1 & 3 \\ 1 & 5 - \lambda & 1 \\ 3 & 1 & 1 - \lambda \end{vmatrix}$$

$$= (1 - \lambda) \cdot [(5 - \lambda) \cdot (1 - \lambda) - 1] - (1 - \lambda) + 3 + 3 \cdot [1 - 3 \cdot (5 - \lambda)]$$

$$= (1 - \lambda) \cdot (\lambda^2 - 6 \cdot \lambda + 3) + 9 \cdot \lambda - 39 = -\lambda^3 + 7 \cdot \lambda^2 - 36$$

$$= -(\lambda + 2) \cdot (\lambda^2 - 9 \cdot \lambda + 18) = 0.$$

Solving this equation, we will obtain three roots, $\lambda_1 = -2$, $\lambda_2 = 6$ and $\lambda_3 = 3$. For each λ, find the eigenvector associated with it.

1. Let $\lambda = -2$. Then,

$$A - \lambda_1 I = \begin{bmatrix} 3 & 1 & 3 \\ 1 & 7 & 1 \\ 3 & 1 & 3 \end{bmatrix}. \qquad (5.36)$$

Assuming that $b = (x, y, z)$, we will obtain the system of equations

$$\begin{cases} 3x + y + 3z = 0, \\ x + 7y + z = 0, \\ 3x + y + 3z = 0. \end{cases} \tag{5.37}$$

Write the matrix that corresponds to this system:

$$\begin{bmatrix} 3 & 1 & 3 \\ 1 & 7 & 1 \\ 3 & 1 & 3 \end{bmatrix}. \tag{5.38}$$

Add to the third row of this matrix the second one, multiplied by (-1):

$$\begin{bmatrix} 3 & 1 & 3 \\ 1 & 7 & 1 \\ 0 & 0 & 0 \end{bmatrix}. \tag{5.39}$$

Drop the zero row and exchange places of the second and the first rows. Then, we have

$$\begin{bmatrix} 1 & 7 & 1 \\ 3 & 1 & 3 \end{bmatrix}. \tag{5.40}$$

Bring the matrix to echelon form; add to the second row the first one, multiplied by -3. We obtain

$$\begin{bmatrix} 1 & 7 & 1 \\ 0 & -20 & 0 \end{bmatrix}. \tag{5.41}$$

Proceed to the equations and write

$$\begin{cases} x + 7y + z = 0, \\ y = 0, \end{cases} \tag{5.42}$$

or $x + z = 0$.

As a free variable, select z. Then, assume that $z = 1$, and then $x = -1$. Thus, $b_1 = (-1, 0, 1)$.

2. For $\lambda = 6$, we obtain the system:

$$\begin{cases} -5x + y + 3z = 0, \\ x - y + z = 0, \\ 3x + y - 5z = 0. \end{cases} \tag{5.43}$$

In order to find the vector b_2, write the matrix of this system and its transformations:

$$\begin{bmatrix} -5 & 1 & 3 \\ 1 & -1 & 1 \\ 3 & 1 & -5 \end{bmatrix} \rightarrow \begin{bmatrix} 1 & -1 & 1 \\ -5 & 1 & 3 \\ 3 & 1 & -5 \end{bmatrix} \rightarrow \begin{bmatrix} 1 & -1 & 1 \\ 0 & -4 & 8 \\ 0 & 4 & -8 \end{bmatrix} \rightarrow \begin{bmatrix} 1 & -1 & 1 \\ 0 & 1 & -2 \end{bmatrix}.$$

Proceed to the system of equations:

$$\begin{cases} x - y = -z, \\ y = 2z. \end{cases} \tag{5.44}$$

Assuming that $z = 1$, we find $y = 2$ and $x = 1$.
Thus, $b_2 = (1, 2, 1)$.
3. For $\lambda = 3$, we have the system:

$$\begin{cases} -2x + y + 3z = 0, \\ x + 2y + z = 0, \\ 3x + y - 2z = 0. \end{cases} \tag{5.45}$$

In order to find the vector b_3, we perform similar equivalent transformations:

$$\begin{bmatrix} -2 & 1 & 3 \\ 1 & 2 & 1 \\ 3 & 1 & -2 \end{bmatrix} \rightarrow \begin{bmatrix} 1 & 2 & 1 \\ -2 & 1 & 3 \\ 3 & 1 & -2 \end{bmatrix} \rightarrow \begin{bmatrix} 1 & 2 & 1 \\ 0 & 5 & 5 \\ 0 & -5 & -5 \end{bmatrix} \rightarrow \begin{bmatrix} 1 & 2 & 1 \\ 0 & 1 & 1 \end{bmatrix}.$$

Proceed to the system of equations:

$$\begin{cases} x + 2y = -z, \\ y = -z. \end{cases} \tag{5.46}$$

Assuming that $z = 1$, we find $y = -1$ and $x = 1$.

Thus, the matrix A has the following eigenvalues and eigenvectors corresponding to them:

$$\lambda_1 = -2, \ b_1 = (-1, 0, 1), \tag{5.47}$$

$$\lambda_2 = 6, \ b_2 = (1, 2, 1), \tag{5.48}$$

$$\lambda_3 = 3, \ b_3 = (1, -1, 1). \tag{5.49}$$

$\square$

Note In physics, the characteristic equation is sometimes referred to as the **secular equation** since such equations appeared during the analysis of the motion of the Solar system's planets and their satellites over considerable periods of time (referred to as "secular" motions) [2].

Annihilating polynomial for the matrix A is such polynomial $p(x)$ whose value of this matrix is equal to the zero matrix: $p(A) = O$.

Theorem 5.5 (Cayley[5]–Hamilton[6] Theorem) *For any square matrix A, the characteristic polynomial is its annihilating polynomial.*

Example 5.12 Let us illustrate the Cayley–Hamilton theorem with the help of the characteristic polynomial of the matrix $A = \begin{bmatrix} 1 & -10 \\ -6 & 5 \end{bmatrix}$.

Indeed, $p(\lambda) = \det |A - \lambda I| = \begin{vmatrix} 1 - \lambda & -10 \\ -6 & 5 - \lambda \end{vmatrix} = \lambda^2 - 6\lambda - 55.$

Check whether the equality $p(A) = O$ is valid:

$$p(A) = \begin{bmatrix} 1 & -10 \\ -6 & 5 \end{bmatrix}^2 - 6 \begin{bmatrix} 1 & -10 \\ -6 & 5 \end{bmatrix} - 55 \begin{bmatrix} 1 & 0 \\ 0 & 1 \end{bmatrix}$$

$$= \begin{bmatrix} 61 & -60 \\ -36 & 85 \end{bmatrix} + \begin{bmatrix} -6 & 60 \\ 36 & -30 \end{bmatrix} + \begin{bmatrix} -55 & 0 \\ 0 & -55 \end{bmatrix} = \begin{bmatrix} 0 & 0 \\ 0 & 0 \end{bmatrix}.$$

Then, $p(\lambda)$ is the annihilating polynomial for the matrix A. $\square$

Recall that the similar matrices A and B are bound by the relation $B = P^{-1}AP$ for some nonsingular matrix P. Selecting P composed of the columns equal to the eigenvectors A (written in random order), we obtain the diagonal matrix B. This is the point of the procedure of **diagonalization** of the initial matrix.

---

[5]Arthur Cayley (1821–1895), English mathematician.
[6]William Rowan Hamilton (1805–1865), Irish mathematician and physicist.

Note that the sufficient condition for the possibility of diagonalization is the presence of different eigenvalues of the initial matrix, and their number should coincide with its order.

Review Questions

1. Define n-dimensional arithmetic vector.
2. Enumerate the basic operations of vectors.
3. What is n-dimensional arithmetic space?
4. How is a linear combination of a system of vectors constructed?
5. What system of vectors is referred to as linearly dependent?
6. Define basis of a vector space.
7. Formulate the criterion that an arbitrary system of vectors is the basis.
8. Explain how a scalar product of vectors can be introduced into a vector space.
9. Enumerate the basic properties of a scalar product.
10. How is the norm of a vector found?
11. What is the Gram–Schmidt orthogonalization procedure used for?
12. What two matrices are called similar?
13. Enumerate the properties of similarity of matrices.
14. Define the concepts of "eigenvector" and "eigenvalue" of a matrix.
15. How can one, knowing the elements of the matrix, set up its characteristic equation?
16. Formulate the Cayley–Hamilton theorem.
17. What is the sufficient condition of diagonalization of a matrix?

Problems

5.1. Find out whether the vectors p, q and r form a basis in a three-dimensional vector space. If they do, expand the vector x in this basis.

$$(1) \quad p = \begin{bmatrix} 2 \\ 1 \\ 0 \end{bmatrix}, q = \begin{bmatrix} 1 \\ 0 \\ 1 \end{bmatrix}, r = \begin{bmatrix} 4 \\ 2 \\ 1 \end{bmatrix}, x = \begin{bmatrix} 3 \\ 1 \\ 3 \end{bmatrix}. \tag{5.50}$$

$$(2) \quad p = \begin{bmatrix} 5 \\ 1 \\ 0 \end{bmatrix}, q = \begin{bmatrix} 2 \\ -1 \\ 3 \end{bmatrix}, r = \begin{bmatrix} 1 \\ 0 \\ -1 \end{bmatrix}, x = \begin{bmatrix} 13 \\ 2 \\ 7 \end{bmatrix}. \tag{5.51}$$

$$
(3) \quad p = \begin{bmatrix} 4 \\ 1 \\ 1 \end{bmatrix}, q = \begin{bmatrix} 2 \\ 0 \\ -3 \end{bmatrix}, r = \begin{bmatrix} -1 \\ 2 \\ 1 \end{bmatrix}, x = \begin{bmatrix} -9 \\ 5 \\ 5 \end{bmatrix}. \tag{5.52}
$$

5.2. The vectors e_1, e_2, e_3 and e_4 are specified by their coordinates in some basis. Show that the vectors e_1, e_2, e_3 and e_4 themselves form a basis, and find the coordinates of the vector x in this basis:

$$
e_1 = \begin{bmatrix} 1 \\ 2 \\ -1 \\ -2 \end{bmatrix}, e_2 = \begin{bmatrix} 2 \\ 3 \\ 0 \\ -1 \end{bmatrix}, e_3 = \begin{bmatrix} 1 \\ 2 \\ 1 \\ 4 \end{bmatrix}, e_4 = \begin{bmatrix} 1 \\ 3 \\ -1 \\ 0 \end{bmatrix}, x = \begin{bmatrix} 7 \\ 14 \\ -1 \\ 2 \end{bmatrix}. \tag{5.53}
$$

5.3. Check that the vectors $a = (1, 2, 3)$, $b = (-3, -2, 3)$ and $c = (0, -2, -2)$ are linearly independent and thus form a basis. Is it orthogonal? Is it orthonormal? If the answers are negative, then use the Gram–Schmidt orthogonalization process to construct an orthonormal basis.

5.4. Check that the vectors $a = (3, -1, 1, 2)$, $b = (-3, 1, 0, -2)$, $c = (0, -2, 2, -2)$ and $d = (1, 4, 2, -7)$ form a basis in $\mathbb{R}^4$. Is it orthogonal? Is it orthonormal? If the answers are negative, then use the Gram–Schmidt orthogonalization process to construct an orthonormal basis.

5.5. Prove that the set of vectors $\{(i, 2-i, 5), (1, 2+i, -i), (1, i, -1)\}$ is a basis in the vector space $\mathbb{C}^3$. What are the coordinates of the vectors $(1, 0, 0)$, $(1, 1, 0)$ and $(1, 1, 1)$ in this basis?

5.6. Assume that the vectors $v_1 = [a_1, a_2]^T$ and $v_2 = [b_1, b_2]^T$ are linearly independent. What can you say about the linear dependence or independence of the vectors $w_1 = [a_1, b_1]^T$ and $w_2 = [a_2, b_2]^T$?

5.7. Prove that the set $\mathcal{M} = \{M_{pq}\}$ of all matrices with p rows and q columns with real elements forms a vector space relative to the operations of matrix addition and matrix multiplication by a number.

5.8. Prove the **Cauchy[7]–Bunyakovsky[8] inequality** (also referred to as the **Cauchy–Schwarz[9] inequality**):

For arbitrary vectors $x, y \in \mathbb{R}^n$, the following relation is valid:

$$
(x \cdot y)^2 \leqslant (x \cdot x)(y \cdot y), \tag{5.54}
$$

---

[7] Augustin-Louis Cauchy (1789–1857), French mathematician and mechanician.

[8] Viktor Yakovlevich Bunyakovsky (1804–1889), Russian mathematician and mathematics historian.

[9] Karl Hermann Amandus Schwarz (1843–1921), German mathematician.

and the equality is valid if and only if the vectors x and y differ in the scalar factor, i.e. are proportional.

5.9. Prove the **Pythagorean**[10] **theorem**:

If the vectors x, $y \in \mathbb{R}^n$ are orthogonal, then the equality

$$\|x + y\|^2 = \|x\|^2 + \|y\|^2 \tag{5.55}$$

is valid.

5.10. Check the validity of the identity

$$\|x + y\|^2 + \|x - y\|^2 \equiv 2(\|x\|^2 + \|y\|^2) \tag{5.56}$$

for arbitrary elements of the n-dimensional vector space $\mathbb{R}^n$. What is the geometric sense of this identity in the spaces $\mathbb{R}^2$ and $\mathbb{R}^3$?

5.11. It is known that the equalities $\|x\| = 6$, $\|x + y\| = 10$ and $\|x - y\| = 12$ are valid. What is the variable $\|x\|$?

5.12. Find the maximum number of linearly independent vectors

(1) on the plane;
(2) in the three-dimensional space;
(3) in $\mathbb{R}^n$.

5.13. Check that the system of vectors

$$[1, 1, 1, \ldots, 1]^T, [0, 1, 1, \ldots, 1]^T, [0, 0, 1, \ldots, 1]^T, \ldots, [0, 0, 0, \ldots, 1]^T$$

forms a basis in $\mathbb{R}^n$.

5.14. Is the system of vectors

$$[1, 1, 1, \ldots, 1]^T, \quad [1, 2, 3, \ldots, n]^T, \quad [1, 2^2, 3^2, \ldots, n^2]^T, \quad \ldots,$$
$$[1, 2^{n-1}, 3^{n-1}, \ldots, n^{n-1}]^T$$

a basis in $\mathbb{R}^n$?

5.15. Show that the matrices $\begin{bmatrix} 0 & 0 \\ a & 0 \end{bmatrix}$ and $\begin{bmatrix} 0 & a \\ 0 & 0 \end{bmatrix}$, where $a \in \mathbb{R}$, are similar.

5.16. Find whether the matrices A_1 and A_2 are similar:

(1) $A_1 = \begin{bmatrix} 1 & -1 \\ 0 & 0 \end{bmatrix}$ and $A_2 = \begin{bmatrix} 0 & 0 \\ 1 & 1 \end{bmatrix}$;

---

[10]Pythagoras of Samos, Πυθαγόρας ὁ Σάμιος (about 570 B.C.—about 495 B.C.), Ancient Greek philosopher and mathematician.

$$(2) \quad A_1 = \begin{bmatrix} 1 & 0 & 0 \\ 0 & 1 & 0 \\ 0 & 0 & 0 \end{bmatrix} \text{ and } A_2 = \begin{bmatrix} 1 & 0 & 1 \\ 1 & 0 & 0 \\ 0 & 1 & 0 \end{bmatrix}.$$

5.17. Is it true that the traces of similar matrices coincide?

5.18. Find the eigenvalues and eigenvectors of the matrix A.

$$(1) \quad A = \begin{bmatrix} 4 & -5 & 2 \\ 5 & -7 & 3 \\ 6 & -9 & 4 \end{bmatrix}; \quad (2) \quad A = \begin{bmatrix} 3 & 1 & 0 \\ -4 & -1 & 0 \\ 4 & -8 & -2 \end{bmatrix}; \tag{5.57}$$

$$(3) \quad A = \begin{bmatrix} 2 & -1 & 2 \\ 5 & -3 & 3 \\ -1 & 0 & -2 \end{bmatrix}; \quad (4) \quad A = \begin{bmatrix} 0 & 1 & 0 \\ -4 & 4 & 0 \\ -2 & 1 & 2 \end{bmatrix}; \tag{5.58}$$

$$(5) \quad A = \begin{bmatrix} 1 & -3 & 3 \\ -2 & -6 & 13 \\ -1 & -4 & 8 \end{bmatrix}. \tag{5.59}$$

5.19. Diagonalize the matrix, i.e. bring the matrix to diagonal form:

$$A = \begin{bmatrix} 4 & 15 & -3 \\ 8 & -3 & 3 \\ 0 & -15 & 7 \end{bmatrix}.$$

5.20. Bring the following matrices to diagonal form:

(1)

$$A = \begin{bmatrix} 4 & 1 & 4 \\ 6 & 3 & 6 \\ -11 & -5 & -11 \end{bmatrix};$$

(2)

$$A = \begin{bmatrix} -23 & -16 & -28 \\ 58 & 39 & 64 \\ -11 & -7 & -10 \end{bmatrix}.$$

5.21. Bring the matrix that depends on the real parameter a to diagonal form:

$$A = \begin{bmatrix} a & -1 & -1 \\ -1 & a & 1 \\ 1 & 1 & a \end{bmatrix}.$$

*__5.22.__ Write the characteristic equation for the matrix Ω of size $n \times n$:

$$\Omega = \begin{bmatrix} 0 & 1 & 0 & \ldots & 0 & 0 \\ 0 & 0 & 1 & \ldots & 0 & 0 \\ & & \ldots & \ldots & & \\ 0 & 0 & 0 & \ldots & 0 & 1 \\ \omega & 0 & 0 & \ldots & 0 & 0 \end{bmatrix}.$$

*__5.23.__ Using the Cayley–Hamilton theorem, compute the n-th power of the matrix Ω, found in the previous problem.

*__5.24.__ Find the value of the limit:

$$\lim_{n \to \infty} \begin{bmatrix} 1 & \varphi/n \\ -\varphi/n & 1 \end{bmatrix}^n. \tag{5.60}$$

5.25. Bring the complex matrix to diagonal form:

$$Z = \begin{bmatrix} 1 - 2i & 2i & 2 \\ 0 & i & 0 \\ i & -2 & 0 \end{bmatrix}.$$

5.26. It is known that two out of three eigenvalues of the matrix

$$Y = \begin{bmatrix} 910 & 1013 + 57i & -1013 + 57i \\ 57 - 899i & 68 - 1070i & 57 + 1013i \\ -57 - 899i & 57 - 1013i & 68 + 1070i \end{bmatrix}$$

are equal to $11 + 57i$ and $11 - 57i$. Without solving the characteristic equation, find the third eigenvalue.

5.27. Prove that the eigenvalues of Hermitian operator are real.

*__5.28.__ Prove that all eigenvalues of the unitary matrix lie in a complex plane, on a unit circle with the centre at the origin of coordinates.

5.29. Compute the eigenvalues and eigenvectors of the Pauli matrices σ_1, σ_2 and σ_3 (see page 185).

Answers and Solutions

5.3 *Solution.*

Write the matrix A, composed of the coordinates of the vectors $\boldsymbol{a}$, $\boldsymbol{b}$ and $\boldsymbol{c}$:

$$A = \begin{bmatrix} 1 & -3 & 0 \\ 2 & -2 & -2 \\ 3 & 3 & -2 \end{bmatrix}.$$

Since $\det A = 16 \neq 0$, then the vectors are linearly independent and form a basis. The basis is neither orthogonal, because, for example, $(\boldsymbol{a} \cdot \boldsymbol{b}) = 2 \neq 0$, nor orthonormal, because $\|\boldsymbol{a}\| = \sqrt{14} \neq 1$.

In order to construct an orthonormal basis, apply the Gram–Schmidt algorithm:

$$\boldsymbol{t}_1 = \boldsymbol{a} = \begin{bmatrix} 1 \\ 2 \\ 3 \end{bmatrix}, \quad \boldsymbol{q}_1 = \frac{\boldsymbol{t}_1}{\|\boldsymbol{t}_1\|} = \frac{1}{\sqrt{14}}(1, 2, 3);$$

$$\boldsymbol{t}_2 = \boldsymbol{b} - (\boldsymbol{b} \cdot \boldsymbol{q}_1)\boldsymbol{q}_1$$

$$= \begin{bmatrix} -3 \\ -2 \\ 3 \end{bmatrix} - \frac{1}{\sqrt{14}}(1 \cdot (-3) + 2 \cdot (-2) + 3 \cdot 3)\frac{1}{\sqrt{14}}\begin{bmatrix} 1 \\ 2 \\ 3 \end{bmatrix} = \frac{2}{7}\begin{bmatrix} -11 \\ -8 \\ 9 \end{bmatrix},$$

$$\boldsymbol{q}_2 = \frac{\boldsymbol{t}_2}{\|\boldsymbol{t}_2\|} = \frac{1}{\sqrt{266}}\begin{bmatrix} -11 \\ -8 \\ 9 \end{bmatrix};$$

$$\boldsymbol{t}_3 = \boldsymbol{c} - (\boldsymbol{c} \cdot \boldsymbol{q}_1)\boldsymbol{q}_1 - (\boldsymbol{c} \cdot \boldsymbol{q}_2)\boldsymbol{q}_2$$

$$= \begin{bmatrix} 0 \\ -2 \\ -2 \end{bmatrix} - \frac{1}{\sqrt{14}}(0 \cdot 1 - 2 \cdot 2 - 2 \cdot 3)\frac{1}{\sqrt{14}}\begin{bmatrix} 1 \\ 2 \\ 3 \end{bmatrix}$$

$$-\frac{1}{\sqrt{266}}(0\cdot(-11)-2\cdot(-8)-2\cdot9)\frac{1}{\sqrt{266}}\begin{bmatrix}-11\\-8\\9\end{bmatrix}=\frac{4}{19}\begin{bmatrix}3\\-3\\1\end{bmatrix};$$

$$q_3=\frac{t_3}{\|t_3\|}=\frac{1}{\sqrt{19}}\begin{bmatrix}3\\-3\\1\end{bmatrix}.$$

As a result, we obtain the orthonormal basis:

$$q_1=\frac{1}{\sqrt{14}}\begin{bmatrix}1\\2\\3\end{bmatrix},\quad q_2=\frac{1}{\sqrt{266}}\begin{bmatrix}-11\\-8\\9\end{bmatrix},\quad q_3=\frac{1}{\sqrt{19}}\begin{bmatrix}3\\-3\\1\end{bmatrix}.$$

5.4 *Solution.*

Write the matrix A, composed of the coordinates of the specified vectors:

$$A=\begin{bmatrix}3 & -3 & 0 & 1\\-1 & 1 & -2 & 4\\1 & 0 & 2 & 2\\2 & -2 & -2 & -7\end{bmatrix}.$$

Its determinant is equal to $\det A=-72\neq0$.

It is clear that $\operatorname{rk}A=4$; then, according to the basic minor theorem, the vectors $a=(3,-1,1,2)$, $b=(-3,1,0,-2)$, $c=(0,-2,2,-2)$ and $d=(1,4,2,-7)$ form a basis in the arithmetical space $\mathbb{R}^4$.

This basis is neither orthogonal nor orthonormal, because, for example, $(a\cdot b)=-14\neq0$ and $(a\cdot a)=15\neq1$.

In order to construct an orthonormal basis (q_1,q_2,q_3,q_4), apply the Gram–Schmidt algorithm:

$$t_1=a=(3,-1,1,2),\quad q_1=\frac{t_1}{\|t_1\|}=\frac{1}{\sqrt{15}}(3,-1,1,2);$$

$$t_2=b-(b\cdot q_1)q_1$$

$$=\begin{bmatrix}3\\-1\\1\\2\end{bmatrix}-\frac{1}{\sqrt{15}}((-3)\cdot3+1\cdot(-1)+0\cdot1-2\cdot2)\frac{1}{\sqrt{15}}\begin{bmatrix}3\\-1\\1\\2\end{bmatrix}=\frac{1}{15}\begin{bmatrix}-3\\1\\14\\-2\end{bmatrix},$$

$$q_2 = \frac{t_2}{\|t_2\|} = \frac{1}{\sqrt{210}} \begin{bmatrix} -3 \\ 1 \\ 14 \\ -2 \end{bmatrix};$$

$$t_3 = c - (c \cdot q_1)q_1 - (c \cdot q_2)q_2 = \begin{bmatrix} 0 \\ -2 \\ 2 \\ -2 \end{bmatrix} - 0 \cdot q_1 - \frac{1}{7} \begin{bmatrix} -3 \\ 1 \\ 14 \\ -2 \end{bmatrix} = \frac{1}{7} \begin{bmatrix} 3 \\ -15 \\ 0 \\ -12 \end{bmatrix},$$

$$q_3 = \frac{t_3}{\|t_3\|} = \frac{1}{\sqrt{42}} \begin{bmatrix} 1 \\ -5 \\ 0 \\ 4 \end{bmatrix};$$

$$t_4 = d - (d \cdot q_1)q_1 - (d \cdot q_2)q_2 - (d \cdot q_3)q_3$$

$$= \begin{bmatrix} 1 \\ 4 \\ 2 \\ -7 \end{bmatrix} - \frac{3}{124} \begin{bmatrix} 1 \\ -5 \\ 0 \\ -4 \end{bmatrix} - \frac{43}{210} \begin{bmatrix} -3 \\ 1 \\ 14 \\ -2 \end{bmatrix} + \frac{13}{15} \begin{bmatrix} 3 \\ -1 \\ 1 \\ 2 \end{bmatrix} = \begin{bmatrix} 4 \\ 4 \\ 0 \\ -4 \end{bmatrix},$$

$$q_4 = \frac{t_4}{\|t_4\|} = \frac{1}{\sqrt{3}} \begin{bmatrix} 1 \\ 1 \\ 0 \\ -1 \end{bmatrix}.$$

As a result, enumerate the vectors of the orthonormal basis:

$$q_1 = \frac{1}{\sqrt{15}} \begin{bmatrix} 3 \\ -1 \\ 1 \\ 2 \end{bmatrix}, \quad q_2 = \frac{1}{\sqrt{210}} \begin{bmatrix} -3 \\ 1 \\ 14 \\ -2 \end{bmatrix},$$

$$q_3 = \frac{1}{\sqrt{42}} \begin{bmatrix} 1 \\ -5 \\ 0 \\ -4 \end{bmatrix}, \quad q_4 = \frac{1}{4\sqrt{3}} \begin{bmatrix} 4 \\ 4 \\ 0 \\ -4 \end{bmatrix}.$$

5.5 *Solution.*

Construct a matrix A of the coordinates of the specified vectors:

$$A = \begin{bmatrix} i & 1 & 1 \\ 2-i & 2+i & i \\ 5 & -i & -1 \end{bmatrix}.$$

Its determinant is $\det A = -8 - 6i \neq 0$, and therefore, the set of vectors forms a basis in the space $\mathbb{C}^3$.

Find the coordinates of the vectors $a = (1, 0, 0)$, $b = (1, 1, 0)$ and $c = (1, 1, 1)$ in this basis.

The vector a in the basis (e_1, e_2, e_3) has the coordinates (a_1, a_2, a_3) that satisfy the system of equations, which in matrix notation has the form:

$$\begin{bmatrix} i & 1 & 1 \\ 2-i & 2+i & i \\ 5 & -i & -1 \end{bmatrix} \begin{bmatrix} a_1 \\ a_2 \\ a_3 \end{bmatrix} = \begin{bmatrix} 1 \\ 0 \\ 0 \end{bmatrix}.$$

In order to solve the obtained system, let us use Cramer's rule, according to which, for $i \in \{1, 2, 3\}$, the equalities $a_i = \dfrac{\Delta_i}{\Delta}$ are valid.

$$\Delta = -8 - 6i,$$

$$\Delta_1 = \begin{vmatrix} 1 & 1 & 1 \\ 0 & 2+i & i \\ 0 & -i & -1 \end{vmatrix} = -3 - i,$$

$$\Delta_2 = \begin{vmatrix} i & 1 & 1 \\ 2-i & 0 & i \\ 5 & 0 & -1 \end{vmatrix} = 2 + 4i,$$

$$\Delta_3 = \begin{vmatrix} i & 1 & 1 \\ 2-i & 2+i & 0 \\ 5 & -i & 0 \end{vmatrix} = -11 - 7i.$$

Hence, we obtain the following coordinates:

$$\boldsymbol{a} = (a_1, a_2, a_3), \text{ where } a_1 = \frac{-3-i}{-8-6i} = \frac{1}{10}(3-i), a_2 = -\frac{1}{10}(4+2i) \text{ and}$$

$$a_3 = \frac{1}{10}(13-i).$$

Write the system of equations for the coordinates of the second vector $\boldsymbol{b}$:

$$\begin{bmatrix} i & 1 & 1 \\ 2-i & 2+i & i \\ 5 & -i & -1 \end{bmatrix} \begin{bmatrix} b_1 \\ b_2 \\ b_3 \end{bmatrix} = \begin{bmatrix} 1 \\ 1 \\ 0 \end{bmatrix}.$$

$$\Delta_1 = \begin{vmatrix} 1 & 1 & 1 \\ 1 & 2+i & i \\ 0 & -i & -1 \end{vmatrix} = -2 - 2i,$$

$$\Delta_2 = \begin{vmatrix} i & 1 & 1 \\ 2-i & 1 & i \\ 5 & 0 & -1 \end{vmatrix} = -3 + 3i,$$

$$\Delta_3 = \begin{vmatrix} i & 1 & 1 \\ 2-i & 2+i & 1 \\ 5 & -i & 0 \end{vmatrix} = -7 - 7i.$$

The values of the coordinates are $b_1 = \frac{1}{25}(7+i)$, $b_2 = \frac{1}{50}(3-21i)$ and $b_3 = \frac{1}{50}(49+7i)$.

Finally, the coordinates of the vector $\boldsymbol{c}$ satisfy the following system of equations in matrix form:

$$\begin{bmatrix} i & 1 & 1 \\ 2-i & 2+i & i \\ 5 & -i & -1 \end{bmatrix} \begin{bmatrix} a_1 \\ a_2 \\ a_3 \end{bmatrix} = \begin{bmatrix} 1 \\ 1 \\ 1 \end{bmatrix}.$$

$$\Delta_1 = \begin{vmatrix} 1 & 1 & 1 \\ 1 & 2+i & i \\ 1 & -i & -1 \end{vmatrix} = -4 - 2i,$$

$$\Delta_2 = \begin{vmatrix} i & 1 & 1 \\ 2-i & 1 & i \\ 5 & 1 & -1 \end{vmatrix} = 2i,$$

$$\Delta_3 = \begin{vmatrix} i & 1 & 1 \\ 2-i & 2+i & 1 \\ 5 & -i & 1 \end{vmatrix} = -10 - 4i.$$

The coordinates of the vector c are equal to $c_1 = \dfrac{1}{25}(11-2i)$, $c_2 = -\dfrac{1}{25}(3+4i)$ and $c_3 = \dfrac{1}{25}(26 - 7i)$.

5.6 Solution.

The criterion, i.e. the necessary and sufficient condition of the linear independence of the system of vectors v_1 and v_2 is the determinant being not equal to zero, which determinant is composed of their coordinates:

$$\begin{vmatrix} a_1 & b_1 \\ a_2 & b_2 \end{vmatrix} \neq 0.$$

As is known, transposition of a matrix does not change its determinant. Therefore, there exists the inequality

$$\begin{vmatrix} a_1 & a_2 \\ b_1 & b_2 \end{vmatrix} \neq 0,$$

and the vectors $w_1 = [a_1, b_1]^T$ and $w_2 = [a_2, b_2]^T$ are independent.

5.7 Proof.

Each matrix from the set $\mathcal{M}$ can be presented as a numerical sequence of length $p \times q$. Indeed, for this, it is enough to write the matrix elements row by row into a vector, or, in other words, into a one-dimensional array of size $p \times q$. Since the vectors of the same size form an arithmetical space, then the set $\mathcal{M} = \{M_{pq}\}$ also forms a vector space relative to the operations of matrix addition and matrix multiplication by a number.

5.8 *Proof.*

Let us introduce for consideration the vector $z = x - \lambda y$, where λ is some real number.

Based on the fourth property of scalar product, $(z, z) \geqslant 0$, or

$$((x - \lambda y) \cdot (x - \lambda y)) = (x \cdot x) - 2\lambda(x \cdot y) + \lambda^2(y \cdot y) \geqslant 0. \qquad (5.61)$$

This inequality should be valid for any $\lambda \in \mathbb{R}$. Note that the left side of the inequality (5.61) is a quadratic trinomial. The necessary and sufficient condition of its non-negativity is non-positivity of the discriminant:

$$(x \cdot y)^2 - (x \cdot x)(y \cdot y) \leqslant 0.$$

The obtained inequality, as it is easy to see, after transferring the summand $(x \cdot x)(y \cdot y)$ to the right side, coincides with the Cauchy–Bunyakovsky inequality.

The equals sign will occur if and only if $z \equiv 0$, i.e. x and y are proportional and differ in the scalar factor.

5.9 *Proof.*

From the definition of vector length and the properties of scalar product, follow the equalities

$$\|x + y\|^2 = ((x + y) \cdot (x + y)) = (x \cdot x) + (x \cdot y) + (y \cdot x) + (y \cdot y) = \|x\|^2 + \|y\|^2.$$

Thus, the Pythagorean theorem is proved for all orthogonal vectors $x, y \in \mathbb{R}^n$.

5.10 *Solution.*

Transforming the squares of norms in the left side of the identity, we obtain

$$\|x + y\|^2 = ((x + y) \cdot (x + y)) = \|x\|^2 + 2(x \cdot y) + \|y\|^2, \qquad (5.62)$$

$$\|x - y\|^2 = ((x - y) \cdot (x - y)) = \|x\|^2 - 2(x \cdot y) + \|y\|^2. \qquad (5.63)$$

From these relations, follow the equality (5.56). The geometric sense of this equality consists in that the sum of the squares of the parallelogram's diagonals is equal to the sum of the squares of the sides.

5.11 *Solution.*

Express the variable $\|y\|$, using the identity (5.56) from the previous problem:

$$\|y\| = \sqrt{(\|x + y\|^2 + \|x - y\|^2)/2 - \|x\|^2}.$$

Having substituted the numeric data, we obtain $\|y\| = \sqrt{(100 + 144)/2 - 36} = \sqrt{86}$.

5.12 *Answer.*

(1) The maximum number of linearly independent vectors equals to 2;
(2) 3;
(3) n.

5.13 *Solution.*

The matrix of the system of vectors under consideration has a lower triangular

form:
$$
\begin{bmatrix}
1\,0\ldots0\ 0 \\
1\,1\ldots0\ 0 \\
\cdots\cdots\cdots \\
1\,1\ldots1\ 0 \\
1\,1\ldots1\ 1
\end{bmatrix}
$$

Its determinant is equal to the product of the elements located on the main diagonal. Therefore, the vectors are linearly independent and form a basis in the vector space $\mathbb{R}^n$.

5.14 *Answer:* It is. This is easily seen from the Vandermonde determinant being not equal to zero in this case (see (2.74) on page 67).

5.15 *Solution.*

It is easy to verify that the equality is valid:

$$
\begin{bmatrix} 0\ a \\ 0\ 0 \end{bmatrix} = P^{-1} \begin{bmatrix} 0\ 0 \\ a\ 0 \end{bmatrix} P,
$$

where $P = \begin{bmatrix} 0\ 1 \\ 1\ 0 \end{bmatrix}$. Therefore, by definition of similarity relation (5.27), the matrices mentioned in the statement of the problem are similar.

5.16 *Answer:*

(1) Yes, $A_2 = \begin{bmatrix} 0\ 1 \\ -1\ 0 \end{bmatrix}^{-1} A_1 \begin{bmatrix} 0\ 1 \\ -1\ 0 \end{bmatrix}$;

(2) No, since the equality of the determinants of these matrices is not fulfilled.

5.17 *Solution.*

Yes, as follows from Problem **1.44**, for any matrices A and the invertible matrix P, the equality

$$
\operatorname{tr}(P^{-1}AP) = \operatorname{tr}(AP^{-1}P) = \operatorname{tr} A
$$

is fulfilled.

5.18 *Solution.*

(1) As is known, the eigenvector of the matrix A is such vector u, in which multiplication of A by u results in the vector λu, where $\lambda \in \mathbb{R}$ is the eigenvalue. Write an equation of the form $Au = \lambda u$:

$$
\begin{bmatrix} 4 & -5 & 2 \\ 5 & -7 & 3 \\ 6 & -9 & 4 \end{bmatrix} \begin{bmatrix} u_1 \\ u_2 \\ u_3 \end{bmatrix} = \lambda \begin{bmatrix} u_1 \\ u_2 \\ u_3 \end{bmatrix},
$$

or in the coordinates of the vector u:

$$
\begin{cases} 4u_1 - 5u_2 + 2u_3 = \lambda u_1, \\ 5u_1 - 7u_2 + 3u_3 = \lambda u_2, \\ 6u_1 - 9u_2 + 4u_3 = \lambda u_3, \end{cases}
$$

$$
\begin{cases} (4 - \lambda)u_1 - 5u_2 + 2u_3 = 0, \\ 5u_1 - (7 + \lambda)u_2 + 3u_3 = 0, \\ 6u_1 - 9u_2 + (4 - \lambda)u_3 = 0. \end{cases}
$$

Note that the eigenvector cannot be zero by definition:

$$
\begin{bmatrix} u_1 \\ u_2 \\ u_3 \end{bmatrix} \neq \begin{bmatrix} 0 \\ 0 \\ 0 \end{bmatrix}.
$$

Therefore, the equations are linearly dependent and the determinant of the system matrix is equal to zero:

$$
\begin{vmatrix} (4 - \lambda) & 5u_2 & 2u_3 \\ 5u_1 & -(7 + \lambda)u_2 & 3u_3 \\ 6u_1 & -9u_2 & (4 - \lambda)u_3 \end{vmatrix} = 0.
$$

Having computed the determinant, we obtain the characteristic equation: $\lambda^2 - \lambda^3 = 0$.

The eigenvalues are $\lambda_{1,2} = 0$ and $\lambda_3 = 1$, so zero is an eigenvalue of multiplicity two.

Find the respective eigenvectors:

Let $\lambda = 0$.

$$\begin{cases} 4u_1 - 5u_2 + 2u_3 = 0, \\ 5u_1 - 7u_2 + 3u_3 = 0, \\ 6u_1 - 9u_2 + 4u_3 = 0. \end{cases}$$

Having solved this system, we obtain

$$u_1 = 0, u_2 = 2u_1, u_3 = 3u_1, \text{ or } \begin{bmatrix} u_1 \\ u_2 \\ u_3 \end{bmatrix} = \begin{bmatrix} u_1 \\ 2u_1 \\ 3u_1 \end{bmatrix}, u_1 \in \mathbb{R}.$$

The eigenvector is $\begin{bmatrix} 1 \\ 2 \\ 3 \end{bmatrix}$.

Let $\lambda = 1$.

$$\begin{cases} 3u_1 - 5u_2 + 2u_3 = 0, \\ 5u_1 - 8u_2 + 3u_3 = 0, \\ 6u_1 - 9u_2 + 3u_3 = 0. \end{cases}$$

Having solved this system, we obtain

$$u_1 = u_2, 0u_2 = 0u_2, u_3 = u_2 \text{ or } \begin{bmatrix} u_1 \\ u_2 \\ u_3 \end{bmatrix} = \begin{bmatrix} u_2 \\ u_2 \\ u_2 \end{bmatrix}, u_2 \in \mathbb{R}. \text{ The eigenvector}$$

is $\begin{bmatrix} 1 \\ 1 \\ 1 \end{bmatrix}$.

(2) By analogy with the solution from the previous item, write an equation of the form $Au = \lambda u$. Then, we obtain the characteristic equation:

$$-(\lambda + 2)(\lambda - 1)^2 = 0.$$

The eigenvalues are equal to $\lambda_1 = -2$ and $\lambda_{2,3} = 1$.

Let us find the eigenvectors.
For $\lambda = -2$,

$$\begin{cases} 5u_1 + u_2 = 0, \\ -4u_1 + u_2 = 0, \\ 4u_1 - 8u_2 = 0, \end{cases}$$

$$u_1 = 0, u_2 = 0, u_3 \in \mathbb{R}.$$

The eigenvector is $\begin{bmatrix} 0 \\ 0 \\ 1 \end{bmatrix}$.

For $\lambda_{2,3} = 1$,

$$\begin{cases} 2u_1 + u_2 = 0, \\ -4u_1 - 2u_2 = 0, \\ 4u_1 - 8u_2 - 3u_3 = 0, \end{cases}$$

$$u_2 = -2u_1, u_3 = \frac{20}{3}u_1,$$

$$\begin{bmatrix} u_1 \\ u_2 \\ u_3 \end{bmatrix} = \begin{bmatrix} u_1 \\ -2u_1 \\ \dfrac{20}{3}u_1 \end{bmatrix}, u_1 \in \mathbb{R}.$$

The eigenvector is $\begin{bmatrix} 3 \\ -6 \\ 20 \end{bmatrix}$.

(3) The characteristic equation: $-\lambda^3 - 3\lambda^2 - 3\lambda - 1 = 0$.
The eigenvalue is multiple of three: $\lambda = -1$.

$$\begin{cases} 3u_1 - u_2 + 2u_3 = 0, \\ 5u_1 - 2u_2 + 3u_3 = 0, \\ -u_1 - u_3 = 0. \end{cases}$$

Find non-trivial solutions: $u_1 = -u_3$ and $u_2 = -u_3$.

The eigenvector is $\begin{bmatrix} -1 \\ -1 \\ 1 \end{bmatrix}$.

(4) The characteristic equation has the form: $-(\lambda - 2)^3 = 0$.

The eigenvalue is $\lambda = 2$.

Find the eigenvector for $\lambda = 2$:

$$\begin{cases} -2u_1 + u_2 = 0, \\ -4u_1 + 2u_2 = 0, \\ -2u_1 + u_2 = 0, \end{cases}$$

then $u = c_1[1/2, 1, 0]^T + c_2[0, 0, 1]^T$.

The eigenvectors: $[1, 2, 0]^T$, $[0, 0, 1]^T$.

(5) The characteristic equation: $-\lambda^3 + 3\lambda^2 - 3\lambda + 1 = 0$.

The eigenvalue $\lambda = 1$ is multiple of three.

$$\begin{cases} -3u_2 + 3u_3 = 0, \\ -2u_1 - 7u_2 + 13u_3 = 0, \\ -u_1 - 4u_2 + 7u_3 = 0, \end{cases}$$

and therefore, $u_2 = u_3$ and $u_1 = 3u_3$.

$$\begin{bmatrix} 3u_3 \\ u_3 \\ u_3 \end{bmatrix}, u_3 \in \mathbb{R}.$$

The eigenvector is $\begin{bmatrix} 3 \\ 1 \\ 1 \end{bmatrix}$.

5.19 *Solution.*

Compose the characteristic equation:

$$\det \begin{bmatrix} 4 - \lambda & 15 & -3 \\ 8 & -3 - \lambda & 3 \\ 0 & -15 & 7 - \lambda \end{bmatrix} = 0.$$

Find the characteristic polynomial, expanding the determinant in the first column:

$$p(\lambda) = (4 - \lambda)((-3 - \lambda)(7 - \lambda) + 45) - 8(15(7 - \lambda) - 45),$$
$$p(\lambda) = \lambda^3 - 8\lambda^2 - 80\lambda + 384.$$

Find the eigenvalues: $\lambda_1 = 12$, $\lambda_2 = -8$ and $\lambda_3 = 4$.
Compose the eigenvectors.

(1) For λ_1, $\begin{cases} -8x_1 + 15x_2 - 3x_3 = 0, \\ 8x_1 - 15x_2 + 3x_3 = 0, \\ -15x_2 - 5x_3 = 0 \end{cases}$ $\Rightarrow a_1 = c(3, 1, -3).$

(2) For λ_2,

$$\begin{cases} 12x_1 + 15x_2 - 3x_3 = 0, \\ 8x_1 + 5x_2 + 3x_3 = 0, \\ -15x_2 + 15x_3 = 0 \end{cases} \Rightarrow a_2 = c(-1, 1, 1).$$

(3) For λ_3,

$$\begin{cases} 15x_2 - 3x_3 = 0, \\ 8x_1 - 7x_2 + 3x_3 = 0, \\ -15x_2 + 3x_3 = 0 \end{cases} \Rightarrow a_3 = c(-1, 1, 5).$$

Write the transformation matrix and the matrix inverse of it:

$$P = \begin{bmatrix} 3 & -1 & -1 \\ 1 & 1 & 1 \\ -3 & 1 & 5 \end{bmatrix}, \quad P^{-1} = \frac{1}{4} \begin{bmatrix} 1 & 1 & 0 \\ -2 & 3 & -1 \\ 1 & 0 & 1 \end{bmatrix}.$$

As a result, the diagonalized matrix is equal to

$$B = P^{-1}AP = \begin{bmatrix} 12 & 0 & 0 \\ 0 & -8 & 0 \\ 0 & 0 & 4 \end{bmatrix}, \text{ where } P = \begin{bmatrix} -3 & -1 & -1 \\ -1 & 1 & 1 \\ 3 & 1 & 5 \end{bmatrix}.$$

5.20 *Solution.*

(1) The characteristic equation has the form:

$$|A - \lambda I| = \begin{bmatrix} 4 - \lambda & 1 & 4 \\ 6 & 3 - \lambda & 6 \\ -11 & -5 & -11 - \lambda \end{bmatrix} = 0.$$

Compute the characteristic polynomial:

$$p(\lambda) = (4 - \lambda)((3 - \lambda)(-11 - \lambda) + 30) - (6(-11 - \lambda) + 30)$$

$$+ 4(-30 - (3 - \lambda) \cdot (-11)),$$

$$p(\lambda) = -\lambda^3 - 4\lambda^2 - 3\lambda.$$

The eigenvalues are $\lambda_1 = -3$, $\lambda_2 = -1$ and $\lambda_3 = 0$.
Compose the eigenvectors.
For λ_1,

$$\begin{cases} 7x_1 + x_2 + 4x_3 = 0, \\ 6x_1 + 6x_2 + 6x_3 = 0, \\ -11x_1 - 5x_2 - 8x_3 = 0 \end{cases} \Rightarrow a_1 = c(1, 1, -2).$$

For λ_2,

$$\begin{cases} 5x_1 + x_2 + 4x_3 = 0, \\ 6x_1 + 4x_2 + 6x_3 = 0, \\ -11x_1 - 5x_2 - 10x_3 = 0 \end{cases} \Rightarrow a_2 = c(5, 3, -7).$$

For λ_3, $\begin{cases} 4x_1 + x_2 + 4x_3 = 0, \\ 6x_1 + 3x_2 + 6x_3 = 0, \\ -11x_1 - 5x_2 - 11x_3 = 0 \end{cases} \Rightarrow a_3 = c(-1, 0, 1).$

The transformation matrix is $P = \begin{bmatrix} 1 & 5 & -1 \\ 1 & 3 & 0 \\ -2 & -7 & 1 \end{bmatrix}$,

the matrix inverse of it has the form $P^{-1} = \begin{bmatrix} -3 & -2 & -3 \\ 1 & 1 & 1 \\ 1 & 3 & 2 \end{bmatrix}$.

The diagonalized matrix:

$$B = P^{-1}AP = \begin{bmatrix} -3 & 0 & 0 \\ 0 & -1 & 0 \\ 0 & 0 & 0 \end{bmatrix}, \text{ where } P = \begin{bmatrix} 1 & 5 & -1 \\ 1 & 3 & 0 \\ -2 & -7 & 1 \end{bmatrix}.$$

(2) Write the characteristic equation:

$$A - \lambda I = \begin{bmatrix} -23 - \lambda & -16 & -28 \\ 58 & 39 - \lambda & 64 \\ -11 & -7 & -10 - \lambda \end{bmatrix} = 0,$$

$$p(\lambda) = (-23 - \lambda)((39 - \lambda)(-10 - \lambda) + 64 \cdot 7) + 16(58(-10 - \lambda)$$
$$+ 11 \cdot 64) - 28(-7 \cdot 58 + 11(39 - \lambda)),$$

$$p(\lambda) = -\lambda^3 + 6\lambda^2 - 11\lambda + 6.$$

The eigenvalues are $\lambda_1 = 3$, $\lambda_2 = 2$ and $\lambda_3 = 1$.
Compose the eigenvectors.
For λ_1,

$$\begin{cases} -26x_1 - 16x_2 - 28x_3 = 0, \\ 58x_1 + 36x_2 + 64x_3 = 0, \\ -11x_1 - 7x_2 - 13x_3 = 0 \end{cases} \Rightarrow a_1 = c(2, -5, 1).$$

For λ_2,

$$\begin{cases} -25x_1 - 16x_2 - 28x_3 = 0, \\ 58x_1 + 37x_2 + 64x_3 = 0, \\ -11x_1 - 7x_2 - 12x_3 = 0 \end{cases} \Rightarrow a_2 = c(4, -8, 1).$$

For λ_3,

$$\begin{cases} -24x_1 - 16x_2 - 28x_3 = 0, \\ 58x_1 + 38x_2 + 64x_3 = 0, \\ -11x_1 - 7x_2 - 11x_3 = 0 \end{cases} \Rightarrow a_3 = c(5, -11, 2).$$

The transformation matrix is $P = \begin{bmatrix} 2 & 4 & 5 \\ -5 & -8 & -11 \\ 1 & 1 & 2 \end{bmatrix}$, and the matrix inverse of it is $P^{-1} = \begin{bmatrix} -5 & -3 & -4 \\ -1 & -1 & -3 \\ 3 & 2 & 4 \end{bmatrix}$.

Write the diagonalized matrix:

$$B = P^{-1}AP = \begin{bmatrix} 3 & 0 & 0 \\ 0 & 2 & 0 \\ 0 & 0 & 1 \end{bmatrix}, \text{ where } P = \begin{bmatrix} 2 & 4 & 5 \\ -5 & -8 & -11 \\ 1 & 1 & 2 \end{bmatrix}.$$

5.21 *Answer:*

The characteristic polynomial has the form $p(\lambda) = \lambda^3 - 3a\lambda^2 + 3a^2\lambda - \lambda + (a - a^3)$; its roots are equal to $\lambda_1 = a - 1$, $\lambda_2 = a$ and $\lambda_3 = a + 1$.

The diagonalized matrix is equal to

$$B = P^{-1}AP = \begin{bmatrix} a+1 & 0 & 0 \\ 0 & a & 0 \\ 0 & 0 & a-1 \end{bmatrix}, \text{ where } P = \begin{bmatrix} -1 & 1 & 0 \\ 1 & -1 & -1 \\ 0 & 1 & 1 \end{bmatrix}.$$

5.22 *Solution.*

Write the left side of the characteristic equation $\det(\Omega - \lambda I) = 0$, expanding the determinant in the first column:

$$\det(\Omega - \lambda I) = \begin{vmatrix} -\lambda & 1 & 0 \dots & 0 & 0 \\ 0 & -\lambda & 1 \dots & 0 & 0 \\ \multicolumn{5}{c}{\dots\dots\dots\dots\dots\dots} \\ 0 & 0 & 0 \dots & -\lambda & 1 \\ \omega & 0 & 0 \dots & 0 & -\lambda \end{vmatrix}$$

$$= \begin{vmatrix} -\lambda & 1 & 0 \dots & 0 \\ 0 & -\lambda & 1 \dots & 0 \\ \multicolumn{4}{c}{\dots\dots\dots\dots} \\ 0 & 0 & 0 \dots & 1 \\ \omega & 0 & 0 \dots & -\lambda \end{vmatrix} - (-1)^n \begin{vmatrix} 1 & 0 \dots & 0 & 0 \\ -\lambda & 1 \dots & 0 & 0 \\ \multicolumn{4}{c}{\dots\dots\dots\dots} \\ 0 & 0 \dots & 1 & 0 \\ 0 & 0 \dots & -\lambda & 1 \end{vmatrix}$$

$$= -\lambda(-\lambda)^{n-1} - (-1)^n\omega = (-\lambda)^n - (-1)^n\omega.$$

As a result, the characteristic equation for the matrix Ω has the form $(-1)^n(\lambda^n - \omega) = 0$ or $(\lambda^n - \omega) = 0$.

5.23 *Solution.*

According to the Cayley–Hamilton theorem, when the matrix is substituted into its characteristic equation, an identity is obtained. As is shown in Problem **5.22**, the characteristic equation for Ω has the form $(\lambda^n - \omega) = 0$. Then, the equality is valid:

$$\Omega^n - \omega I = O, \text{ or } \Omega^n = \omega I,$$

where, as usual, O is the zero matrix of size $n \times n$ and I is the identity matrix of the same size.

5.24 *Solution.*

Denote $A = \begin{bmatrix} 1 & \varphi/n \\ -\varphi/n & 1 \end{bmatrix}$.

The eigenvalues of this matrix are equal to $\lambda_{1,2} = 1 \pm i\varphi/n$; the eigenvectors corresponding to them are equal to $[1, i]^T$ and $[1, -i]^T$.

Compute the power of A^n, having brought the matrix to diagonal form A' first:

$$A' = \begin{bmatrix} 1 & 1 \\ i & -i \end{bmatrix}^{-1} \begin{bmatrix} 1 & \varphi/n \\ -\varphi/n & 1 \end{bmatrix} \begin{bmatrix} 1 & 1 \\ i & -i \end{bmatrix} = \begin{bmatrix} 1 + i\varphi/n & 0 \\ 0 & 1 - i\varphi/n \end{bmatrix},$$

and therefore, according to the theorem on the power of a special form matrix on page 56,

$$A^n = \begin{bmatrix} 1 & 1 \\ i & -i \end{bmatrix} (A')^n \begin{bmatrix} 1 & 1 \\ i & -i \end{bmatrix}$$

$$= \begin{bmatrix} 1 & 1 \\ i & -i \end{bmatrix}^{-1} \begin{bmatrix} (1 + i\varphi/n)^n & 0 \\ 0 & (1 - i\varphi/n)^n \end{bmatrix} \begin{bmatrix} 1 & 1 \\ i & -i \end{bmatrix}^{-1}.$$

After simple computations, we obtain

$$A^n = \frac{1}{2} \begin{bmatrix} (1 + i\varphi/n)^n + (1 - i\varphi/n)^n & i(-(1 + i\varphi/n)^n + (1 - i\varphi/n)^n) \\ i((1 + i\varphi/n)^n - (1 - i\varphi/n)^n) & (1 + i\varphi/n)^n + (1 - i\varphi/n)^n \end{bmatrix}.$$

Using the relation known from mathematical analysis [76]

$$\lim_{n \to \infty} (1 + t/n)^n = e^t,$$

which is valid for all $t \in \mathbb{C}$, perform the limit operation:

$$\lim_{n \to \infty} A^n = \frac{1}{2} \begin{bmatrix} e^{i\varphi/n} + e^{-i\varphi/n} & -i(e^{i\varphi/n} - e^{-i\varphi/n}) \\ -i(e^{-i\varphi/n} - e^{i\varphi/n}) & e^{i\varphi/n} + e^{-i\varphi/n} \end{bmatrix}.$$

Representing the exponent of the imaginary number by Euler's formula $\cos z + i \sin z = e^{iz}$, we arrive at the final answer:

$$\lim_{n \to \infty} \begin{bmatrix} 1 & \varphi/n \\ -\varphi/n & 1 \end{bmatrix}^n = \begin{bmatrix} \cos \varphi & \sin \varphi \\ -\sin \varphi & \cos \varphi \end{bmatrix}.$$

5.25 *Answer:*

$$Z' = P^{-1}ZP = \begin{bmatrix} -2i & 0 & 0 \\ 0 & i & 0 \\ 0 & 0 & 1 \end{bmatrix}, \text{ where } P = \begin{bmatrix} -2 & -2i & -i \\ 0 & 1-i & 0 \\ 1 & 2 & 1 \end{bmatrix}.$$

5.26 *Solution.*

As is known, the traces of identical matrices coincide (see Problem **5.17**). Therefore, the sum of the eigenvalues of the matrix is equal to the trace of this matrix:

$$\mathrm{tr}\, Y = \lambda_1 + \lambda_2 + \lambda_3 = \sum_{i=1}^{3} y_{ii} = 910 + (68 - 1070i) + (68 + 1070i) = 1046,$$

and the third eigenvalue is equal to

$$\lambda_3 = 1046 - (\lambda_1 + \lambda_2) = 1046 - ((11 + 57i) + (11 - 57i)) = 1024.$$

5.27 *Proof.*

Let A be an arbitrary Hermitian matrix of size $n \times n$ that has the eigenvector b, which is satisfied by the eigenvalue λ_0. This means that the equality is fulfilled

$$Ab = \lambda_0 b. \tag{5.64}$$

Consider the expression $b^H Ab$. It is equal to a real number, since according to the theorem on Hermitian conjugation of product on page 181,

$$(b^H Ab)^H = b^H A^H (b^H)^H = b^H Ab. \tag{5.65}$$

According to (5.64), the equality is fulfilled

$$b^H A b = \lambda_0 b^H b = \lambda_0(|b_1|^2 + |b_2|^2 + \cdots + |b_n|^2), \tag{5.66}$$

where $b_1, b_2, \ldots, b_n$ are the components of the vector b.

Comparing (5.65) and (5.66), we obtain that $\lambda_0 \in \mathbb{R}$.

5.28 *Proof.*

Let U be an arbitrary unitary matrix that has the eigenvector b, which is satisfied by the eigenvalue λ_0. This means that the equality is fulfilled

$$U b = \lambda_0 b. \tag{5.67}$$

Consider the expression $(U b)^H (U b)$. By virtue of (5.67), this expression can be presented in the form:

$$(U b)^H (U b) = (\lambda_0 b)^H (\lambda_0 b) = (\lambda_0^* b^H)(\lambda_0 b) = \lambda_0^* \lambda_0 b^H b = |\lambda_0|^2 b^H b. \tag{5.68}$$

Note that the transformations used property (3) of the Hermitian conjugate operation (see page 181).

On the other hand, relying on the theorem on Hermitian conjugation of product on page 181 and on the property of unitarity of $U^H U = I$, we obtain

$$(U b)^H (U b) = (b^H U^H)(U b) = b^H (U^H U) b = b^H I b = b^H b. \tag{5.69}$$

Comparing (5.68) and (5.69), we come to the conclusion: $|\lambda_0|^2 = 1$.

Therefore, the complex number λ_0 is located on a complex plane, at a distance of $\rho = 1$ from the origin of coordinates. The locus of all such points λ_0 is the unit circle with the centre at the origin of coordinates, which is what we set out to prove.

5.29 *Answer:*

for σ_1, $\lambda_{1,2} = \pm 1$, the eigenvectors are $b_{1,2} = \dfrac{1}{\sqrt{2}} \begin{bmatrix} 1 \\ \pm 1 \end{bmatrix}$;

for σ_2, $\lambda_{1,2} = \pm 1$, the eigenvectors are $b_{1,2} = \dfrac{1}{\sqrt{2}} \begin{bmatrix} 1 \\ \pm i \end{bmatrix}$;

for σ_3, $\lambda_{1,2} = \pm 1$, the eigenvectors are $b_1 = \begin{bmatrix} 1 \\ 0 \end{bmatrix}$ and $b_2 = \begin{bmatrix} 0 \\ 1 \end{bmatrix}$.

Chapter 6
Vectors in a Three-Dimensional Space

Geometrical vector is a directed segment $\overrightarrow{AB}$ with the beginning at the point A and the end at the point B. Hereinafter, the word "geometrical" in this definition will be omitted for brevity.

Zero vector, or **null vector**, is a vector whose beginning and end coincide.

The length of the segment AB is called the modulus or **magnitude** of the vector $\overrightarrow{AB}$ and is denoted by $|\overrightarrow{AB}|$.

The vectors lying on parallel lines are referred to as **collinear**.

Unit vector is a vector whose modulus is equal to one.

The vector $\boldsymbol{b}$ is the product of the number α and the vector $\boldsymbol{a}$, if the following conditions are met:

(1) $|\boldsymbol{b}| = \mathrm{abs}(\alpha)|\boldsymbol{a}|$;
(2) directions of the vectors $\boldsymbol{a}$ and $\boldsymbol{b}$ coincide if $\alpha > 0$, and these vectors are oppositely directed if $\alpha < 0$.

The product of zero and $\boldsymbol{a}$ is equal to zero vector by definition.

The two vectors $\overrightarrow{AB}$ and $\overrightarrow{CD}$ are considered to be **equal**, and if they are collinear, they have the same moduli and are unidirectional.

A vector

$$\boldsymbol{\varepsilon} = \frac{\overrightarrow{AB}}{|\overrightarrow{AB}|} \tag{6.1}$$

is called the **normalized vector** of the vector $\overrightarrow{AB}$. It is a unit vector whose direction coincides with that of the vector $\overrightarrow{AB}$.

When determining the **sum** of vectors, the law of parallelogram should be used (see Fig. 6.1).

© Springer Nature Switzerland AG 2021
S. Kurgalin, S. Borzunov, *Algebra and Geometry with Python*,
https://doi.org/10.1007/978-3-030-61541-3_6

Fig. 6.1 Determine the
vector c as the sums of the
vectors a and b (law of
parallelogram)

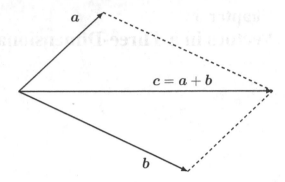

Let the line L and the point A be specified in the space. Let us draw, through the
point A, the plane π, which is orthogonal to the line L. The point of intersection of
the plane π and the line L is called the **projection** of the point A on the line L.

Consider the vector $\overrightarrow{AB}$ and the line L in space. Let A' and B' be projections
of the points A and B on the line L, respectively. Then, the vector $\overrightarrow{A'B'}$ is called a
projection of the vector $\overrightarrow{AB}$ on the line L and is denoted by $\mathrm{Pr}_L \overrightarrow{AB}$.

The **numeric projection** of the vector $\overrightarrow{AB}$ on the line L is equal to the modulus
$|\overrightarrow{AB}|$, multiplied by the cosine of the angle α between the vector $\overrightarrow{AB}$ and the line L,
i.e. $\mathrm{Pr}_L \overrightarrow{AB} = |\overrightarrow{AB}| \cos \alpha$.

Numeric projections of vectors have the following properties:

$$\mathrm{Pr}_L (a + b) = \mathrm{Pr}_L a + \mathrm{Pr}_L b, \tag{6.2}$$

$$\mathrm{Pr}_L (\alpha a) = \alpha \, \mathrm{Pr}_L a. \tag{6.3}$$

6.1 Cartesian Coordinate System

Cartesian[1] coordinate system is a system that consists of the reference point O, the
mutually perpendicular axes Ox, Oy, Oz, that intersect at the point O, and a scale
unit segment.

Let A be an arbitrary point in space. The vector $\overrightarrow{OA}$ is called a **position vector**
of the point A. The numeric projection of the vector $\overrightarrow{OA}$ on the axis Ox is denoted
by x and is called an **abscissa**, on the axis Oy by y and is called an **ordinate**, and
on the axis Oz by z and is called an **applicate** of the point A [19, 23].

The respective coordinates of the equal vectors coincide.

---

[1]René Descartes (1596–1650), French philosopher, mathematician, physicist and physiologist.

Fig. 6.2 Cartesian coordinate system

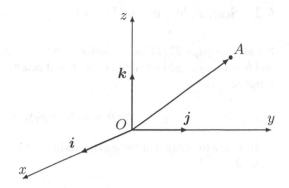

The real numbers x, y, z are coordinates of the point A and the position vector $\overrightarrow{OA}$, which can be written as $A(x, y, z)$ and $\overrightarrow{OA} = (x, y, z)$.

Normalized vectors (basis vectors) of Cartesian coordinate system are the unit vectors i, j, k (see Fig. 6.2). In coordinate representation, they have the following form: $i = (1, 0, 0)$, $j = (0, 1, 0)$ and $k = (0, 0, 1)$.

An arbitrary vector $\overrightarrow{AB} = (x, y, z)$ can be represented in the form of a **basis expansion** of Cartesian coordinate system:

$$\overrightarrow{AB} = xi + yj + zk. \tag{6.4}$$

The linear operations on vectors, i.e. addition of vectors and multiplication of a vector by a number, are said to be performed component-wise or coordinate-wise. This means that if $a = (a_x, a_y, a_z)$, $b = (b_x, b_y, b_z)$ and $c \in \mathbb{R}$, then

$$a + b = (a_x + b_x, a_y + b_y, a_z + b_z), \quad ca = (ca_x, ca_y, ca_z). \tag{6.5}$$

With the help of the normalized vectors of the coordinate system, the obtained equalities can be written as

$$a+b = (a_x+b_x)i+(a_y+b_y)j+(a_z+b_z)k, \quad ca = ca_xi+ca_yj+ca_zk. \tag{6.6}$$

Let us illustrate the use of the introduced definitions by the following example.

Example 6.1 For the vectors $v_1 = (-2, -1, 7)$, $v_2 = (0, 4, -6)$ and the scalar $t = 3$, we have

$$v_1 + v_2 = (-2 + 0, -1 + 4, 7 + (-6)) = (-2, 3, 1) = -2i + 3j + k,$$

$$tv_1 = (3 \cdot (-2), 3 \cdot (-1), 3 \cdot 7) = (-6, -3, 21) = -6i - 3j + 21k.$$

$\square$

6.2   Scalar Product of Vectors

Recall (see page 223) that the scalar product of the two vectors $a = (x_a, y_a, z_a)$ and $b = (x_b, y_b, z_b)$ is denoted by $a \cdot b$ and determined through their coordinates as follows:

$$a \cdot b = x_a x_b + y_a y_b + z_a z_b. \tag{6.7}$$

It is easy to verify that for scalar products of basis vectors the series of equalities is valid:

$$i \cdot j = i \cdot k = j \cdot k = 0 \quad \text{and} \quad i \cdot i = j \cdot j = k \cdot k = 1. \tag{6.8}$$

Theorem 6.1 *The scalar product of the two vectors a and b is equal to the product of the moduli of these vectors by the cosine of the angle α between them.*

The concepts of projection of a vector on a line and scalar product of vectors are closely connected. Indeed, since the projection of the vector a on the line, containing the vector b, is equal to $\mathrm{Pr}_b\, a = |a| \cos \alpha$ and, on the other hand, $\mathrm{Pr}_a\, b = |b| \cos \alpha$, then we can write

$$a \cdot b = |a||b| \cos \alpha = |a|\, \mathrm{Pr}_a\, b = |b|\, \mathrm{Pr}_b\, a. \tag{6.9}$$

If $a \cdot b = 0$, but $a \neq 0$ and $b \neq 0$, then such vectors are referred to as **orthogonal**, since the angle between them is equal to $\pi/2$. Recall (see page 224) that the notation of the form $a \perp b$ is used to denote orthogonality of vectors.

Example 6.2 Compute the scalar product of the vectors $a = (3, 2, 1)$ and $b = (0, 2, 1)$.

Solution

$$a \cdot b = 3 \cdot 0 + 2 \cdot 2 + 1 \cdot 1 = 5. \tag{6.10}$$

□

The length $|a|$ of the vector $a = (x_a, y_a, z_a)$ is computed by the formula:

$$|a| = \sqrt{x_a^2 + y_a^2 + z_a^2}. \tag{6.11}$$

Example 6.3 Find the length of the vector $a = (5, -3, -1)$.

Solution The length $|a|$ is equal to $\sqrt{5^2 + (-3)^2 + (-1)^2} = \sqrt{25 + 9 + 1} = \sqrt{35}$.

□

The **distance** between the points $A(x_a, y_a, z_a)$ and $B(x_b, y_b, z_b)$ is computed by the formula:

$$AB = \sqrt{(x_a - x_b)^2 + (y_a - y_b)^2 + (z_a - z_b)^2}. \tag{6.12}$$

Example 6.4 Find the distance between the points $A(5, 3, 2)$ and $B(0, 3, 2)$.

Solution

$$AB = \sqrt{(5 - 0)^2 + (3 - 3)^2 + (2 - 2)^2} = 5. \tag{6.13}$$

$\square$

Find the angle between two arbitrary vectors of a three-dimensional vector space. Let the vectors $a = (x_a, y_a, z_a)$ and $b = (x_b, y_b, z_b)$ be given. Represent the scalar product $a \cdot b$ in two ways, namely by the formulae (6.7) and (6.9):

$$a \cdot b = x_a x_b + y_a y_b + z_a z_b, \tag{6.14}$$

$$a \cdot b = |a||b| \cos \alpha. \tag{6.15}$$

Hence, we can conclude that the cosine of the angle α is equal to

$$\cos \alpha = \frac{x_a x_b + y_a y_b + z_a z_b}{\sqrt{x_a^2 + y_a^2 + z_a^2}\sqrt{x_b^2 + y_b^2 + z_b^2}}. \tag{6.16}$$

It is clear that

$$\alpha = \arccos\left(\frac{x_a x_b + y_a y_b + z_a z_b}{\sqrt{x_a^2 + y_a^2 + z_a^2}\sqrt{x_b^2 + y_b^2 + z_b^2}}\right). \tag{6.17}$$

Example 6.5 Find the angle between the vectors $a = (1, 2, 3)$ and $b = (0, 2, 1)$.

Solution Using the formula (6.17), we obtain

$$\alpha = \arccos\left(\frac{1 \cdot 0 + 2 \cdot 2 + 3 \cdot 1}{\sqrt{1 + 4 + 9} \cdot \sqrt{0 + 4 + 1}}\right) = \arccos \frac{7}{\sqrt{70}}. \tag{6.18}$$

$\square$

Example 6.6 Let two vectors be given: $a = (5, 4, 1)$ and $b = (2, -2, -2)$. Are these vectors collinear or mutually orthogonal?

Solution If the vectors are collinear, then there exists such a real number λ that the condition $a = \lambda b$ is satisfied. Hence, it follows:

$$\frac{x_a}{x_b} = \frac{y_a}{y_b} = \frac{z_a}{z_b}. \tag{6.19}$$

But since there exist two inequalities $\frac{5}{2} \neq \frac{4}{-2} \neq \frac{1}{-2}$, then the vectors a and b are non-collinear.

In order to check the mutual orthogonality of the vectors, find the scalar product of $a \cdot b$:

$$a \cdot b = 5 \cdot 2 - 4 \cdot 2 - 1 \cdot 2 = 0. \tag{6.20}$$

Then, the vectors a and b are mutually orthogonal.   □

6.3   Vector Product of Vectors

A **vector product**, or **cross product**, of two vectors specified in Cartesian coordinate system as $a = (x_a, y_a, z_a)$ and $b = (x_b, y_b, z_b)$ is a vector denoted by $a \times b$, or $[a, b]$, and determined according to the rule:

$$a \times b = \begin{vmatrix} i & j & k \\ x_a & y_a & z_a \\ x_b & y_b & z_b \end{vmatrix} = (y_a \cdot z_b - z_a \cdot y_b)i + (z_a \cdot x_b - x_a \cdot z_b)j + (x_a \cdot y_b - x_b \cdot y_a)k.$$

$$\tag{6.21}$$

6.3.1   Properties of the Vector Product

For the arbitrary vectors of the three-dimensional vector space $\mathbb{R}^3$, the following properties are valid.

Property 1

$$|a \times b| = |a||b| \sin \alpha, \tag{6.22}$$

where α is the angle between a and b.

Proof Prove that the equality is valid:

$$|a \times b|^2 = |a|^2|b|^2 \sin^2 \alpha. \tag{6.23}$$

Express $|a \times b|^2$ in terms of the coordinates of the initial vectors:

$$|a \times b|^2 = (y_a z_b - z_a y_b)^2 + (z_a x_b - x_a z_b)^2 + (x_a y_b - x_b y_a)^2 = y_a^2 z_b^2$$
$$+ z_a^2 y_b^2 + z_a^2 x_b^2 + x_a^2 z_b^2 + x_a^2 y_b^2 + x_b^2 y_a^2 - 2 y_a z_b z_a y_b$$
$$- 2 z_a x_b x_a z_b - 2 x_a y_b x_b y_a.$$

On the other hand, the right side of the equality (6.23) can be represented in the form:

$$|a|^2 |b|^2 \sin^2 \alpha = |a|^2 |b|^2 (1 - \cos^2 \alpha) = (x_a^2 + y_a^2 + z_a^2)(x_b^2 + y_b^2 + z_b^2)$$
$$- (x_a x_b + y_a y_b + z_a z_b)^2 = x_a^2 x_b^2 + x_a^2 y_b^2 + x_a^2 z_b^2 + y_a^2 x_b^2 + y_a^2 y_b^2$$
$$+ y_a^2 z_b^2 + z_a^2 x_b^2$$
$$+ z_a^2 y_b^2 + z_a^2 z_b^2 - x_a^2 x_b^2 - y_a^2 y_b^2 - z_a^2 z_b^2 - 2 x_a x_b y_a y_b$$
$$- 2 x_a x_b z_a z_b - 2 y_a y_b z_a z_b$$
$$= y_a^2 z_b^2 + z_a^2 y_b^2 + z_a^2 x_b^2 + x_a^2 z_b^2 + x_a^2 y_b^2 + x_b^2 y_a^2 - 2 y_a z_b z_a y_b$$
$$- 2 z_a x_b x_a z_b - 2 x_a y_b x_b y_a. \tag{6.24}$$

From the relation $|a \times b|^2 = |a|^2 |b|^2 \sin^2 \alpha$, we finally arrive at the conclusion that $|a \times b| = |a||b| \sin \alpha$.

Hence, it follows: if a and b are collinear, then $a \times b = 0$ (since $\sin 0 = 0$ and $\sin \pi = 0$).

Property 2 If the vectors a and b are non-collinear, then the vector $c = a \times b$ is orthogonal to each of the vectors a and b.

Proof Expand the scalar product of the vectors with regard to their coordinates:

$$(a \times b) \cdot a = y_a z_b x_a - z_a y_b x_a - x_a z_b y_a + z_a x_b y_a + x_a y_b z_a - y_a x_b z_a = 0. \tag{6.25}$$

Similar to formula (6.25), we obtain that $(a \times b) \cdot b = 0$.

Therefore, the vector c is orthogonal to both the vector a and the vector b.

Property 3

$$a \times b = -b \times a. \tag{6.26}$$

Property 4

$$a \times (b + c) = a \times b + a \times c, \tag{6.27}$$
$$(a + b) \times c = a \times c + b \times c. \tag{6.28}$$

Fig. 6.3 To the computation
of the area of the triangle
ABC (Example 6.8)

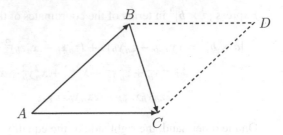

Example 6.7 The vectors $a = (3, -1, -2)$ and $b = (1, 2, -1)$ are given. Find the
coordinates of the vector $(2a - b) \times (2a + b)$.

Solution Determine the coordinates of the new vectors: $2a - b = (5, -4, -3)$ and
$2a + b = (7, 0, -5)$. Then, the sought-for vector will be equal to

$$\begin{vmatrix} i & j & k \\ 5 & -4 & -3 \\ 7 & 0 & -5 \end{vmatrix} = 20i + 4j + 28k. \tag{6.29}$$

We obtain the answer: $(20, 4, 28)$.   □

Example 6.8 The points $A(1, 2, 0)$, $B(3, 0, -3)$ and $C(5, 2, 6)$ are given. Compute
the area $S_{\triangle ABC}$ of the triangle ABC (see Fig. 6.3).

The square of the triangle ABC is a half of the square of parallelogram, formed
by the vectors $\overrightarrow{AB} = (2, -2, -3)$ and $\overrightarrow{AC} = (4, 0, 6)$. Therefore,

$$S_{\triangle ABC} = \frac{1}{2} S_{ABDC} = \frac{1}{2} |\overrightarrow{AB} \times \overrightarrow{AC}|. \tag{6.30}$$

Then,

$$\overrightarrow{AB} \times \overrightarrow{AC} = \begin{vmatrix} i & j & k \\ 2 & -2 & -3 \\ 4 & 0 & 6 \end{vmatrix} = -12i - 24j + 8k. \tag{6.31}$$

Hence,

$$S_{\triangle ABC} = \frac{1}{2} \sqrt{12^2 + 24^2 + 8^2} = 14. \tag{6.32}$$

6.4 Scalar Triple Product

The **scalar triple product**, or **mixed product**, of the vectors a, b and c is the number (a, b, c), resulting from computation of the expression $(a \times b) \cdot c$. First, the vectors a and b are multiplied vectorially, and then, the resulting vector is multiplied by c scalarly.

It is easy to verify the formula that expresses the scalar triple product in terms of their coordinates:

$$(a, b, c) \equiv ((a \times b) \cdot c) = (y_a z_b - z_a y_b)x_c + (z_a x_b - x_a z_b)y_c + (x_a y_b - y_a x_b)z_c$$

$$= \begin{vmatrix} x_a & y_a & z_a \\ x_b & y_b & z_b \\ x_c & y_c & z_c \end{vmatrix}. \tag{6.33}$$

For designation of the triple scalar product $(a \times b) \cdot c$, the notation (abc) is also used.

6.4.1 Properties of Scalar Triple Product

Property 1

$$(a \times b) \cdot c = a \cdot (b \times c) = (c \times a) \cdot b. \tag{6.34}$$

Property 2

$$\mathrm{abs}((a \times b) \cdot c) = V_n, \tag{6.35}$$

where V_n is the volume of the parallelepiped formed by the vectors a, b and c (see Fig. 6.4).

Proof By the definition of numeric projection of the vector c on the line, specified by the vector $a \times b$, the equality is valid:

$$(a \times b) \cdot c = |a \times b| \, \mathrm{Pr}_{a \times b} c, \tag{6.36}$$

where $|a \times b| = S$ the area of the parallelogram lying in the base of the parallelepiped.

The vector $a \times b$ is orthogonal to the base of the parallelepiped, and, therefore, $\mathrm{Pr}_{a \times b} c$ coincides with the height of the parallelepiped h.

Fig. 6.4 To the computation
of the volume of the
parallelepiped

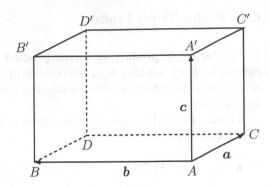

Thus,

$$\text{abs}((a \times b) \cdot c) = S \cdot h, \tag{6.37}$$

as we set out to prove.

Note Since the tetrahedron $ABCA'$ forms one-sixth part of the volume of the parallelepiped, its volume is equal to

$$V_{ABCA'} = \frac{1}{6}V_n = \frac{1}{6}\text{abs}((a \times b) \cdot c). \tag{6.38}$$

Three vectors are called **coplanar** if all of them are parallel to the same plane.

Property 3 The vectors a, b and c are coplanar if and only if $(a, b, c) = 0$.

Example 6.9 Let the vertices of the tetrahedron be given $A(2, -1, 1)$, $B(5, 5, 4)$, $C(3, 2, -1)$ and $D(4, 1, 3)$. Compute its volume.

Solution Let us make the vectors $\overrightarrow{AB} = (3, 6, 3)$, $\overrightarrow{AC} = (1, 3, -2)$ and $\overrightarrow{AD} = (2, 2, 2)$. Then, the volume of the tetrahedron $ABCD$ is equal to

$$V_{ABCD} = \frac{1}{6}\text{abs}(\overrightarrow{AB} \times \overrightarrow{AC}) \cdot \overrightarrow{AD} = \frac{1}{6}\text{abs}\det \begin{bmatrix} 3 & 6 & 3 \\ 1 & 3 & -2 \\ 2 & 2 & 2 \end{bmatrix} \tag{6.39}$$

$$= \frac{1}{6} \cdot 3 \cdot 2 \cdot \text{abs}\det \begin{bmatrix} 1 & 2 & 1 \\ 1 & 3 & -2 \\ 1 & 1 & 1 \end{bmatrix} = \text{abs}(-3) = 3. \tag{6.40}$$

$\square$

Example 6.10 Find whether the following vectors are coplanar $a = (2, -1, 2)$, $b = (1, 2, -3)$ and $c = (3, -4, 7)$.

Solution Find the scalar triple product of the given vectors:

$$(a, b, c) = \begin{vmatrix} 2 & -1 & 2 \\ 1 & 2 & -3 \\ 3 & -4 & 7 \end{vmatrix} = 2 \cdot (14 - 12) + 1 \cdot (7 + 9) + 2 \cdot (-4 - 6) = 0. \quad (6.41)$$

Therefore, the vectors a, b and c are coplanar.

6.5   Vector Triple Product

Having three vectors, for example, a, b and c, apply the vector product operation first to b and c, and then vectorially multiply a and $b \times c$. As a result, we obtain the **vector triple product** $a \times (b \times c)$.

Theorem 6.2 *For arbitrary vectors of a three-dimensional vector space a, b and c, the identity is valid:*

$$a \times (b \times c) \equiv b(a \cdot c) - c(a \cdot b). \quad (6.42)$$

For proof see Problem **6.12**.

Note The relation (6.42) is also referred to as **Lagrange's identity**.

Consequence. *For the operation of vector triple product, the **Jacobi identity** is valid:*

$$a \times (b \times c) + c \times (a \times b) + b \times (c \times a) \equiv 0. \quad (6.43)$$

Proof For each of the three summands of the sum, use the expansion (6.42):

$$a \times (b \times c) = b(a \cdot c) - c(a \cdot b), \quad (6.44)$$

$$b \times (c \times a) = c(b \cdot a) - a(b \cdot c), \quad (6.45)$$

$$c \times (a \times b) = a(c \cdot b) - b(c \cdot a). \quad (6.46)$$

Computation of the sum of the three vector products (6.44)–(6.46) after collecting similar summands results in zero. The consequence is proved.

Review Questions

1. Define geometric vector.
2. What are zero vector and unit vector?
3. What vectors are called collinear?
4. What is the condition of equality of vectors?
5. Formulate the rule of parallelogram for addition of vectors.
6. How is the projection of the vector b onto the vector a determined?
7. What is the Cartesian coordinate system?
8. Enumerate the normalizing vectors of the Cartesian coordinate system. How can one use them to expand an arbitrary vector in the Cartesian basis?
9. Define scalar product of vectors.
10. What vectors are called orthogonal?
11. How do we find the distance between the two points specified by their Cartesian coordinates?
12. Define vector product of vectors.
13. Enumerate the properties of a vector product.
14. Define scalar triple product of vectors.
15. Enumerate the properties of a scalar triple product.
16. What three vectors are called coplanar?
17. What is vector triple product?
18. Write Lagrange's identity for the vector triple product.
19. Write the Jacobi identity.

Problems

6.1. Compute the scalar and vector products of the vectors $c_1 = 2a - b$ and $c_2 = -a + 3b$, if:

(a) $a = (-2, 1, 1)$, $b = (3, -2, 4)$;
(b) $a = (2, 1, -2)$, $b = (-1, 0, -2)$.

6.2. There are given vertices of the triangle ABC. Compute its area and the cosine of the inner angle at the vertex B:

(a) $A(2, 1, 0)$, $B(3, 0, 3)$, $C(2, -3, 7)$;
(b) $A(4, -3, 2)$, $B(-1, 4, 3)$, $C(6, 3, -2)$.

6.3. Find whether vectors a, b, c are coplanar:

(a) $a = (1, 1, 1)$, $b = (2, 3, 0)$, $c = (3, -1, -1)$;
(b) $a = (-1, 0, -2)$, $b = (0, 0, -1)$, $c = (-1, 0, 3)$.

6.4. Prove that the points $A(1, -1, 1)$, $B(1, 3, 1)$, $C(4, 3, 1)$ and $D(4, -1, 1)$ are vertices of a rectangle. Compute the length of its diagonals.

6.5. Compute the coordinates of the vector c that is orthogonal to the vectors $a = 2j - k$ and $b = -i + 2j - 3k$ and forms an obtuse angle with the axis Oy, if $|c| = \sqrt{7}$.

6.6. Find the angle between the vectors $a + b$ and $a - b$, if $a = 3i - j + 2k$ and $b = i + j - k$.

6.7. The vectors a and b form the angle $\pi/3$. Find the length of the vector $a - 2b$, if $|a| = 2$, $|b| = 1$.

6.8. For what value of the real parameter d are the vectors $a = (12, 2, d)$ and $b = (-3, 17d, -1)$ orthogonal?

6.9. For what value of the real parameter $\varkappa$ will the vectors of the three-dimensional Euclidean vector space $t_1 = a - 10b$ and $t_2 = a + \varkappa b$ be orthogonal, if $|a| = 5$, $|b| = 3$, and the angle φ between the vectors a and b is equal to $\dfrac{\pi}{6}$?

6.10. For what value of the real parameter $\varkappa$ will the vectors of the three-dimensional Euclidean vector space $t_1 = 2a + \varkappa b$ and $t_2 = b - 2a$ be orthogonal, if $|a| = 1$, $|b| = 3/2$, and the angle φ between the vectors a and b is equal to $\dfrac{2\pi}{3}$?

6.11. The vectors a and b have, in Cartesian basis, the coordinates $a = (a_1, a_2, 0)$ and $b = (b_1, b_2, 0)$. Find the sine of the angle between these vectors.

6.12. Prove the theorem on vector triple product (6.42).

6.13. Prove the identities valid for the arbitrary vectors a, b, c and d:

(1) $(a \times b) \cdot (c \times d) \equiv (a \cdot c)(b \cdot d) - (a \cdot d)(b \cdot c)$;

(2) $(a \times b) \times (c \times d) \equiv c(abd) - d(abc)$;

(3) $a \times (b \times (c \times d)) \equiv (b \cdot d)(a \times c) - (b \cdot c)(a \times d)$;

(4) $(((a \times b) \times (b \times c))((b \times c) \times (c \times a))((c \times a) \times (a \times b))) \equiv (abc)^4$.

6.14. Prove the identity valid for the arbitrary vectors a, b, c, d, e, f:

$$(a \times b) \cdot ((c \times d) \times (e \times f)) \equiv (abd)(cef) - (abc)(def).$$

∗6.15. Simplify the vector expression that depends on the natural parameter n:

$$f_n = \underbrace{(a \times \cdots \times (a \times (a \times b))\dots)}_{n \text{ products}}.$$

6.16. Prove that for $p_1, p_2, p_3, p_4, p_5 \in \mathbb{R}^3$ the equality is fulfilled:

$$(p_1 p_2 p_3)(p_1 p_4 p_5) = \begin{vmatrix} (p_1 p_2 p_4) & (p_1 p_2 p_5) \\ (p_1 p_3 p_4) & (p_1 p_3 p_5) \end{vmatrix}.$$

∗6.17. Solve the system of linear equations, represented in vector form, relative to the unknown variables x_1, x_2, x_3:

$$\alpha \cdot x = \gamma,$$

$$\alpha \times x + \beta = 0,$$

where $x = (x_1, x_2, x_3)$, and the vectors $\alpha \neq 0$, β and the scalar γ do not depend on x_1, x_2, x_3.

∗6.18. Solve the system of linear equations, represented in vector form, relative to the unknown variables x_1, x_2, x_3:

$$\alpha \cdot x = c_1,$$

$$\beta \cdot x = c_2,$$

$$\gamma \cdot x = c_3,$$

where $x = (x_1, x_2, x_3)$, and the vectors α, β, γ and the constants c_1, c_2, c_3 do not depend on x_1, x_2, x_3 and $(\alpha, \beta, \gamma) \neq 0$.

∗6.19. Solve the system of equations relative to the unknown vectors x and y:

$$\begin{cases} \pi \times x + \rho \times y = \sigma, \\ \rho \times x - \pi \times y = \tau, \end{cases}$$

where $\pi, \rho, \sigma, \tau \in \mathbb{R}^n$, and the vectors π and ρ are not equal to the zero vector simultaneously.

Answers and Solutions

6.1 *Solution.*

(a) Write the vectors c_1 and c_2 in coordinate form:

$$c_1 = 2(-2, 1, 1) - (3, -2, 4) = (-7, 4, -2), c_2 = -(-2, 1, 1)$$

$$+3(3, -2, 4) = (11, -7, 11).$$

The scalar product is equal to $c_1 \cdot c_2 = -7 \cdot 11 + 4 \cdot (-7) - 2 \cdot 11 = -127$.

The vector product is equal to

$$
c_1 \times c_2 = \begin{vmatrix} i & j & k \\ -7 & 4 & -2 \\ 11 & -7 & 11 \end{vmatrix} = 30i + 55j + 5k = (30, 55, 5).
$$

(b) Write the vectors c_1 and c_2: $c_1 = (5, 2, -2)$, $c_2 = (-5, -1, -4)$.
Their scalar product is equal to $c_1 \cdot c_2 = 5 \cdot (-5) + 2 \cdot (-1) - 2 \cdot (-4) = -19$.
The vector product:

$$
c_1 \times c_2 = \begin{vmatrix} i & j & k \\ 5 & 2 & -2 \\ -5 & -1 & -4 \end{vmatrix} = -10i + 30j + 5k = (-10, 30, 5).
$$

6.2 *Solution.*

(a) Compute the coordinates of the vectors $\overrightarrow{BA}$ and $\overrightarrow{BC}$:

$$
\overrightarrow{BA} = (2 - 3, 1 - 0, 0 - 3) = (-1, 1, -3),
$$
$$
\overrightarrow{BC} = (2 - 3, -3 - 0, 7 - 3) = (-1, -3, 4).
$$

Then, find the cosine of the angle φ at the vertex B:

$$
\cos\varphi = \frac{\overrightarrow{BA} \cdot \overrightarrow{BC}}{|\overrightarrow{BA}| \cdot |\overrightarrow{BC}|} = \frac{-1 \cdot (-1) + 1 \cdot (-3) + (-3) \cdot 4}{\sqrt{(-1)^2 + 1^2 + (-3)^2} \cdot \sqrt{(-1)^2 + (-3)^2 + 4^2}}
$$
$$
= -\frac{14}{\sqrt{286}}.
$$

The area can be computed in two ways.
The first method
The product of the vectors $\overrightarrow{BA}$ and $\overrightarrow{BC}$ is determined as

$$
\overrightarrow{BA} \times \overrightarrow{BC} = \begin{vmatrix} i & j & k \\ -1 & 1 & -3 \\ -1 & -3 & 4 \end{vmatrix} = (4 - 9)i - (-4 - 3)j + (3 + 1)k = -5i + 7j + 4k.
$$

Substitute the obtained values of the coordinates into the formula for the area of the triangle:

$$
S = \frac{1}{2}|\overrightarrow{BA} \times \overrightarrow{BC}| = \frac{1}{2}\sqrt{(-5)^2 + 7^2 + 4^2} = \frac{3}{2}\sqrt{10}.
$$

The second method

According to the fundamental trigonometric identity (see Appendix B, formula (B.1)), we have

$$\sin\varphi = \sqrt{1 - \cos^2\varphi} = \sqrt{1 - \frac{14^2}{286}} = \frac{3\sqrt{10}}{\sqrt{286}}.$$

Substitute the coordinate values into the area formula:

$$S = \frac{1}{2}|\overrightarrow{BA}| \cdot |\overrightarrow{BC}| \cdot \sin\varphi$$

$$= \frac{1}{2}\sqrt{(-1)^2 + 1^2 + (-3)^2} \cdot \sqrt{(-1)^2 + (-3)^2 + 4^2} \cdot \frac{3\sqrt{10}}{\sqrt{286}} = \frac{3}{2}\sqrt{10}.$$

Obviously, both methods of computing the area of the triangle result in the same answer:

$$\cos\varphi = -\frac{14}{\sqrt{286}}, \quad S = \frac{3}{2}\sqrt{10}.$$

(b) The coordinates of the vectors $\overrightarrow{BA}$ and $\overrightarrow{BC}$ are equal to

$$\overrightarrow{BA} = (5, -7, -1), \quad \overrightarrow{BC} = (7, -1, -5).$$

Compute the cosine of the angle φ between these vectors:

$$\cos\varphi = \frac{(\overrightarrow{BA} \cdot \overrightarrow{BC})}{|\overrightarrow{BA}| \cdot |\overrightarrow{BC}|} = \frac{5 \cdot 7 + (-7) \cdot (-1) + (-5) \cdot (-1)}{\sqrt{5^2 + (-7)^2 + (-1)^2} \cdot \sqrt{7^2 + (-1)^2 + (-5)^2}} = \frac{47}{75}.$$

Then, compute the vector product of $\overrightarrow{BA}$ and $\overrightarrow{BC}$:

$$\overrightarrow{BA} \times \overrightarrow{BC} = \begin{vmatrix} i & j & k \\ 5 & -7 & -1 \\ 6 & -1 & -5 \end{vmatrix} = 34i + 18j + 44k.$$

Substitute the coordinate values into the area formula:

$$S = \frac{1}{2}|\overrightarrow{BA} \times \overrightarrow{BC}| = \frac{1}{2} \cdot \sqrt{34^2 + 18^2 + 44^2} = \sqrt{854}.$$

We finally obtain

$$\cos \varphi = \frac{47}{75}, \quad S = \sqrt{854}.$$

6.3 *Solution.*

(a) Since the determinant composed of the vector coordinates is not equal to zero:

$$\begin{vmatrix} 1 & 1 & 1 \\ 2 & 3 & 0 \\ 3 & -1 & -1 \end{vmatrix} = -12,$$

then the vectors a, b and c are not coplanar.

(b) The determinant composed of the vector coordinates is equal to zero:

$$\begin{vmatrix} -1 & 0 & -2 \\ 0 & 0 & 1 \\ -1 & 0 & 3 \end{vmatrix} = 0,$$

and, therefore, these vectors are coplanar.

6.4 *Solution.*

Note that the points A, B, C and D lie in the same plane, since the applicate of all these points is equal to $z = 1$.

Let us prove that these four points are vertices of a parallelogram. We have to prove that

$$|\overrightarrow{AB}| = |\overrightarrow{CD}|, \quad |\overrightarrow{BC}| = |\overrightarrow{AD}|.$$

Find the coordinates of the vectors introduced for consideration:

$$\overrightarrow{AB} = (1 - 1, 3 - (-1), 1 - 1) = (0, 4, 0),$$
$$\overrightarrow{CD} = (4 - 4, -1 - 3, 1 - 1) = (0, -4, 0),$$
$$\overrightarrow{BC} = (4 - 1, 3 - 3, 1 - 1) = (3, 0, 0),$$
$$\overrightarrow{AD} = (4 - 1, -1 - (-1), 1 - 1) = (3, 0, 0).$$

Therefore,

$$
\begin{cases}
|\overrightarrow{AB}| = \sqrt{0^2 + 4^2 + 0^2} = 4, \\
|\overrightarrow{CD}| = \sqrt{0^2 + 4^2 + 0^2} = 4, \\
|\overrightarrow{BC}| = \sqrt{3^2 + 0^2 + 0^2} = 3, \\
|\overrightarrow{AD}| = \sqrt{3^2 + 0^2 + 0^2} = 3;
\end{cases}
\Rightarrow
\quad
\begin{aligned}
|\overrightarrow{AB}| &= |\overrightarrow{CD}|, \\
|\overrightarrow{BC}| &= |\overrightarrow{AD}|.
\end{aligned}
$$

If one of the parallelogram angles is equal to $\dfrac{\pi}{2}$, then all other angles are also equal to $\dfrac{\pi}{2}$.

Show that the scalar product of the vectors $\overrightarrow{AB}$ and $\overrightarrow{AD}$ is equal to zero:

$$
\overrightarrow{AB} \cdot \overrightarrow{AD} = 0 \cdot 3 + 4 \cdot 0 + 0 \cdot 0 = 0 \Rightarrow \overrightarrow{AB} \perp \overrightarrow{AD}.
$$

Therefore, $ABCD$ is a rectangle. Find its diagonals with the help of the Pythagorean theorem:

$$
AC = \sqrt{AB^2 + BC^2} = \sqrt{4^2 + 3^2} = 5.
$$

It is clear that $AC = BD$, since the diagonals of the rectangle are equal.

As a result, we obtain $AC = BD = 5$.

6.5 *Solution.*

Let $c = (c_x, c_y, c_z)$, where c_x, c_y, c_z are the unknown coordinates of the vector. The condition of orthogonality of a and c has the form $a \cdot c = 0$, or $2c_y - c_z = 0$. Then, the condition of orthogonality of b and c is $b \cdot c = 0$, or $-c_x + 2c_y - 3c_z = 0$. Since $|c| = \sqrt{7}$ and the length of the vector is equal to the square root of the sum of squares of its coordinates, $\sqrt{c_x^2 + c_y^2 + c_z^2} = \sqrt{7}$, which can be written as $c_x^2 + c_y^2 + c_z^2 = 7$.

We obtain the system of three equations relative to the variables c_x, c_y, c_z:

$$
\begin{cases}
2c_y - c_z = 0, \\
-c_x + 2c_y - 3c_z = 0, \\
c_x^2 + c_y^2 + c_z^2 = 7.
\end{cases}
$$

As the independent variable select c_y and express through it two other variables of the system: $c_z = 2c_y$ and $c_x = 2c_y - 3c_z = 2c_y - 3(2c_y) = -4c_y$. Therefore, $c = (-4c_y, c_y, 2c_y) = c_y(-4, 1, 2)$, where $c_y \in \mathbb{R}$.

The length of the vector c is equal to $|c| = \sqrt{7}$, and, therefore,

$$\mathrm{abs}(c_y)\sqrt{(-4)^2 + 1^2 + 2^2} = \sqrt{7}, \mathrm{abs}(c_y) = \frac{\sqrt{7}}{\sqrt{21}} = \frac{1}{\sqrt{3}}, \text{ and } c_y = \pm\frac{1}{\sqrt{3}}.$$

According to the problem statement, the vector c forms an obtuse angle with the axis Oy, i.e. $\cos\varphi < 0$, where φ is the angle between the vectors c and j. Since

$$\cos\varphi = \frac{c \cdot j}{|c||j|} = \frac{c_y}{\sqrt{7}} < 0,$$

$$c_y = -\frac{1}{\sqrt{3}}.$$

We finally obtain $c = \dfrac{1}{\sqrt{3}}(4, -1, -2)$.

6.6 *Solution.*

Write the vectors a and b in coordinate form:

$$a = (3, -1, 2),$$
$$b = (1, 1, -1).$$

Then, the sum and the difference of these vectors are

$$a + b = (3 + 1, -1 + 1, 2 + (-1)) = (4, 0, 1),$$
$$a - b = (3 - 1, -1 - 1, 2 - (-1)) = (2, -2, 3).$$

Use the formula of the cosine of the angle α between the vectors:

$$\cos\alpha = \frac{(a \cdot b)}{|a| \cdot |b|}.$$

After simple computations, we obtain

$$\cos\alpha = \frac{(a + b) \cdot (a - b)}{|a + b| \cdot |a - b|} = \frac{4 \cdot 2 + 0 \cdot (-2) + 1 \cdot 3}{\sqrt{4^2 + 0^2 + 1^2} \cdot \sqrt{2^2 + (-2)^2 + 3^2}} = \frac{11}{17}.$$

Therefore, $\alpha = \arccos\dfrac{11}{17}$.

6.7 *Solution.*

Compute $|a - 2b|^2$:

$$|a-2b|^2 = (a-2b)\cdot(a-2b) = a\cdot a - 4a\cdot b + 4b\cdot b = |a|^2 - 4|a||b|\cos\varphi + 4|b|^2,$$

where $\varphi = \pi/3$ is the angle between the vectors a and b. Having substituted numeric values, we obtain $|a - 2b|^2 = 1$, and, therefore, $|a - 2b| = 1$.

6.8 *Solution.*

As is known, the necessary and sufficient condition of orthogonality of two vectors is that their scalar product is equal to zero: $a\cdot b = 0$. Substitute the coordinate values from the problem statement:

$$12\cdot(-3) + 2\cdot 17d - 1\cdot d = 0,$$
$$d = \frac{12}{11}.$$

Therefore, the vectors a and b are orthogonal for the parameter value $d = \dfrac{12}{11}$.

6.9 *Solution.*

In order for the vectors to be orthogonal, their scalar product must be equal to zero: $t_1\cdot t_2 = 0$,

$$(a - 10b)\cdot(a + \varkappa b) = 0,$$

$$a\cdot a + \varkappa a\cdot b - 10a\cdot b - 10\varkappa b\cdot b = 0.$$

Then, we find $\varkappa = \dfrac{10\sqrt{3} - 10}{3\sqrt{3} - 36}$.

6.10 *Solution.*

The necessary condition of the vector orthogonality: $t_1\cdot t_2 = 0$, or

$$(2a + \varkappa b)\cdot(b - 2a) = 0,$$

$$2ab - 4|a|^2 + \varkappa|b|^2 - 2\varkappa ab = 0.$$

Substitute the numeric values from the problem statement: $-\dfrac{3}{2} - 4 + \dfrac{9\varkappa}{4} + \dfrac{3\varkappa}{2} = 0$, and, therefore, $\varkappa = \dfrac{22}{15}$.

6.11 *Answer:* $\sin\varphi = \dfrac{a_1 b_2 - a_2 b_1}{\sqrt{a_1^2 + b_1^2}\sqrt{a_2^2 + b_2^2}}$.

6.15 *Answer:*

$$f_n = (-1)^{\lfloor n/2 \rfloor} a^{n-2} \begin{cases} a(a \times b), & \text{if } n \text{ is odd,} \\ a^2 b - (a \cdot b)a, & \text{if } n \text{ is even.} \end{cases}$$

6.17 *Answer:*
$$x = \frac{\gamma}{\alpha^2}\alpha + \frac{1}{a^2}\alpha \times \beta.$$

6.18 *Answer:*
$$x = \frac{1}{(\alpha, \beta, \gamma)}\Big(c_1(\beta \times \gamma) + c_2(\gamma \times \alpha) + c_3(\alpha \times \beta)\Big).$$

6.19 *Solution.*
Multiply the first equation of the system by the vector ρ scalarly on the right; multiply the second equation by the vector π scalarly on the right. Subtracting one equation from the other, we find

$$x = -\frac{\pi \times \sigma + \rho \times \tau}{\pi^2 + \rho^2} + \gamma_1\pi + \gamma_2\rho,$$

where γ_1, γ_2 are real numbers.

Having substituted the obtained expression into the first equation of the system, we find y:

$$y = \frac{\pi \times \tau - \rho \times \sigma}{\pi^2 + \rho^2} + \gamma_2\pi - \gamma_1\rho.$$

Chapter 7
Equation of a Straight Line on a Plane

7.1 Slope-Intercept Form of the Equation of a Straight Line

Consider Cartesian coordinate system on a plane.

Let the straight line L intersect the axis Oy at the point B with the coordinates $(0, b)$ and forms with the axis Ox the angle α (see Fig. 7.1). For definiteness, we will assume that the angle $\alpha < \dfrac{\pi}{2}$.

On the line, take an arbitrary point A with the coordinates (x, y). Then, from the point A, drop a perpendicular to the axis Ox, and from the point B, a perpendicular to the axis Oy. Consider the obtained triangle ABC. It is obvious that $BC = x$, $AC = y - b$, $\angle ABC = \alpha$. Since $AC = BC \tan \alpha$, we obtain $y - b = x \cdot \tan \alpha$ or $y = \tan \alpha \cdot x + b$.

Denote the tangent of the angle α by k. The variable $k = \tan \alpha$ is referred to as the **slope** of the straight line on the plane. As a result, we arrive at the equation of a straight line of the form:

$$y = kx + b. \tag{7.1}$$

It is called the **slope-intercept form of the equation of a straight line**.

Note For the lines of the form $y = \mathrm{const}$, the slope is equal to zero; for the lines of the form $x = \mathrm{const}$, the slope is undefined.

© Springer Nature Switzerland AG 2021
S. Kurgalin, S. Borzunov, *Algebra and Geometry with Python*,
https://doi.org/10.1007/978-3-030-61541-3_7

Fig. 7.1 The line L on the plane xOy

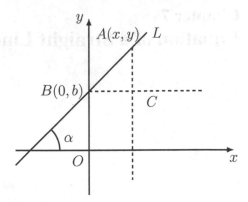

Fig. 7.2 Construction of n—a normal vector to the line

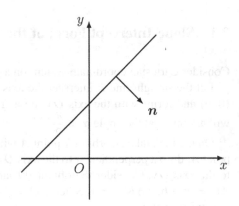

7.2   General Equation of a Straight Line

General equation of a straight line on a plane has the form:

$$Ax + By + C = 0. \tag{7.2}$$

The real numbers A, B and C are called **coefficients of the straight line equation**.

The variables A and B cannot simultaneously be equal to zero, because, in this case, if $C = 0$, then all points on the plane will satisfy this equation. However, if $C \neq 0$, then none of the points on the plain satisfies this equation.

The vector $n = \begin{bmatrix} A \\ B \end{bmatrix}$ is called a **normal vector** of a line or a **normal**. The normal vector is orthogonal to the respective line (see Fig. 7.2).

Let the inequality $B \neq 0$ be valid. In this case, the summand By can be rearranged to the right, and both parts of the equation can be divided by $-B \neq 0$.

As a result, we obtain

$$Ax + C = -By \qquad (7.3)$$

or

$$-\frac{A}{B}x - \frac{C}{B} = y. \qquad (7.4)$$

Let us introduce notations $-\dfrac{A}{B} = k$ and $-\dfrac{C}{B} = b$; then, we arrive at the equation $y = kx + b$, which is a slope-intercept form of the equation of a straight line (see Eq. (7.1)).

7.3 Slope-Intercept Form of the Equation of a Straight Line Through a Given Point

Consider the slope-intercept form of the equation of a straight line:

$$y = kx + b. \qquad (7.5)$$

Let this straight line pass through a point with the coordinates (x_0, y_0). Substitute these coordinates into the equation:

$$y_0 = kx_0 + b. \qquad (7.6)$$

Subtract from the Eq. (7.5) the Eq. (7.6). We obtain the sought **slope-intercept form of the equation of a straight line passing through the given point**:

$$(y - y_0) = k(x - x_0). \qquad (7.7)$$

Example 7.1 Find the slope-intercept form of the equation of a straight line $k = 2$ through $T(1, 5)$.

Solution Use the formula (7.7). Then, we have

$$(y - 5) = 2(x - 1), \qquad (7.8)$$

or

$$y = 2x + 3. \qquad (7.9)$$

$\square$

7.4   Equation of a Straight Line Through Two Given Points

Find the equation of a straight line through two given points $T_1(x_1, y_1)$ and $T_2(x_2, y_2)$ subject to the condition that $x_1 \neq x_2$ and $y_1 \neq y_2$. For this purpose, write the equation of a straight line in the form (7.7), assuming that is passes through the point T_1:

$$y - y_1 = k(x - x_1). \tag{7.10}$$

Since this line also passes through the point T_2, then we will substitute its coordinates into the Eq. (7.10):

$$y_2 - y_1 = k(x_2 - x_1). \tag{7.11}$$

Divide the Eq. (7.10) by Eq. (7.11). We obtain

$$\frac{y - y_1}{y_2 - y_1} = \frac{x - x_1}{x_2 - x_1}. \tag{7.12}$$

This is the **equation of a straight line through two given points**.

Note The formula (7.12) is not applicable in the case of equality of the abscissas or equality of the ordinates of the initial points T_1 and T_2. If $x_1 = x_2$, then the equation of the line $T_1 T_2$ has the form $x = x_1$. If the condition $y_1 = y_2$ is fulfilled, then the equation of this line has the form $y = y_1$.

Example 7.2 Find the equation of the line through $T_1(a, b)$ and $T_2(b, a)$, where $a, b \in \mathbb{R}$, and $a \neq b$.

Solution Use the formula (7.12). Substitute into it the coordinates of the points $x_1 = a$, $y_1 = b$, $x_2 = b$, $y_2 = a$:

$$\frac{y - b}{a - b} = \frac{x - a}{b - a}. \tag{7.13}$$

After simple transformations, we obtain the equation of the straight line:

$$y - b = -(x - a), \tag{7.14}$$

$$x + y - a - b = 0. \tag{7.15}$$

$\square$

7.5 Angle Between Two Straight Lines

Consider two straight lines specified by the equations $y = k_1x + b_1$ and $y = k_2x + b_2$ (see Fig. 7.3).

By the angle α between the lines, we will understand the angle by which one of these lines should be turned around their intersection point, anticlockwise, until the first superposition on the other line.

From Fig. 7.3, it is seen that the angle α between the lines is equal to $\alpha_1 - \alpha_2$. And the equalities $\tan\alpha_1 = k_1$ and $\tan\alpha_2 = k_2$ are fulfilled. In this case, based on the formula of tangent of a difference of two arguments (see formula (B.14) in Appendix B), we can write

$$\tan\alpha = \tan(\alpha_1 - \alpha_2) = \frac{\tan\alpha_1 - \tan\alpha_2}{1 + \tan\alpha_1 \cdot \tan\alpha_2} = \frac{k_1 - k_2}{1 + k_1k_2}. \tag{7.16}$$

Hence,

$$\alpha = \arctan\frac{k_1 - k_2}{1 + k_1k_2}. \tag{7.17}$$

Having exchanged places of the parameters k_1 and k_2, we obtain the tangent of the **adjacent** angle $\widetilde{\varphi} = \pi - \varphi$.

From the obtained formula (7.17) follow two consequences:

(a) The straight lines with the slopes k_1 and k_2 are orthogonal if the condition $1 + k_1k_2 = 0$ is fulfilled, which is equivalent to $k_2 = -\dfrac{1}{k_1}$.

(b) The straight lines are parallel if $k_1 = k_2$.

Consider the straight lines specified by the equations in general form:

$$A_1x + B_1y + C_1 = 0 \quad \text{and} \quad A_2x + B_2y + C_2 = 0. \tag{7.18}$$

Fig. 7.3 Definition of the angle α between two straight lines

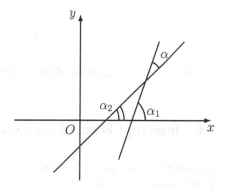

Then, $k_1 = -\dfrac{A_1}{B_1}$ and $k_2 = -\dfrac{A_2}{B_2}$. Therefore,

$$\tan \alpha = \frac{A_2 B_1 - A_1 B_2}{A_1 A_2 + B_1 B_2}. \tag{7.19}$$

From (7.19) directly follows that the lines specified by the Eqs. (7.18) are orthogonal at $A_1 A_2 + B_1 B_2 = 0$ and parallel at $A_2 B_1 - A_1 B_2 = 0$.

Example 7.3 A straight line $2x - 5y + 1 = 0$ is given. Set up the equation of a straight line that passes through the point $T_0(3, 3)$:

(a) parallel to this line;
(b) perpendicular to this line.

Solution

(a) Find the slope $k_0 = -\dfrac{A}{B} = -\dfrac{2}{(-5)} = \dfrac{2}{5}$.

Write the equation of the straight line that passes through the given point with the specified slope $k_1 = k_0$:

$$y - y_0 = k_1(x - x_0), \tag{7.20}$$

$$y - 3 = \frac{2}{5}(x - 3). \tag{7.21}$$

Thus, the sought straight line has the form: $2x - 5y + 9 = 0$.

(b) Find the slope of the line that is perpendicular to the given one: $k_2 = -\dfrac{1}{k_0} = -\dfrac{5}{2}$.

Write the equation of the straight line that passes through the given point with the specified slope k_2:

$$y - y_0 = k_2(x - x_0), \tag{7.22}$$

$$y - 3 = -\frac{5}{2}(x - 3). \tag{7.23}$$

We obtain the equation of the straight line: $5x + 2y - 21 = 0$. □

7.6　Intercept Form of the Equation of a Straight Line

Consider the equation of the line $Ax + By + C = 0$, where the variables A, B and C are not equal to zero.

Fig. 7.4 The line that passes through the points $(a, 0)$ and $(0, b)$

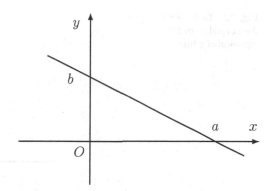

Let us transform it as follows:

$$-\frac{A}{C}x - \frac{B}{C}y = 1. \tag{7.24}$$

Let us introduce notations: $a = -\frac{C}{A}$ and $b = -\frac{C}{B}$.
As a result, we obtain the equation:

$$\frac{x}{a} + \frac{y}{b} = 1, \tag{7.25}$$

which is the **intercept form of the equation of a straight line**.

It is obvious that the given line passes through the points with the coordinates $(a, 0)$ and $(0, b)$. It is shown in Fig. 7.4.

Thus, this line cuts off segments of length abs(a) and abs(b) on the coordinate axes.

7.7 Normal (Symmetric) Form of the Equation of a Line

Consider an arbitrary straight line L. Let us draw, through the origin of coordinates O, a line, perpendicular to L, and denote by the letter P the intersection point of these lines.

On the line OP, take the unit vector $\boldsymbol{n}$, whose direction coincides with the direction of the vector $\overrightarrow{OP}$.

Assume that $p = |\overrightarrow{OP}|$, and the angle θ is the angle between the vector $\boldsymbol{n}$ and the axis Ox (see Fig. 7.5).

Since $\boldsymbol{n}$ is a unit vector, its coordinates are equal to the projections of this vector on the coordinate axes:

$$\boldsymbol{n} = (\cos\theta, \ \sin\theta). \tag{7.26}$$

Fig. 7.5 To the derivation of the normal form of the equation of a line

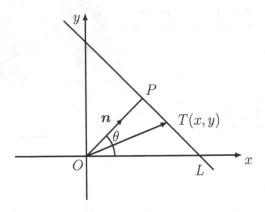

The arbitrary point $T(x, y)$ lies on the considered line L if and only if the projection of the vector $\overrightarrow{OT}$ on the axis determined by the vector $\boldsymbol{n}$ is equal to p:

$$\Pr_{\boldsymbol{n}} \overrightarrow{OT} = p. \tag{7.27}$$

As is well known,

$$\Pr_{\boldsymbol{n}} \overrightarrow{OT} = \frac{\overrightarrow{OT} \cdot \boldsymbol{n}}{|\boldsymbol{n}|} = \overrightarrow{OT} \cdot \boldsymbol{n}. \tag{7.28}$$

Bearing in mind that $\overrightarrow{OT} = (x, y)$, and the vector $\boldsymbol{n}$ is determined by the equality (7.26), we obtain the following expression for their scalar product:

$$\overrightarrow{OT} \cdot \boldsymbol{n} = x \cos \theta + y \sin \theta. \tag{7.29}$$

From the above reasoning, it follows that the point $T(x, y)$ lies on the line L if and only if the coordinates of this point satisfy the relation:

$$x \cos \theta + y \sin \theta - p = 0. \tag{7.30}$$

This equation is called the **normal form of the equation of the line** L or the **Hesse[1] normal form**.

Let the straight line L be specified by the general equation:

$$Ax + By + C = 0, \tag{7.31}$$

where $C \neq 0$.

---

[1] Ludwig Otto Hesse (1811–1874), German mathematician.

In order to transform the general equation of a line into a normal form, multiply both sides of the equation by the so-called **normalizing factor**:

$$\mu = -\frac{1}{\sqrt{A^2 + B^2}}\mathrm{sgn}(C), \tag{7.32}$$

where $\mathrm{sgn}(C)$ is the sign of the coefficient C determined by the rule:

$$\mathrm{sgn}(C) = \begin{cases} +1, & \text{if } C > 0, \\ 0, & \text{if } C = 0, \\ -1, & \text{if } C < 0. \end{cases} \tag{7.33}$$

As a result, the new coefficients at x and y (namely, μA and μB) will satisfy the condition:

$$(\mu A)^2 + (\mu B)^2 = 1. \tag{7.34}$$

If we select the angle θ so that $\cos\theta = \mu A$ and $\sin\theta = \mu B$, while $p = -\mathrm{abs}(\mu C)$, then we will obtain the normal form of the equation of a line.

Note If the condition $C = 0$ is fulfilled, then the line passes through the origin of coordinates, and the normalizing factor can be taken with an arbitrary sign: $\mu = \pm\frac{1}{\sqrt{A^2 + B^2}}$.

Let us introduce the concept of **deviation** of an arbitrary point $T(x, y)$ from the given line L. Let the number d denote the distance from the point T to this line.

We will call the number $+d$ the deviation δ of the point T from the line L in the event when the point T and the origin of coordinates O lie on the opposite sides of the line L and the number $-d$ in the event when the points T and O lie on the same side of L.

Show that the left side of the normal form of the equation of the line is equal to the deviation of the point $T(x, y)$ from this line.

Let Q be the projection of the point T on the axis determined by the vector $\boldsymbol{n}$. The deviation δ of the point T from the line L is equal to PQ.

From Fig. 7.6, it is seen that

$$\delta = PQ = OQ - OP = OQ - p. \tag{7.35}$$

But $OQ = \mathrm{Pr}_n \overrightarrow{OT} = x\cos\theta + y\sin\theta$. So,

$$OQ = x\cos\theta + y\sin\theta. \tag{7.36}$$

Fig. 7.6 The deviation
$\delta = PQ$ of the point T from
the line L

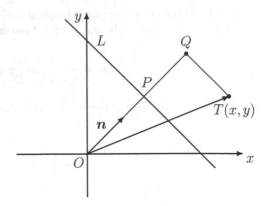

Correlating the obtained formulae (7.35) and (7.36), we obtain

$$\delta = x\cos\theta + y\sin\theta - p. \tag{7.37}$$

This brings us to the following rule: in order to find the deviation δ of the point
$T(x_0, y_0)$ from the line L, we should substitute to the left side of the normal form
of the equation of the line the coordinates x_0 and y_0 of the point T instead of x and
y. The distance from the point T to the line L is equal to the absolute deviation.

Example 7.4 Let us compute the distance from the point $T(5, 4)$ to the line through
$A(1, -2)$ and $B(0, 3)$.

Solution Write the equation of a straight line through the points A and B:

$$\frac{x-1}{-1} = \frac{y+2}{5} \quad \text{or} \quad 5x + y - 3 = 0. \tag{7.38}$$

Having multiplied the resulting equality by $\mu = \dfrac{1}{\sqrt{26}}$, we bring the equation to
the normal form:

$$\frac{5x}{\sqrt{26}} + \frac{y}{\sqrt{26}} - \frac{3}{\sqrt{26}} = 0. \tag{7.39}$$

Then, the distance d from the point T to the line is equal to

$$d = \text{abs}\left(\frac{5\cdot 5}{\sqrt{26}} + \frac{4}{\sqrt{26}} - \frac{3}{\sqrt{26}}\right) = \frac{26}{\sqrt{26}} = \sqrt{26}. \tag{7.40}$$

□

7.8 Line Segments

Line segment M_1M_2 is a part of a straight line between its two points $M_1(x_1, y_1)$ and $M_2(x_2, y_2)$. The points $M_1(x_1, y_1)$ and $M_2(x_2, y_2)$ are the **endpoints** of the segment [8].

A set of points belonging to the segment is specified as follows:

$$M_1M_2 = \{(x, y): x = (1-t)x_1 + tx_2, \ y = (1-t)y_1 + ty_2, t \in [0, 1]\}. \quad (7.41)$$

There exists an equivalent notation of M_1M_2:

$$M_1M_2 = \{(x, y): x = x_1 + (x_2 - x_1)t, \ y = y_1 + (y_2 - y_1)t, t \in [0, 1]\}. \quad (7.42)$$

The variable $0 \leqslant t \leqslant 1$ in the formulas (7.41) and (7.42) is referred to as the **parameter** of the segment.

Example 7.5 Let us write a program in Python that determines, by the segment endpoint coordinates, in which coordinate quadrants it is located. For example, the segment L_1L_2 that connects the points $L_1(-1, -2)$ and $L_2(4, 1)$, lies in the I, III and IV quadrants. Another example: the segment that connects the points $M_1(-1, -2)$ and $M_2(-1, -2)$ entirely belongs to the II quadrant (see Fig. 7.7, which shows the segments L_1L_2 and M_1M_2 with the numbers of each quadrant).

Solution In order to represent a Cartesian plane point in the computer memory, introduce a class `Point`, which contains two fields: `x` and `y`, the abscissa and

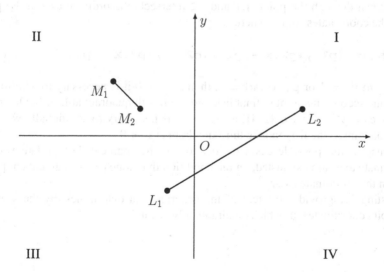

Fig. 7.7 To the Example 7.5. The segment L_1L_2 lies in the I, III and IV quadrants, while the segment M_1M_2 entirely belongs to the II quadrant

the ordinate of the point. Thus, the segment is determined by the boundary points; denote them by P_1 and P_2.

The main computing work is performed by the function get_quadrants (p1, p2). It returns the list that contains the numbers of the quadrants where the segment $P_1 P_2$ is located.

The auxiliary function get_quadrant (p) is used to determine the number of the quadrant to which the only argument belongs, namely the point p. This function returns an integer number from the set {0, 1, 2, 3, 4}. The variable get_quadrant (p) is equal to zero if and only if p lies on the Ox or Oy axis and therefore does not belong to any of the plane quadrants.

Execution of the function get_quadrant (p1, p2) begins with checking whether the points p1 and p2 lie in the adjacent quadrants, i.e. those that form the unordered pairs 1–2, 2–3, 3–4, 4–1. During this check, the variables p1_quad and p2_quad will be assigned the numbers of the quarters to which points p1 and p1 belong, respectively.

Since the numbers of the adjacent quadrants differ by one modulo two, the False value of the boolean variable

is_adjacent = abs(p1_quad - p2_quad) % 2 == 1

is a sufficient condition of adjacency.

Then, the following operations are executed. If the points p1 and p2 lie in the adjacent quadrants, then into the final list are written the values p1_quad and p2_quad, following which the function get_quadrants () terminates.

Otherwise, the equality of the numbers p1_quad and p2_quad is checked. If it is valid, then the entire segment lies in the quadrant number p1_quad, and the function get_quadrants () terminates.

The last case of the opposite quadrants remains, i.e. of the pairs 1–3 or 2–4. The line drawn through the points p1 and p2 intersects the ordinate axis at the point with the coordinates (0, b), where

$$b = (p1.y*p2.x - p1.x*p2.y) / (p2.x - p1.x). \qquad (7.43)$$

If p1_quad = 1 or p1_quad = 3, then at b > 0 it is necessary to additionally write the second quadrant to final list, and the fourth quadrant at b < 0. Otherwise (in the case p1_quad $\in$ {2, 4}) at b > 0, it is necessary to additionally write the first quadrant to the list, and the third quadrant at b < 0.

Thus, all the possible cases of location of the segment $P_1 P_2$ relative to the coordinate axes are exhausted, on the condition that none of the segment endpoints lies on the coordinate axis.

Listing 7.1 provides the text of the program that determines, by the segment endpoint coordinates, in which quadrants it is located.

Listing 7.1

```
1  class Point:
2      def __init__(self, x, y):
3          self.x = x
4          self.y = y
5
6
7  def get_quadrant(p):
8      if p.x > 0 and p.y > 0:
9          return 1
10     elif p.x < 0 and p.y > 0:
11         return 2
12     elif p.x < 0 and p.y < 0:
13         return 3
14     elif p.x > 0 and p.y < 0:
15         return 4
16     else:
17         return 0
18
19
20 def get_quadrants(p1, p2):
21     p1_quad = get_quadrant(p1)
22     p2_quad = get_quadrant(p2)
23
24     is_adjacent = abs(p1_quad - p2_quad) % 2 == 1
25
26     if is_adjacent:
27         return [p1_quad, p2_quad]
28     elif p1_quad == p2_quad:
29         return [p1_quad]
30     else:
31         b = (p1.y * p2.x - p1.x * p2.y) \
32             / (p2.x - p1.x)
33
34         if b == 0:
35             return [p1_quad, p2_quad]
36         elif p1_quad == 1 or p2_quad == 3:
37             quadrant = 2 if b > 0 else 4
38             return [p1_quad, p2_quad, quadrant]
39         else:
40             quadrant = 1 if b > 0 else 3
41             return [p1_quad, p2_quad, quadrant]
```

The most general case is when P_1 or P_2 can belong to the coordinate axes and is discussed in Problem **7.30**. The solution of this problem includes the function `get_quadrants_general()`, which is free from the mentioned constraint.

□

Review Questions

1. How is the slope of a straight line on a plane determined?
2. Write the equation of a straight line with a slope.
3. What is the form of the general equation of a straight line on a plane?
4. What is the normal to the line?
5. What does the equation with a slope for a straight line through a specified point look like?
6. Write the equation of a straight line through two specified points.
7. How is the angle between two lines computed?
8. Write intercept form of the equation of a straight line.
9. Define the deviation of an arbitrary point from a given line.
10. For solution of what problem is it convenient to use the normal form of the equation of a line?
11. How can the set of points of the segment $M_1 M_2$ be specified with the help of a parameter?

Problems

7.1. Find the intersection point of the lines $2x - 3y + 4 = 0$ and $4x + y - 6 = 0$.

7.2. The sides of the triangle lie on the lines $5x - y + 12 = 0$, $x + y + 3 = 0$ and $4x + 3y - 6 = 0$. Find the coordinates of the vertices of this triangle.

7.3. The coordinates of the vertices of a triangle are $(5, -4)$, $(6, -6)$ and $(-15, 4)$. Find the equations of its sides.

7.4. Show that the area of a triangle with the vertices (x_1, y_1), (x_2, y_2) and (x_3, y_3) is equal to

$$S = \frac{1}{2} \text{abs} \begin{vmatrix} x_1 & y_1 & 1 \\ x_2 & y_2 & 1 \\ x_3 & y_3 & 1 \end{vmatrix}. \tag{7.44}$$

7.5. The sides of the triangle lie on the lines $x + y + 1 = 0$, $x + 2y - 3 = 0$ and $4x - 3y - 2 = 0$. Compute the area of this triangle.

7.6. Median of a triangle is the segment that connects its vertex with the midpoint of the opposite side. Set up the equations of the lines on which the medians of the triangle ABC lie, if $A(1, 2)$, $B(4, -3)$, $C(6, 6)$.

7.7. Compute the distance from the point $T(1, 7)$ to the line through $A(-3, -20)$ and $B(4, 17)$.

7.8. Compute the distance from the origin of coordinates to the line given by the equation $\dfrac{x - x_0}{x_0} + \dfrac{y - y_0}{y_0} = 0$, where x_0, y_0 are real numbers not equal to zero.

7.9. One of the sides of the square lies on the line $x + 3y + 10 = 0$. Find the area of this square if the coordinates of one of its vertices are $(-4, -4)$.

7.10. At what point do the lines specified by the equations $x/a + y/b = 1$ and $x/b + y/a = 1$, where $a, b \neq 0$, intersect?

7.11. Find the angle between the lines $3x + 5y - 10 = 0$ and $-2x + y + 4 = 0$.

7.12. Find the values of the parameters λ and μ at which the lines $\lambda x + 6y - 2 = 0$ and $2x + 3y - \mu = 0$:

(1) have exactly one common point,
(2) coincide,
(3) are parallel.

7.13. Compute the distance between the parallel lines specified by the equations $Ax + By + C = 0$ and $Ax + By + C' = 0$, where $C \neq C'$.

*7.14. On what condition do the lines $A_1 x + B_1 y + C_1 = 0$, $A_2 x + B_2 y + C_2 = 0$, $\ldots$, $A_n x + B_n y + C_n = 0$ intersect at one point?

7.15. The line L passes through the point $T(x_0, y_0)$ at the angle α the abscissa axis. Write the equation of the line L^* that passes through the same point T_0 at the angle $\Delta\alpha$ to the line L.

*7.16. The sides of a triangle are specified by the equation $A_i x + B_i y + C_i = 0$, where $i = 1, 2, 3$. Find the equation of

(a) median,
(b) altitude,
(c) bisector,

drawn to the third side.

7.17. Compute the area of a triangle intercepted by the line $Ax + By + C = 0$ from the quadrantal angle.

7.18. Find the equation of the line that passes through the point $T(x_0, y_0)$ and intercepts from the quadrantal angle a triangle with an area equal to S. The variables x_0 and y_0 are positive.

7.19. Assume that some line passing through the point $T(x_0, y_0)$ intercepts from the quadrantal angle a right triangle. What is the least area of this triangle? The variables x_0 and y_0 are positive.

∗7.20. The sides of a triangle are specified by the equation $\alpha_i x + \beta_i y + \gamma_i = 0$, where $i = 1, 2, 3$. Show that the area of this triangle can be calculated by the formula:

$$S = \frac{1}{2} \frac{\Delta^2}{\operatorname{abs}(\Delta_1 \Delta_2 \Delta_3)},$$
                                                                                                                    (7.45)

where $\Delta = \begin{vmatrix} \alpha_1 & \beta_1 & \gamma_1 \\ \alpha_2 & \beta_2 & \gamma_2 \\ \alpha_3 & \beta_3 & \gamma_3 \end{vmatrix}$, Δ_i are the cofactors of the element γ_i, $i \in \{1, 2, 3\}$.

7.21. Prove that the points $T_1(-2, -8)$, $T_2(18, 2)$ and $T_3(3, -11/2)$ lie on the same line.

7.22. For what values of the real parameter a do the points $T_1(0, 1)$, $T_2(a, 2)$ and $T_3(3, a)$ lie on the same line?

7.23. The coordinates of a triangle are given: $A(1, -1)$, $B(2, 4)$, $C(-8, -1)$. Set up the equation of the line that passes through the vertex A parallel to the side BC.

7.24. The coordinates of a triangle are given: $A(-2, 0)$, $B(2, 3)$, $C(-1, -1)$. Set up the equation of the line that passes through the vertex B parallel to the side AC.

7.25. It is known about the point N that it lies on the ordinate axis, and the distance from this point to $N'(-2, -5/2)$ is equal to $d = 2\sqrt{2}$. Find the coordinates of the point N.

7.26. It is known that the area of the triangle is equal to $S = 6$ and its two vertices have the coordinates $(1, 1)$ and $(-2, -3)$. Find the coordinates of the third vertex of the triangle if this vertex lies on the abscissa axis.

7.27. It is known that the area of the triangle is equal to $S = 10$ and its two vertices have the coordinates $(-2, 3)$ and $(-7, -1)$. Find the coordinates of the third vertex of the triangle if this vertex lies on the ordinate axis.

7.28. Find the projection of the point $(2, -13)$ on the line that passes through the points $(0, 2)$ and $(2, -8)$.

7.29. Find the projection of the point (a, a) on the line that passes through the points $(1, 2a)$ and $(2, 3a)$, if a is an arbitrary real number.

7.30. Write a program in Python that determines, by the segment endpoint coordinates, in which coordinate quadrants it is located. In contrast to the solution of the Example 7.5 on page 287, consider the full set of possible cases, including the one when the segment endpoints can belong to the coordinate axes.

Answers and Solutions

7.1 *Answer:* $(1, 2)$.

7.2 *Answer:* $(-5/2, -1/2)$, $(-30/19, 78/19)$, $(15, -18)$.

7.3 *Solution.*
Let us use the equation of a straight line through two given points (7.12):

$$\frac{y - y_1}{y_2 - y_1} = \frac{x - x_1}{x_2 - x_1}.$$

Substituting the coordinates of the points from the problem statement, we obtain the equation of the triangle sides:

$$2x + y - 6 = 0, \quad 2x + 5y + 10 = 0, \quad 10x + 21y + 66 = 0.$$

7.4 *Solution.*
Denote the triangle vertices (x_1, y_1), (x_2, y_2) and (x_3, y_3) by A_1, A_2 and A_3, respectively.

As is well known (see page 262), the area of an arbitrary triangle can be represented as half of the modulus of the vector product $\overrightarrow{A_1 A_2} \times \overrightarrow{A_1 A_3}$:

$$S = \frac{1}{2} |\overrightarrow{A_1 A_2} \times \overrightarrow{A_1 A_3}| = \frac{1}{2} \begin{vmatrix} x_2 - x_1 & y_2 - y_1 \\ x_3 - x_1 & y_3 - y_1 \end{vmatrix}.$$

This expression can be rewritten in the equivalent form:

$$S = \frac{1}{2} \text{abs} \begin{vmatrix} x_1 & y_1 & 1 \\ x_2 & y_2 & 1 \\ x_3 & y_3 & 1 \end{vmatrix}.$$

Thus, the formula (7.44) is proved. It implies that *the necessary and sufficient condition of the three points belonging to one line is that the respective third-order determinant is equal to zero.*

7.5 *Solution.*
Solving the systems of equations:

$$\begin{cases} x + y + 1 = 0, \\ x + 2y - 3 = 0; \end{cases} \quad \begin{cases} x + y + 1 = 0, \\ 4x - 3y - 2 = 0; \end{cases} \quad \begin{cases} x + 2y - 3 = 0, \\ 4x - 3y - 2 = 0, \end{cases}$$

we find the coordinates of the vertices of the triangle: $(-5, 4)$, $(-1/7, -6/7)$, $(13/11, 10/11)$.

As is shown in Problem **7.4**, the area of an arbitrary triangle $A_1 A_2 A_3$, whose vertices have the coordinates (x_1, y_1), (x_2, y_2) and (x_3, y_3), respectively, is equal to

$$S = \frac{1}{2}\text{abs} \begin{vmatrix} x_1 & y_1 & 1 \\ x_2 & y_2 & 1 \\ x_3 & y_3 & 1 \end{vmatrix}.$$

In our case,

$$S = \frac{1}{2}\text{abs} \begin{vmatrix} -5 & 4 & 1 \\ -1/7 & -6/7 & 1 \\ 13/11 & 10/11 & 1 \end{vmatrix} = \frac{1}{2}\text{abs}\left(\frac{1156}{77}\right) = \frac{578}{77}.$$

Note. See Problem **7.20** for a general solution.

7.6 *Solution.*

The abscissa and the ordinate of the midpoint of the segment with the endpoints (x_1, y_1) and (x_2, y_2) are determined by the formulae $x_m = (x_1 + x_2)/2$ and $x_m = (y_1 + y_2)/2$, respectively.

Find the midpoints of the sides: $(5/2, -1/2)$, $(7/2, 4)$, $(5, 3/2)$.

Then, we apply the formula (7.12) and obtain the following equations of the medians of the triangle ABC:

$$x + 8y - 17 = 0, \quad 14x + y - 53 = 0, \quad 13x - 7y - 36 = 0.$$

7.7 *Solution.*

The general equation of a straight line through $(-3, -20)$ and $(4, 17)$ has the form $37x - 7y - 29 = 0$.

We obtain the normal equation of this line. The normalizing multiplier (7.32) is equal to

$$\mu = -\frac{1}{\sqrt{37^2 + 7^2}}\text{sgn}(-29) = \frac{1}{\sqrt{1418}}.$$

Thus, the equation in normal form is

$$\frac{37}{\sqrt{1418}}x - \frac{7}{\sqrt{1418}}y - \frac{29}{\sqrt{1418}} = 0,$$

and the deviation of the point $T(1, 7)$ from the line is equal to

$$\delta = \frac{37}{\sqrt{1418}} \cdot 1 - \frac{7}{\sqrt{1418}} \cdot 7 - \frac{29}{\sqrt{1418}} = -\frac{41}{\sqrt{1418}}.$$

Therefore, the sought distance is equal to $\dfrac{41}{\sqrt{1418}}$.

7.8 *Solution.*

Bring the equation of a line to normal form.

The normalizing multiplier is equal to

$$\mu = -\frac{1}{\sqrt{x_0^2 + y_0^2}} \operatorname{sgn}(-2x_0 y_0) = \frac{1}{\sqrt{x_0^2 + y_0^2}} \operatorname{sgn}(x_0 y_0),$$

and, therefore, the equation of a line can be written in the form:

$$\left(\frac{x_0}{\sqrt{x_0^2 + y_0^2}} x + \frac{y_0}{\sqrt{x_0^2 + y_0^2}} y - 2 \frac{x_0 y_0}{\sqrt{x_0^2 + y_0^2}} \right) \operatorname{sgn}(x_0 y_0) = 0.$$

Compute the deviation from the origin of coordinates $(0, 0)$:

$$\delta = -2 \frac{x_0 y_0}{\sqrt{x_0^2 + y_0^2}} \operatorname{sgn}(x_0 y_0).$$

The distance from the origin of coordinates to the line is equal to the absolute value of the deviation: $\dfrac{2 \operatorname{abs}(x_0 y_0)}{\sqrt{x_0^2 + y_0^2}}$.

7.9 *Solution.*

The deviation from the point $(-4, -4)$ to the line $-\dfrac{1}{\sqrt{10}} x - \dfrac{3}{\sqrt{10}} y - \dfrac{10}{\sqrt{10}} = 0$

is equal to $\delta = -3\sqrt{\dfrac{2}{5}}$.

Hence, the distance from the point to this line is equal to $d = 3\sqrt{\dfrac{2}{5}}$; it coincides with the length of the side of the square.

The area of the square is $S = d^2 = 18/5$.

7.10 *Solution.*

The system of equations:

$$\begin{cases} x/a + y/b = 1, \\ x/b + y/a = 1, \end{cases}$$

for $a, b \neq 0$ has the unique solution $x = y = ab/(a + b)$.

Therefore, the lines intersect at the point $\left(\dfrac{ab}{a + b}, \dfrac{ab}{a + b} \right)$.

7.11 *Solution.*

Take the formula (7.19) in order to find the tangent of the angle between the lines specified in the general form. Substitute into it the values $A_1 = 3$, $B_1 = 5$, $A_2 = -2$ and $B_2 = 1$, and we obtain

$$\tan \alpha = \frac{A_2 B_1 - A_1 B_2}{A_1 A_2 + B_1 B_2} = 13.$$

Therefore, the angle between the lines is $\alpha = \arctan 13$.

7.12 *Solution.*

Consider the system of equations:

$$\begin{cases} \lambda x + 6y = 2, \\ 2x + 3y = \mu. \end{cases}$$

Using the bordering minor method (see page 64), we find the ranks of the system matrix and the augmented matrix:

$$\mathrm{rk} \begin{bmatrix} \lambda & 6 \\ 2 & 3 \end{bmatrix} = \begin{cases} 1, & \text{if } \lambda = 4, \\ 2, & \text{if } \lambda \neq 4; \end{cases}$$

$$\mathrm{rk} \begin{bmatrix} \lambda & 6 & 2 \\ 2 & 3 & \mu \end{bmatrix} = \mathrm{rk} \begin{bmatrix} 1 & 6 & 2 \\ 0 & 0 & \mu - 4 \end{bmatrix} = \begin{cases} 1, & \text{if } \mu = 1, \\ 2, & \text{if } \mu \neq 1. \end{cases}$$

Therefore:

(a) at $\lambda \neq 4$, $\mu \neq 1$, the system has the unique solution, and the lines intersect exactly at one point;

(b) at $\lambda = 4$, $\mu = 1$, the system has an infinite set of solutions, and the lines coincide;

(c) at $\lambda = 4$, $\mu \neq 1$, the system has no solution, and the lines are parallel.

7.13 *Solution.*
Find the deviation of each line from the origin of coordinates:

$$\delta_1 = -\frac{1}{\sqrt{A^2 + B^2}} C\operatorname{sgn}(C), \quad \delta_2 = -\frac{1}{\sqrt{A^2 + B^2}} C'\operatorname{sgn}(C').$$

Then, as is easy to see, the distance d between the parallel planes is equal to the absolute value of the difference of the deviations:

$$d = \operatorname{abs}(\delta_1 - \delta_2) = \operatorname{abs}\left(-\frac{1}{\sqrt{A^2 + B^2}} C\operatorname{sgn}(C) + \frac{1}{\sqrt{A^2 + B^2}} C'\operatorname{sgn}(C') \right)$$

$$= \frac{\operatorname{abs}(C' - C)}{\sqrt{A^2 + B^2}}.$$

7.14 *Answer:* the criterion of intersection of n lines at one point is the equality of the ranks of the two matrices:

$$\begin{bmatrix} A_1 & B_1 \\ A_2 & B_2 \\ \cdots \cdots \\ A_n & B_n \end{bmatrix} \quad \text{and} \quad \begin{bmatrix} A_1 & B_1 & C_1 \\ A_2 & B_2 & C_2 \\ \cdots \cdots \cdots \\ A_n & B_n & C_n \end{bmatrix}.$$

7.15 *Solution.*
The equation of the line L^* that passes through the point $T(x_0, y_0)$ has the form $y - y_0 = k(x - x_0)$. In this equation, k is the unknown slope.

In order to find the value of k, let us use its property: $k = \tan \alpha^*$, where α^* is the angle of inclination of L^* relative to the abscissa axis.

Since L^* passes at the angle $\Delta\alpha$ to the line L, then the two options arise: $\alpha^* = \alpha - \Delta\alpha$ and $\alpha^* = \alpha + \Delta\alpha$. Write these equations in the form of a common equality:

$$\alpha^* = \alpha \pm \Delta\alpha.$$

By the formulae of tangent of sum and difference of two angles (B.13) and (B.14) on page 412, we have

$$k = \tan(\alpha \pm \Delta\alpha) = \frac{\tan\alpha \pm \tan\Delta\alpha}{1 \mp \tan\Delta\alpha \tan\alpha}.$$

We finally obtain the equation of the line L^*:

$$y - y_0 = \frac{\tan\alpha \pm \tan\Delta\alpha}{1 \mp \tan\Delta\alpha \tan\alpha}(x - x_0).$$

7.16 *Answer:*

(a) $(A_1x + B_1y + C_1) \begin{vmatrix} A_2 & B_2 \\ A_3 & B_3 \end{vmatrix} = (A_2x + B_2y + C_2) \begin{vmatrix} A_3 & B_3 \\ A_1 & B_1 \end{vmatrix}$;

(b) $(A_1x + B_1y + C_1)(A_2A_3 + B_2B_3) = (A_2x + B_2y + C_2)(A_1A_3 + B_1B_3)$;

(c) $\dfrac{A_1x + B_1y + C_1}{\sqrt{A_1^2 + B_1^2}} = -s\dfrac{A_2x + B_2y + C_2}{\sqrt{A_2^2 + B_2^2}}$,

where $s = \text{sgn}\left(\begin{vmatrix} A_1 & B_1 \\ A_3 & B_3 \end{vmatrix} \begin{vmatrix} A_2 & B_2 \\ A_3 & B_3 \end{vmatrix} \right)$.

7.17 *Solution.*

Denote the intersection points of the line $Ax + By + C = 0$ with the coordinate axes by $P(x_0, 0)$ and $Q(0, y_0)$, where $x_0 = -\dfrac{C}{A}$, $y_0 = -\dfrac{C}{B}$. The coordinates x_0 and y_0 in the absolute value are equal to the lengths of the cathetuses of the right triangle POQ lying on the axes Ox and Oy, respectively. The area of this triangle is equal to the half of the product of the cathetuses:

$$S = \frac{1}{2}x_0y_0 = \frac{1}{2}\left(-\frac{C}{A}\right)\left(-\frac{C}{B}\right) = \frac{C^2}{2AB}.$$

7.18 *Solution.*

Write the equation of the line that passes through the point (x_0, y_0): $y - y_0 = k(x - x_0)$, where $k < \infty$ is the slope.

The case $k \to \infty$ does not require separate consideration, since in that case the line will not intercept the triangle from the quadrantal angle.

Find the coordinates of the points of intersection of the line with the coordinate axes:

with the axis Ox: $y = 0$, $-y_0 = k(x^* - x_0)$, $x^* = x_0 - \dfrac{y_0}{k}$, where $\text{abs}(x^*)$ is the length of the cathetus lying on the axis Ox;

with the axis Oy: $x = 0$, $y^* - y_0 = -kx_0$, $y^* = y_0 - kx_0$, where $\text{abs}(y^*)$ is the length of the cathetus lying on the axis Oy.

The area of such a triangle is equal to the half of the product of the cathetuses:

$$S = \frac{1}{2}\text{abs}(x^*y^*), \; S = \frac{1}{2}\left(x_0 - \frac{y_0}{k}\right)(y_0 - kx_0) = \frac{1}{2}\text{abs}\left(\frac{(y_0 - kx_0)^2}{-k}\right).$$

Let us express the variable k:

$k^2x_0^2 - 2k(x_0y_0 - S) + y_0^2 = 0$. The solution of this square equation leads to two possible values of k:

$$k_{1,2} = \frac{x_0y_0 - S \pm \sqrt{S(S - 2x_0y_0)}}{x_0^2}.$$

We obtain the sought equation of the line:

$$y - y_0 = \left(\frac{x_0 y_0 - S \pm \sqrt{S(S - 2x_0 y_0)}}{x_0^2}\right)(x - x_0).$$

7.19 *Solution.*

As is shown in the previous problem, the area of the triangle S depends on the slope of the line k as follows: $S(k) = \frac{1}{2}\text{abs}\left(\frac{(y_0 - kx_0)^2}{-k}\right)$.

For the points $T(x_0, y_0)$, whose coordinates x_0 and y_0 are positive, the equality $k < 0$ is fulfilled, i.e. in this case the line forms an obtuse angle with the axis Ox.

In order to find the minimal value of the function, let us compute the points at which the derivative $\dfrac{d\,S(k)}{d\,k} = \dfrac{y_0^2 - k^2 x_0^2}{k^2}$ is equal to zero.

The condition $k < 0$ is satisfied by the value $k^* = -\dfrac{y_0}{x_0}$. The point k^* is the minimum point, since the second derivative is $\dfrac{d^2\,S(k^*)}{d\,k} = -\dfrac{2y_0^2}{(k^*)^3} > 0$.

Thus, the least value of the area of the triangle is equal to

$$S_{\min} = S(k^*) = 2x_0 y_0.$$

7.20 *Solution.*

Compute the coordinates of the intersection points of the line pairs. Taking into account that the area of the triangle is expressed in the form of a half of the modulus of the vector form of the vectors that from both sides, we obtain

$$S = \frac{\beta_1\gamma_2 - \beta_2\gamma_1}{\alpha_1\beta_2 - \alpha_2\beta_1}\left(\frac{\alpha_2\gamma_3 - \alpha_3\gamma_2}{\alpha_3\beta_2 - \alpha_2\beta_3} - \frac{\alpha_3\gamma_1 - \alpha_1\gamma_3}{\alpha_1\beta_3 - \alpha_3\beta_1}\right)$$
$$+ \frac{\beta_2\gamma_3 - \beta_3\gamma_2}{\alpha_2\beta_3 - \alpha_3\beta_2}\left(\frac{\alpha_3\gamma_1 - \alpha_1\gamma_3}{\alpha_1\beta_3 - \alpha_3\beta_1} - \frac{\alpha_1\gamma_2 - \alpha_2\gamma_1}{\alpha_2\beta_1 - \alpha_1\beta_2}\right)$$
$$+ \frac{\beta_3\gamma_1 - \beta_1\gamma_3}{\alpha_3\beta_1 - \alpha_1\beta_3}\left(\frac{\alpha_1\gamma_2 - \alpha_2\gamma_1}{\alpha_2\beta_1 - \alpha_1\beta_2} - \frac{\alpha_2\gamma_3 - \alpha_3\gamma_2}{\alpha_3\beta_2 - \alpha_2\beta_3}\right).$$

Reduce the fractions to a common denominator. After simple but somewhat cumbersome algebraic transformations, we arrive at the formula:

$$S = \frac{(\alpha_1(\beta_2\gamma_3 - \beta_3\gamma_2) + \alpha_2(\beta_3\gamma_1 - \beta_1\gamma_3) + \alpha_3(\beta_1\gamma_2 - \beta_2\gamma_1))^2}{(\alpha_1\beta_2 - \alpha_2\beta_1)(\alpha_2\beta_3 - \alpha_3\beta_2)(\alpha_3\beta_1 - \alpha_1\beta_3)}.$$

As is easy to see, the obtained expression can be presented as

$$S = \frac{1}{2} \frac{\Delta^2}{\mathrm{abs}(\Delta_1 \Delta_2 \Delta_3)},$$

where $\Delta = \begin{vmatrix} \alpha_1 & \beta_1 & \gamma_1 \\ \alpha_2 & \beta_2 & \gamma_2 \\ \alpha_3 & \beta_3 & \gamma_3 \end{vmatrix}$, Δ_i are the cofactors of the elements γ_i, $i \in \{1, 2, 3\}$.

7.21 *Solution.*

Let us draw a line through the points T_1 and T_2 (see the formula (7.12)). The slope of this line is $k = 1/2$. Then, the line $T_1 T_3$ has the slope $k' = 1/2$. Since $k = k'$, then the points T_1, T_2 and T_3 lie on the same line.

7.22 *Solution.*

The first method

Let the equation of the line passing through the three points T_1, T_2 and T_3 have the form $y = kx + b$. Having substituted the coordinates of each of these points into the equation of the line, we obtain the system of relatively unknown variables k and b:

$$\begin{cases} b = 1, \\ ak + b = 2, \\ 3k + b = a. \end{cases}$$

The condition of definiteness of this system (i.e. the uniqueness of its solution) leads to the square equation $a^2 - a - 3 = 0$.

Therefore, the points T_1, T_2 and T_3 lie on the same line at the two values of the parameter a: $a_1 = \frac{1}{2}(1 + \sqrt{13})$ and $a_2 = \frac{1}{2}(1 - \sqrt{13})$.

The second method

Let us use the consequence of the formula (7.44) from Problem **7.4**: the criterion of the three points belonging to one line is that the third-order determinant is equal to zero:

$$\begin{vmatrix} 0 & 1 & 1 \\ a & 2 & 1 \\ 3 & a & 1 \end{vmatrix} = 0.$$

The obtained equation has two roots: $a_{1,2} = \frac{1}{2}(1 \pm \sqrt{13})$.

Note. The necessary and sufficient condition for the three points T_1, T_2 and T_3 to lie on the same line is collinearity of the vectors $\overrightarrow{T_1 T_2}$ and $\overrightarrow{T_1 T_3}$. In view of this, there exists one more method of solving this problem, which is based on checking whether the vector product $\overrightarrow{T_1 T_2} \times \overrightarrow{T_1 T_3}$ is equal to zero. Of course, different solution methods result in the same answer.

7.23 *Solution.*

The slope of the sought line coincides with the slope of the line BC. The equation of the line BC has the form $y = \frac{1}{2}x + 3$ (see the formula (7.12)). Then, use the formula (7.10) for the line that passes through a given point:

$$y - y_1 = k(x - x_1), \tag{7.46}$$

where $k = \frac{1}{2}$, $x_1 = 1$, $y_1 = -1$.

Substituting the numeric values, we obtain the equation of the line that passes through the vertex A parallel to the side BC: $x - 2y - 3 = 0$.

7.24 *Answer:* $x + y - 5 = 0$.

7.25 *Solution.*

Since the point N lies on the ordinate axis, its coordinates are equal to $(0, y)$, where y is an unknown value. The distance d between the points N and N' is equal to

$$d = \sqrt{(0 - (-2))^2 + (y - (-5/2))^2} = \sqrt{4 + (y + 5/2)^2}.$$

Thus, we obtain the equation relative to the variable y:

$$\sqrt{(0 - (-2))^2 + (y - (-5/2))^2} = 2\sqrt{2},$$

which has the solutions $y_1 = -1/2$ and $y_2 = -9/2$. Therefore, the coordinates of the point N are equal to $(0, -1/2)$ or $(0, -9/2)$.

7.26 *Solution.*

By the triangle area formula expressed by the coordinates of its vertices $(1, 1)$, $(-2, -3)$ and $(x, 0)$ (see (7.44) on page 290), we obtain $S = \text{abs}(4x - 1)$.

According to the problem statement, $S = 6$, hence $\text{abs}(4x - 1) = 13/4$.

Therefore, the coordinates of the vertices of the triangle are equal to $(-5/4, 0)$ or $(7/4, 0)$.

7.27 *Solution.*

The coordinates of the third vertex are $(0, y)$.

The area of the triangle is equal to

$$S = \frac{1}{2}\text{abs}\begin{vmatrix} -2 & 3 & 1 \\ -7 & -1 & 1 \\ 0 & y & 1 \end{vmatrix} = \frac{1}{2}\text{abs}(23 - 5y) = 10,$$

hence $y = 13/5$ or $y = 33/5$.

The coordinates of the third vertex of the triangle are equal to $(0, 13/5)$ or $(0, 33/5)$.

7.28 *Solution.*

Let $A(0, 2)$, $B(2, -8)$. The sought projection of the point is $K(2, -13)$.

The projection of the point is the foot of the perpendicular to the line AB. Then, the orthogonality property for the line AB and the perpendicular from the point K: $k_1 \cdot k_2 = -1$ is fulfilled.

Write the system of equations relative to the coefficients k_1 and b_1 for the line AB:

$$\begin{cases} 2 = b_1, \\ 8 = 2k_1 + b_1, \end{cases}$$

From this system, we obtain that $b_1 = 2$, $k_1 = -5$. Then, $k_2 = 1/5$.

Substitute k_2 into the equation of the line for the perpendicular from the point K: $-13 = 2 \cdot 1/5 + b_2$, hence $b_2 = -67/5$.

Compute the coordinates of the point (x_0, y_0), at which the lines intersect: $1/5 x_0 - 67/5 = -5x_0 + 2$, $x_0 = 77/26$, $y_0 = -333/26$.

7.29 *Solution.*

Let $A(1, 2a)$, $B(2, 3a)$, $K(a, a)$.

Projection of a point is a foot of a perpendicular to the line AB. Then, the equality is fulfilled for the line AB and the perpendicular from the point K: $k_1 \cdot k_2 = -1$.

Write the system of equations relative to the coefficients k_1 and b_1 for the line AB:

$$\begin{cases} 2a = k_1 + b_1, \\ 3a = 2k_1 + b_1. \end{cases}$$

From this system, we obtain that $b_1 = a$, $k_1 = a$. Then, $k_2 = -1/a$.

Substitute k_2 into the equation of the line for the perpendicular from the point K: $a = a(-1/a) + b_2$, hence $b_2 = a + 1$.

The coordinate of the projection of the point K satisfies the equation:

$$ax_0 + a = -1/ax_0 + b_2, ax_0 + a = -x_0/a + a + 1, x_0(a + 1/a) = 1, x_0 = \frac{a}{a^2 + 1}.$$

Let us express the ordinate of the projection: $y_0 = -\dfrac{1}{a^2 + 1} + a + 1 = a + \dfrac{a}{a^2 + 1}$.

The sought projection has the coordinates $\left(\dfrac{a}{a^2 + 1}, a + \dfrac{a}{a^2 + 1} \right)$.

7.30 *Solution.*

The suggested solution generally repeats the approach discussed in Example 7.5. Note that in order to study the full set of possible cases of mutual arrangement of the segment endpoints and the coordinate axes, the integer variable `neighbor` is additionally used. It plays an important role for the location of the segment illustrated in Fig. 7.8. Here, one of the segment endpoints, namely P_2, is located on the coordinate axis, and into the variable `neighbor` will be written the number of the quadrant where the segment points from the small neighborhood P_2 lie.

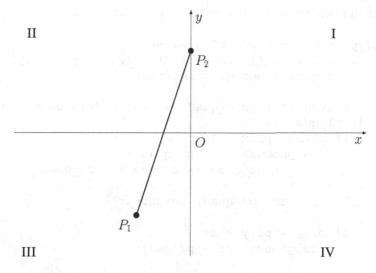

Fig. 7.8 To Problem **7.30**. The segment $P_1 P_2$ lies in the second and third quadrants, get_quadrant(p1) $= 3$, get_quadrant(p2) $= 0$, neighbor $= 2$

After performing all checks, into the answer will be written the numbers of the quadrants within which the segment points fall.

```python
class Point:
    def __init__(self, x, y):
        self.x = x
        self.y = y


def get_quadrant(p):
    if p.x > 0 and p.y > 0:
        return 1
    elif p.x < 0 and p.y > 0:
        return 2
    elif p.x < 0 and p.y < 0:
        return 3
    elif p.x > 0 and p.y < 0:
        return 4
    else:
        return 0


def get_quadrants_general(p1, p2):
    p1_quad = get_quadrant(p1)
    p2_quad = get_quadrant(p2)
```

```python
is_adjacent = abs(p1_quad - p2_quad) % 2 == 1

if p1_quad == 0 and p2_quad == 0:
    mid = Point((p1.x + p2.x) / 2, (p1.y + p2.y) / 2)
    mid_quad = get_quadrant(mid)

    return [] if mid_quad == 0 else [mid_quad]
elif p1_quad == 0:
    if p1.x * p2.x < 0:
        neighbour = 3 - p2_quad \
            if p2_quad <= 2 else 7 - p2_quad

        return [p2_quad, neighbour]

    if p1.y * p2.y < 0:
        neighbour = 5 - p2_quad

        return [p2_quad, neighbour]

    return [p2_quad]
elif p2_quad == 0:
    if p1.x * p2.x < 0:
        neighbour = 3 - p1_quad if p1_quad <= 2 \
            else 7 - p1_quad

        return [p1_quad, neighbour]

    if p1.y * p2.y < 0:
        neighbour = 7 - p1_quad

        return [p1_quad, neighbour]

    return [p1_quad]
elif is_adjacent:
    return [p1_quad, p2_quad]
elif p1_quad == p2_quad:
    return [p1_quad]
else:
    b = (p1.y * p2.x - p1.x * p2.y) \
        / (p2.x - p1.x)

    if b == 0:
        return [p1_quad, p2_quad]
    elif p1_quad == 1 or p2_quad == 3:
        quadrant = 2 if b > 0 else 4
        return [p1_quad, p2_quad, quadrant]
```

```
else:
    quadrant = 1 if b > 0 else 3
    return [p1_quad, p2_quad, quadrant]
```

Chapter 8
Equation of a Plane in Space

8.1 Equation of a Plane That Is Orthogonal to the Specified Vector and Passes Through the Specified Point

Assume that it is known that the plane π is orthogonal to the vector $n = (A, B, C)$ and passes through the point $T_0(x_0, y_0, z_0)$. Take an arbitrary point $T(x, y, z)$ on the plane π. The vector $\overrightarrow{T_0T}$ belongs to the plane π. From the condition of orthogonality of the vector n of the plane π follows that the vector

$$\overrightarrow{T_0T} = (x - x_0, y - y_0, z - z_0) \tag{8.1}$$

is orthogonal to the vector n.

Relying on the property of the scalar product of orthogonal vectors, we can write

$$n \cdot \overrightarrow{T_0T} = 0. \tag{8.2}$$

This equation is called the **vector equation of a plane** [8].

Rewritten in coordinate form

$$A(x - x_0) + B(y - y_0) + C(z - z_0) = 0, \tag{8.3}$$

this equation is called the **equation of a plane that is orthogonal to the vector** $n = (A, B, C)$ **and passes through the point** $T_0(x_0, y_0, z_0)$.

© Springer Nature Switzerland AG 2021
S. Kurgalin, S. Borzunov, *Algebra and Geometry with Python*,
https://doi.org/10.1007/978-3-030-61541-3_8

8.2　General Equation of a Plane

Equation of the first degree

$$Ax + By + Cz + D = 0, \tag{8.4}$$

in which A, B, C and D are arbitrary real constants such that out of the coefficients A, B and C at least one is other than zero, is referred to as the **general equation of a plane**.

A general equation (8.4) is referred to as **complete** if all its coefficients are other than zero. If at least one of the mentioned coefficients is equal to zero, the equation is referred to as **incomplete**.

Consider the possible types of incomplete equations.

(1) If $D = 0$, then the equation $Ax + By + Cz = 0$ determines the plane that passes through the origin of coordinates.

(2) If $A = 0$, then the equation $By + Cz + D = 0$ determines the plane parallel to the axis Ox, since the normal vector of this plane $\boldsymbol{n} = (0, B, C)$ is perpendicular to the axis Ox.

(3) If $B = 0$, then the equation $Ax + Cz + D = 0$ determines the plane, parallel to the axis Oy, since its normal vector $\boldsymbol{n} = (A, 0, C)$ is perpendicular to the axis Oy.

(4) If $C = 0$, then the equation $Ax + By + D = 0$ determines the plane parallel to the axis Oz, since for this axis perpendicular is the normal vector with the coordinates $(A, B, 0)$.

(5) If $A = B = 0$, then the equation $Cz + D = 0$ determines the plane parallel to the coordinate plane xOy.

(6) If $A = 0$ and $C = 0$, then the equation $By + D = 0$ determines the plane, parallel to the coordinate plane xOz.

(7) If $B = 0$ and $C = 0$, then the equation $Ax + D = 0$ determines the plane, parallel to the coordinate plane yOz.

(8) If $A = 0$, $B = 0$ and $D = 0$, then the equation of the plane $Cz = 0$ determines the coordinate plane xOy.

(9) If $A = 0$, $C = 0$ and $D = 0$, then the equation of the plane $By = 0$ determines the coordinate plane xOz.

(10) If $B = 0$, $C = 0$ and $D = 0$, then the equation of the plane $Ax = 0$ determines the coordinate plane yOz.

8.3   Intercept Form of the Equation of a Plane

Consider the general equation of the plane $Ax + By + Cz + D = 0$.

Assume that all the coefficients A, B, C and D are other than zero. Then this equation can be written in the form:

$$\frac{x}{(-D/A)} + \frac{y}{(-D/B)} + \frac{z}{(-D/C)} = 1. \qquad (8.5)$$

Let us introduce the notations: $a = -\dfrac{D}{A}$, $b = -\dfrac{D}{B}$, $c = -\dfrac{D}{C}$. Then Eq. (8.5) will be reduced to the following form:

$$\frac{x}{a} + \frac{y}{b} + \frac{z}{c} = 1. \qquad (8.6)$$

This is the **intercept form of the equation of a plane**.

In Eq. (8.6) the numbers a, b and c have simple geometric meaning: they are equal in absolute value to the lengths of the segments (intercepts) that the plane intercepts on the coordinate axes Ox, Oy and Oz, respectively (see Fig. 8.1). The plane passes through the points $(a, 0, 0)$, $(0, b, 0)$, $(0, 0, c)$.

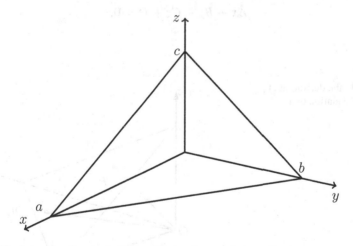

Fig. 8.1 The intercepts that are intercepted by the plane on the coordinate axes

8.4   Normal Equation of a Plane

Consider the plane shown in Fig. 8.2.

Assume that p is the length of the vector $\overrightarrow{OP}$; α, β, γ are the angles between the unit vector $\boldsymbol{n}$ and the coordinate axes; Q is the arbitrary point on the plane with the coordinates (x, y, z).

It is obvious that the projection of the vector $\overrightarrow{OQ}$ on the direction $\boldsymbol{n}$ is equal to p:

$$\text{Pr}_{\boldsymbol{n}}\, \overrightarrow{OQ} = p, \tag{8.7}$$

$$\text{Pr}_{\boldsymbol{n}}\, \overrightarrow{OQ} = \frac{\overrightarrow{OQ} \cdot \boldsymbol{n}}{|\boldsymbol{n}|}, \quad |\boldsymbol{n}| = 1. \tag{8.8}$$

Therefore

$$\overrightarrow{OQ} \cdot \boldsymbol{n} = x \cos\alpha + y \cos\beta + z \cos\gamma = p. \tag{8.9}$$

We have obtained the **normal equation of a plane**. The variables $\cos\alpha$, $\cos\beta$, $\cos\gamma$ are called the **direction cosines** of the vector $\boldsymbol{n}$.

Take the general equation of a plane

$$Ax + By + Cz + D = 0, \tag{8.10}$$

where $D \neq 0$.

Fig. 8.2 To the derivation of the normal equation of a plane

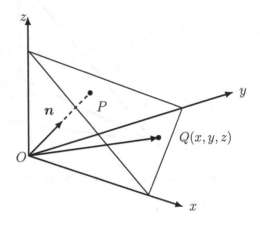

The normalizing factor is calculated by the formula:

$$\mu = -\frac{1}{\sqrt{A^2 + B^2 + C^2}}\mathrm{sgn}(D) \tag{8.11}$$

(cf. the formula (7.32)).

If we multiply (8.10) by the normalizing factor μ, then as a result we obtain the normal equation of the plane:

$$\mu(Ax + By + Cz + D) = 0. \tag{8.12}$$

Note If the condition $D = 0$ is met, then the plane passes through the origin of coordinates, and the normalizing factor can be taken with an arbitrary sign: $\mu = \pm\frac{1}{\sqrt{A^2 + B^2 + C^2}}$.

8.5 Equation of a Plane That Passes Through the Specified Point Parallel to the Two Specified Vectors

Let the plane π be parallel to the vectors $\boldsymbol{a}_1 = (k_1, l_1, m_1)$ and $\boldsymbol{a}_2 = (k_2, l_2, m_2)$ and pass through the point $T_0(x_0, y_0, z_0)$. Further, let $T(x, y, z)$ be an arbitrary point on the plane. Then the vectors $\boldsymbol{a}_1, \boldsymbol{a}_2$ and $\overrightarrow{T_0T}$ are **coplanar**. Therefore,

$$(\overrightarrow{T_0T} \times \boldsymbol{a}_1) \cdot \boldsymbol{a}_2 = 0, \tag{8.13}$$

$$\begin{vmatrix} x - x_0 & y - y_0 & z - z_0 \\ k_1 & l_1 & m_1 \\ k_2 & l_2 & m_2 \end{vmatrix} = 0 \tag{8.14}$$

or

$$(x - x_0)\begin{vmatrix} l_1 & m_1 \\ l_2 & m_2 \end{vmatrix} - (y - y_0)\begin{vmatrix} k_1 & m_1 \\ k_2 & m_2 \end{vmatrix} + (z - z_0)\begin{vmatrix} k_1 & l_1 \\ k_2 & l_2 \end{vmatrix} = 0. \tag{8.15}$$

After expanding the second order determinants and introducing the appropriate notation, we obtain the general equation of the plane.

8.6   Equation of the Plane That Passes Through the Three Specified Points

Given are the following points: $T_1(x_1, y_1, z_1)$, $T_2(x_2, y_2, z_2)$ and $T_3(x_3, y_3, z_3)$. Take an arbitrary point $T(x, y, z)$ and construct the vectors $\overrightarrow{T_1T}$, $\overrightarrow{T_1T_2}$ and $\overrightarrow{T_1T_3}$. They are coplanar, therefore

$$\begin{vmatrix} x - x_1 & y - y_1 & z - z_1 \\ x_2 - x_1 & y_2 - y_1 & z_2 - z_1 \\ x_3 - x_1 & y_3 - y_1 & z_3 - z_1 \end{vmatrix} = 0. \tag{8.16}$$

The obtained equation is the **equation of a plane that passes through the three specified points**.

8.7   Angle Between Two Planes

Consider the planes specified by the equations:

$$A_1x + B_1y + C_1z + D_1 = 0, \tag{8.17}$$

$$A_2x + B_2y + C_2z + D_2 = 0. \tag{8.18}$$

For the given planes, construct the normal vectors $n_1 = (A_1, B_1, C_1)$ and $n_2 = (A_2, B_2, C_2)$. Therefore, the angle ω between the planes will be determined from the relation:

$$\cos \omega = \frac{n_1 \cdot n_2}{|n_1||n_2|} = \frac{A_1A_2 + B_1B_2 + C_1C_2}{\sqrt{A_1^2 + B_1^2 + C_1^2}\sqrt{A_2^2 + B_2^2 + C_2^2}}. \tag{8.19}$$

If $n_1 \cdot n_2 = 0$, then the vectors n_1 and n_2 are orthogonal. Hence,

$A_1A_2 + B_1B_2 + C_1C_2 = 0$ is the **condition of orthogonality of planes**.   (8.20)

But if the equalities $\dfrac{A_1}{A_2} = \dfrac{B_1}{B_2} = \dfrac{C_1}{C_2}$ are valid, then the vectors n_1 and n_2 are collinear, and this will be the **condition of parallelism of planes**.

8.8 Distance from a Point to a Plane

The concept of **deviation of the point** $T_0(x_0, y_0, z_0)$ **from the plane** $Ax + By + Cz + D = 0$ is introduced similarly to the relation (7.37):

$$\delta = -\text{sgn}(D)\frac{Ax_0 + By_0 + Cz_0 + D}{\sqrt{A^2 + B^2 + C^2}}. \tag{8.21}$$

The distance d from the point T_0 to the plane is determined as the absolute value of the deviation $d = \text{abs}(\delta)$.

If the plane is specified in the normal form, then the distance d is determined as

$$d = \text{abs}(\cos \alpha \cdot x_0 + \cos \beta \cdot y_0 + \cos \gamma \cdot z_0 - p), \tag{8.22}$$

where $\cos \alpha$, $\cos \beta$, $\cos \gamma$ are the direction cosines.

8.9 Pencil of Planes

Assume that two planes are specified by the equations:

$$\begin{cases} A_1 x + B_1 y + C_1 z + D_1 = 0, \\ A_2 x + B_2 y + C_2 z + D_2 = 0. \end{cases} \tag{8.23}$$

If these planes are neither parallel nor coincide, then they intersect on some straight line.

It is obvious that for any real constants λ and μ, the plane determined by the equation

$$\lambda(A_1 x + B_1 y + C_1 z + D_1) + \mu(A_2 x + B_2 y + C_2 z + D_2) = 0 \tag{8.24}$$

will also pass through this line, because all the points governed by Eqs. (8.23) satisfy Eq. (8.24) as well. The same equation specifies all the planes that pass through a common line.

A collection of planes passing through the same straight line is called the **pencil of planes**.

Example 8.1 Find the equation of the plane that passes through the point $T(4, -1, 2)$ and a straight line that is the intersection of the planes $x + 3y - z - 5 = 0$ and $-2x + y + z + 4 = 0$.

Solution Write the equation of the pencil of planes and fit λ and μ so that the required plane passes through the point T:

$$\lambda(x + 3y - z - 5) + \mu(-2x + y + z + 4) = 0, \tag{8.25}$$

$$\lambda(4 + 3 \cdot (-1) - 2 - 5) + \mu(-2 \cdot 4 + (-1) + 2 + 4) = 0, \tag{8.26}$$

$$-6\lambda - 3\mu = 0, \tag{8.27}$$

$$2\lambda + \mu = 0. \tag{8.28}$$

For example, take the values $\lambda = 1$ and $\mu = -2$. In this case, we obtain the answer:

$$(x + 3y - z - 5) + (-2)(-2x + y + z + 4) = 0, \text{ or} \tag{8.29}$$

$$5x + y - 3z - 13 = 0. \tag{8.30}$$

$\square$

Review Questions

1. Write the equation of a plane orthogonal to a specified vector and passing through a specified point.
2. Define the general equation of a plane?
3. Write the intercept form of the equation of a plane.
4. For solution of what problem is it convenient to use the normal form of the equation of a plane?
5. How are the directing cosines of the vector $\boldsymbol{a}$ computed?
6. Write the equation of a plane that passes through a specified point parallel to two specified vectors.
7. Write the equation of a plan that passes through three specified points.
8. How is the angle between two planes computed?
9. Define the deviation of an arbitrary point from a given plane.
10. What is pencil of planes?

Problems

8.1. Set up the equation of the plane that passes through the origin of coordinates, if it has the normal vector $\boldsymbol{n} = (1, 2, -3)$.

8.2. Set up the equation of the plane that passes through the point $T(-1, 0, 2)$ and has the normal vector $\boldsymbol{n} = (-3, -2, 0)$.

8.3. Find the equation of the plane that passes through the points $T_1(2, -1, 0)$ and $T_2(-5, 1, 1)$ parallel to the vector $a = (0, -1, -7)$.

8.4. Find the equation of the plane that passes through the point $T_0(2, -1, 0)$ parallel to the vectors $a_1 = (3, 5, -8)$ and $a_2 = (4, 6, -7)$.

8.5. Find the equation of the plane that passes through the point $T_0(1/2, 1/2, 1/2)$ parallel to the vectors $a_1 = (0, 1, -1)$ and $a_2 = (1, 1, 10)$.

8.6. Find the lengths of the segments intercepted by the plane $3x + 4y + 5z - 12 = 0$ on the coordinate axes.

8.7. What is the distance from the point $A(1, 2, 9)$ to the plane $x + y - 2z - 17 = 0$?

8.8. Does the plane $-2x + 2y - z - 1 = 0$ intersect the segment $P_1 P_2$, if the segment endpoint coordinates are as follows: $P_1(-5, -5, -5)$, $P_2(8, 8, 8)$?

8.9. Compute the distance between the parallel planes specified by the equations $Ax + By + Cz + D = 0$ and $Ax + By + Cz + D' = 0$, where $D \neq D'$.

∗8.10. On what condition do the three planes $A_1 x + B_1 y + C_1 z + D_1 = 0$, $A_2 x + B_2 y + C_2 z + D_2 = 0$ and $A_3 x + B_3 y + C_3 z + D_3 = 0$ intersect exactly at one point? Find the coordinates of this point.

8.11. Find the volume of a tetrahedron intercepted by the plane $Ax + By + Cz + D = 0$ from the quadrantal angle.

8.12. On the axis Oz find the points equidistant from the two planes $x - y + z - 10 = 0$ and $x + y - z + 8 = 0$.

∗8.13. Compute the volume of a tetrahedron whose vertices are located at the points with the coordinates (x_1, y_1, z_1), (x_2, y_2, z_2), (x_3, y_3, z_3) and (x_4, y_4, z_4).

Answers and Solutions

8.1 *Solution.*

Use the formula (8.3) that expresses the equation of the plane that passes through the point $A_0(x_0, y_0, z_0)$ perpendicular to the vector $n = (n_1, n_2, n_3)$:

$$n_1(x - x_0) + n_2(y - y_0) + n_3(z - z_0) = 0.$$

Substitute the coordinates of the point $O(0, 0, 0)$: $(x-0)+2(y-0)-3(z-0) = 0$. Hence, the equation of the sought plane has the form $x + 2y - 3z = 0$.

8.2 *Solution.*

Substitute into the formula (8.3) the coordinates of the point $T(-1, 0, 2)$ and the vector $n = (-3, -2, 0)$; we obtain the equation of the plane:

$$(-3)(x - (-1)) + (-2)(y - 0) + 0(z - 2) = 0,$$

or

$$3x + 2y + 3 = 0.$$

8.3 *Solution.*

Select an arbitrary point $T(x, y, z)$ on the sought plane. According to the problem statement, the vectors $\overrightarrow{T_1 T}$, $\overrightarrow{T_1 T_2}$ and a are coplanar, therefore, their scalar triple product is equal to zero:

$$(\overrightarrow{T_1 T}, \overrightarrow{T_1 T_2}, a) = 0.$$

Use the formula (6.33):

$$(\overrightarrow{T_1 T}, \overrightarrow{T_1 T_2}, a) = \begin{vmatrix} x-2 & y+1 & z \\ -7 & 2 & 1 \\ 0 & -1 & 7 \end{vmatrix} = -13x - 49y + 7z - 23 = 0.$$

Then, the equation of the plane that passes through the points T_1 and T_2 parallel to the vector a can be presented in the form $13x + 49y - 7z + 23 = 0$.

8.4 *Solution.*

Consider an arbitrary point $T(x, y, z)$ on the sought plane. The scalar triple product of $(a_1, a_2, \overrightarrow{T_1 T})$ is equal to zero because of coplanarity of these three vectors.

Having computed the scalar triple product, we obtain

$$(a_1, a_2, \overrightarrow{T_0 T}) = \begin{vmatrix} 3 & 5 & -8 \\ 4 & 6 & -7 \\ x-2 & y+1 & z \end{vmatrix} = 13x - 11y - 2z - 37 = 0.$$

Hence, the equation of the plane that passes through the point T_0 parallel to the vectors a_1 and a_2 can be presented in the form $13x - 11y - 2z - 37 = 0$.

8.5 *Solution.*

Using an auxiliary point T with the coordinates $T(x, y, z)$, similarly to the solution of the previous problem, we obtain

$$(a_1, a_2, \overrightarrow{T_0 T}) = 0,$$

$$\begin{vmatrix} 0 & 1 & -1 \\ 1 & 1 & 10 \\ x - \dfrac{1}{2} & y - \dfrac{1}{2} & z - \dfrac{1}{2} \end{vmatrix} = 11x - y - z - \dfrac{9}{2} = 0.$$

As a result, the equation of the plane that passes through the point T_0 parallel to the vectors a_1 and a_2 has the form $22x - 2y - 2z - 9 = 0$.

8.6 *Solution.*

Assume that the general equation of the plane $Ax+By+Cz+D=0$ is specified, and $D \neq 0$. Pass on to the intercept form of the equation of a plane, dividing both sides of the equation by $-D$:

$$-\frac{A}{D}x - \frac{B}{D}y - \frac{C}{D}z = 1.$$

The variables $a = -\dfrac{D}{A}$, $b = -\dfrac{D}{B}$ and $c = -\dfrac{D}{C}$ are modulo equal to the lengths of the segments intercepted by the plane on the coordinate axes Ox, Oy and Oz, respectively (see the formula (8.6)).

Having substituted the values of the coefficients $A = 3$, $B = 4$, $C = 5$, $D = -12$, we obtain $a = 4$, $b = 3$, $c = \dfrac{12}{5}$.

8.7 *Solution.*

Write the equation of a plane in normal form:

$$\frac{1}{\sqrt{1^2 + 1^2 + (-2)^2}}(x + y - 2z - 17) = \frac{1}{\sqrt{6}}(x + y - 2z - 17) = 0.$$

Substitute the coordinates of the point from the problem statement into the normal equation and find the distance from this point to the plane:

$$d = \text{abs}\left(\frac{1}{\sqrt{6}}(1 + 2 - 18 - 17)\right) = \frac{32}{\sqrt{6}}.$$

8.8 *Solution.*

The plane divides the space into two parts: $-2x+2y-z-1 > 0$ and $-2x+2y-z-1 < 0$. If two endpoints of the segment are located in different parts of the space, then it is obvious that the segment intersects the plane. Substitute the coordinates of the points from the problem statement:

$$-2 \cdot (-5) + 2 \cdot (-5) - (-5) - 1 = 4 > 0,$$

$$-2 \cdot 8 + 2 \cdot 8 - 8 - 1 = -9 < 0.$$

Therefore, the points P_1 and P_2 are situated on the opposite sides of the plane, and the segment $P_1 P_2$ intersects the plane $-2x + 2y - z - 1 = 0$.

8.9 *Solution.*

Find the deviation of each plane from the origin of coordinates:

$$\delta_1 = -\frac{1}{\sqrt{A^2 + B^2 + C^2}}D\text{sgn}(D), \quad \delta_2 = -\frac{1}{\sqrt{A^2 + B^2 + C^2}}D'\text{sgn}(D').$$

Then the distance d between the parallel planes is equal to the absolute value of the difference of these deviations:

$$d = \text{abs}(\delta_1 - \delta_2)$$

$$= \text{abs}\left(-\frac{1}{\sqrt{A^2 + B^2 + C^2}}D\text{sgn}(D) + \frac{1}{\sqrt{A^2 + B^2 + C^2}}D'\text{sgn}(D')\right)$$

$$= \frac{\text{abs}(D' - D)}{\sqrt{A^2 + B^2 + C^2}}.$$

8.10 *Answer:*

The condition of three planes intersecting at one point is that the determinant is other than zero:

$$\Delta_1 = \begin{vmatrix} A_1 & B_1 & C_1 \\ A_2 & B_2 & C_2 \\ A_3 & B_3 & C_3 \end{vmatrix} \neq 0.$$

The coordinates (x_P, y_P, z_P) of the intersection point of these planes are

$$x_P = -\frac{1}{\Delta}\begin{vmatrix} D_1 & B_1 & C_1 \\ D_2 & B_2 & C_2 \\ D_3 & B_3 & C_3 \end{vmatrix}, \quad y_P = -\frac{1}{\Delta}\begin{vmatrix} A_1 & D_1 & C_1 \\ A_2 & D_2 & C_2 \\ A_3 & D_3 & C_3 \end{vmatrix}, \quad z_P = -\frac{1}{\Delta}\begin{vmatrix} A_1 & B_1 & D_1 \\ A_2 & B_2 & D_2 \\ A_3 & B_3 & D_3 \end{vmatrix}.$$

Note that if $\Delta = 0$ and at least one second order minor of the matrix $\begin{bmatrix} A_1 & B_1 & C_1 \\ A_2 & B_2 & C_2 \\ A_3 & B_3 & C_3 \end{bmatrix}$ is other than zero, then all planes are parallel to the same line. But if all the second order minors are zeroes, then the planes have the common line.

8.11 *Solution.*

Pass on to the intercept form of the equation of a plane:

$$-\frac{A}{D}x - \frac{B}{D}y - \frac{C}{D}z = 1,$$

$$-\frac{D}{A} = a, \quad -\frac{D}{B} = b, \quad -\frac{D}{C} = c,$$

where a, b and c are modulo equal to the lengths of the segments intercepted on the coordinate axes by the plane and coinciding with the edges of the tetrahedron. The volume of the tetrahedron V_{tetr} and the volume of the parallelepiped V_{par} are bound as follows:

$$V_{\text{tetr}} = \frac{1}{6} V_{\text{par}}, \text{ where } V_{\text{par}} = \text{abs}(abc).$$

Substitute the values of a, b and c:

$$V_{\text{tetr}} = \frac{1}{6} \text{abs} \left(\frac{D^3}{ABC} \right).$$

8.12 *Solution.*

Find the normal equations of the planes:

$$\frac{1}{\sqrt{3}}(x - y + z - 10) = 0,$$
$$-\frac{1}{\sqrt{3}}(x + y - z + 8) = 0.$$

The coordinates of the point located on the axis Oz are $(0, 0, z_0)$, where $z_0 \in \mathbb{R}$. In order to find the distance from this point to the plane, substitute the coordinates into Eqs. (8.22):

$$\begin{cases} \dfrac{\text{abs}(z_0 - 10)}{\sqrt{3}} = d_1, \\ \dfrac{\text{abs}(-z_0 + 8)}{\sqrt{3}} = d_2. \end{cases}$$

Equate the distances:

$$\text{abs}(z_0 - 10) = \text{abs}(-z_0 + 8) \Rightarrow z_0 = 9.$$

So, the problem statement is satisfied by a point with the coordinates $(0, 0, 9)$.

8.13 *Solution.*

Denote the vertices of the tetrahedron (x_i, y_i, z_i) by A_i, respectively ($i = 1, 2, 3, 4$). The volume of the tetrahedron is one sixth part of the absolute value of the scalar triple product $(\overrightarrow{A_1A_2}, \overrightarrow{A_1A_3}, \overrightarrow{A_1A_4})$:

$$V = \frac{1}{6} \text{abs}(\overrightarrow{A_1A_2}, \overrightarrow{A_1A_3}, \overrightarrow{A_1A_4}) = \frac{1}{6} \text{abs} \begin{vmatrix} x_2 - x_1 & y_2 - y_1 & z_2 - z_1 \\ x_3 - x_1 & y_3 - y_1 & z_3 - z_1 \\ x_4 - x_1 & y_4 - y_1 & z_4 - z_1 \end{vmatrix}.$$

Note that this expression can be rewritten in equivalent form:

$$V = \frac{1}{6}\text{abs}\begin{vmatrix} x_1 & y_1 & z_1 & 1 \\ x_2 & y_2 & z_2 & 1 \\ x_3 & y_3 & z_3 & 1 \\ x_4 & y_4 & z_4 & 1 \end{vmatrix}.$$

Chapter 9
Equation of a Line in a Space

9.1 Equation of a Line That Passes Through the Specified Point Parallel to the Specified Vector

Assume that the line L passes through the point $T_0(x_0, y_0, z_0)$ parallel to the vector $a = (k, l, m)$. Take an arbitrary point $T(x, y, z)$ on the line and construct the vector $\overrightarrow{T_0T} = (x - x_0, y - y_0, z - z_0)$, parallel to the line L and collinear with the vector a. Then, we can write the equation

$$\frac{x - x_0}{k} = \frac{y - y_0}{l} = \frac{z - z_0}{m}. \tag{9.1}$$

This relation is referred to as the **canonical equation of a line that passes through the specified point parallel to the specified vector**, and the vector $a = (k, l, m)$ is referred to as the **directing** vector.

From the canonical equation (9.1), we can easily derive the **parametric equation** of such a line:

$$\begin{cases} x - x_0 = ku, \\ y - y_0 = lu, \\ z - z_0 = mu, \end{cases} \tag{9.2}$$

where $u \in \mathbb{R}$ is the **parameter** of the line.

Example 9.1 Let us find the equation of the line that passes through the point $T(3, -1, 0)$ perpendicular to the plane $x - 4y + 7z - 10 = 0$.

© Springer Nature Switzerland AG 2021
S. Kurgalin, S. Borzunov, *Algebra and Geometry with Python*,
https://doi.org/10.1007/978-3-030-61541-3_9

Solution The normal vector of the plane $n = (1, -4, 7)$ will in this case coincide with the directing vector for the line:

$$\frac{x-3}{1} = \frac{y+1}{-4} = \frac{z}{7}.$$ (9.3)

$\square$

Example 9.2 Let us find the canonical equation of the line that is the intersection of the planes determined by the equations $-x - y + z + 2 = 0$ and $2x + 4y - z = 0$.

Solution The line is orthogonal to the normal vectors of each of the planes: $n_1 = (-1, -1, 1)$, $n_2 = (2, 4, -1)$. Therefore, as the directing vector of the line, we can take $a = n_1 \times n_2$.

$$a = \begin{vmatrix} i & j & k \\ -1 & -1 & 1 \\ 2 & 4 & -1 \end{vmatrix} = -3i + j - 2k, \quad a = (-3, 1, -2).$$ (9.4)

As the coordinates of the point that lies on the line, we select any solution of the system

$$\begin{cases} -x - y + z + 2 = 0, \\ 2x + 4y - z = 0. \end{cases}$$ (9.5)

Let $z = 0$, then

$$\begin{cases} -x - y = -2, \\ 2x + 4y = 0, \end{cases} \Rightarrow x = 4, \ y = -2.$$ (9.6)

Therefore, the point $T(4, -2, 0)$ lies on the line.
Therefore, the equation of the sought line has the form:

$$\frac{x-4}{-3} = \frac{y+2}{1} = \frac{z}{-2}.$$ (9.7)

$\square$

9.2 Equation of a Line That Passes Through the Two Specified Points

Assume that the line L passes through the points $T_1(x_1, y_1, z_1)$ and $T_2(x_2, y_2, z_2)$. Then, the vector $a = \overrightarrow{T_1 T_2} = (x_2 - x_1, y_2 - y_1, z_2 - z_1)$ is parallel to the line L, and the **equation of a line that passes through the two specified points** in canonical form looks as follows:

$$\frac{x - x_1}{x_2 - x_1} = \frac{y - y_1}{y_2 - y_1} = \frac{z - z_1}{z_2 - z_1}. \tag{9.8}$$

Example 9.3 The line that passes through two points with the coordinates $(11, 2, -6)$ and $(13, 0, 7)$ is represented by the equation:

$$\frac{x - 11}{2} = \frac{y - 2}{-2} = \frac{z + 6}{13}. \tag{9.9}$$

□

9.3 Angle Between Two Lines

Consider two lines

$$\frac{x - x_1}{k_1} = \frac{y - y_1}{l_1} = \frac{z - z_1}{m_1}, \tag{9.10}$$

$$\frac{x - x_2}{k_2} = \frac{y - y_2}{l_2} = \frac{z - z_2}{m_2}. \tag{9.11}$$

The **angle between two lines** ω will be equal to the angle formed by the directing vectors $a_1 = (k_1, l_1, m_1)$ and $a_2 = (k_2, l_2, m_2)$:

$$\omega = \arccos \frac{(a_1 \cdot a_2)}{|a_1||a_2|} = \arccos \frac{k_1 k_2 + l_1 l_2 + m_1 m_2}{\sqrt{k_1^2 + l_1^2 + m_1^2}\sqrt{k_2^2 + l_2^2 + m_2^2}}. \tag{9.12}$$

Example 9.4 Given are the lines:

$$\frac{x - 4}{6} = \frac{y + 5}{3} = \frac{z + 1}{-1}, \tag{9.13}$$

$$\frac{x - 2}{2} = \frac{y + 6}{1} = \frac{z + 2}{-1}. \tag{9.14}$$

Show that these lines are collinear and find the angle between them.

Solution The necessary and sufficient condition of collinearity of the lines is the equality to zero of the scalar triple product $(a_1, a_2, \overrightarrow{T_1T_2})$, where $a_1 = (6, 3, -1)$, $a_2 = (2, 1, -1)$ are the directing vectors of the lines, and the points T_1 and T_2 have the coordinates $T_1(4, -5, -1)$, $T_2(2, -6, -2)$.

Compute $(a_1, a_2, \overrightarrow{T_1T_2})$:

$$(a_1, a_2, \overrightarrow{T_1T_2}) = \begin{vmatrix} 6 & 3 & -1 \\ 2 & 1 & -1 \\ -2 & -1 & -1 \end{vmatrix} = 0. \tag{9.15}$$

Therefore, the lines are collinear. The angle ω between them is determined from the condition

$$\cos \omega = \frac{6 \cdot 2 + 3 \cdot 1 + (-1) \cdot (-1)}{\sqrt{6^2 + 3^2 + (-1)^2}\sqrt{2^2 + 1^2 + (-1)^2}} = \frac{8}{\sqrt{69}}. \tag{9.16}$$

As a result, we obtain the angle between the two lines

$$\omega = \arccos \frac{8}{\sqrt{69}}. \tag{9.17}$$

$\square$

9.4 Angle Between a Line and a Plane

Assume that the line L and the plane π intersect at some point, and assume that the vectors are given: $n = (A, B, C)$ is the normal vector of plane π, and $a = (k, l, m)$ is the directing vector (see Fig. 9.1).

Fig. 9.1 The angle α between a line and a plane

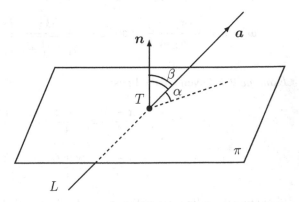

If the line and the plane form the angle α, and the vectors $\boldsymbol{n}$ and $\boldsymbol{a}$ for the angle β, then $\alpha + \beta = \dfrac{\pi}{2}$. Then,

$$\cos \beta = \frac{(\boldsymbol{n} \cdot \boldsymbol{a})}{|\boldsymbol{n}||\boldsymbol{a}|}, \quad \cos \beta = \cos\left(\frac{\pi}{2} - \alpha\right) = \sin \alpha, \tag{9.18}$$

and the **angle α between the line and the plane** is determined from the condition:

$$\sin \alpha = \frac{(\boldsymbol{n} \cdot \boldsymbol{a})}{|\boldsymbol{n}||\boldsymbol{a}|} = \frac{Ak + Bl + Cm}{\sqrt{A^2 + B^2 + C^2}\sqrt{k^2 + l^2 + m^2}}. \tag{9.19}$$

Condition of parallelism of the line L and the plane π (including the belonging of L and π):

$$Ak + Bl + Cm = 0. \tag{9.20}$$

Condition of perpendicularity of the line L and the plane π:

$$\frac{A}{k} = \frac{B}{l} = \frac{C}{m}. \tag{9.21}$$

(Here, the equalities of the form $\dfrac{a}{b} = \dfrac{c}{d}$ are understood in terms of $ad = bc$.)

We obtain the condition of belonging of the line

$$\frac{x - x_1}{k} = \frac{y - y_1}{l} = \frac{z - z_1}{m} \tag{9.22}$$

to the plane $Ax + By + Cz + D = 0$.

For this, it is necessary and sufficient that the point $T_1(x_1, y_1, z_1)$ should lie on the plane, and the vectors $\boldsymbol{n} = (A, B, C)$ and $\boldsymbol{a} = (k, l, m)$ should be perpendicular to each other. Therefore, the **condition of the line's belonging to the plane** consists in the fulfilment of the equalities:

$$\begin{cases} Ax_1 + By_1 + Cz_1 + D = 0, \\ Ak + Bl + Cm = 0. \end{cases} \tag{9.23}$$

9.5   Condition of Two Lines' Belonging to a Plane

The two lines:

$$L_1 : \frac{x - x_1}{k_1} = \frac{y - y_1}{l_1} = \frac{z - z_1}{m_1}, \text{ passing through the point } T_1(x_1, y_1, z_1), \text{ and}$$

$$L_2 : \frac{x - x_2}{k_2} = \frac{y - y_2}{l_2} = \frac{z - z_2}{m_2}, \text{ passing through the point } T_2(x_2, y_2, z_2),$$

in space can

1. intersect;
2. be parallel;
3. be skew.

In the first two cases, they lie in the same plane. The two lines that do not intersect and are not parallel are called **skew lines**.

The necessary and sufficient **condition of belonging of the lines** L_1 **and** L_2 **to the same plane** consists in coplanarity of the vectors $a_1 = (k_1, l_1, m_1)$, $a_2 = (k_2, l_2, m_2)$ and $\overrightarrow{T_1 T_2} = (x_2 - x_1, y_2 - y_1, z_2 - z_1)$, i.e. the determinant's equality to zero must be valid:

$$\begin{vmatrix} x_2 - x_1 & y_2 - y_1 & z_2 - z_1 \\ k_1 & l_1 & m_1 \\ k_2 & l_2 & m_2 \end{vmatrix} = 0. \tag{9.24}$$

Condition of parallelism of two lines:

$$\frac{k_1}{k_2} = \frac{l_1}{l_2} = \frac{m_1}{m_2}. \tag{9.25}$$

For **intersection of lines**, it is sufficient that at least one of the equalities (9.25) is violated and the condition (9.24) is valid.

Example 9.5 Find the planes that pass through the line

$$\frac{x - 4}{-8} = \frac{y - 8}{-1} = \frac{z + 1}{3} \tag{9.26}$$

and are orthogonal to the plane $-3x + y + z + 3 = 0$.

Solution The directing vector of the given line $a = (-8, -1, 3)$ is the normal vector to the plane $n = (-3, 1, 1)$.

Fig. 9.2 To Example 9.5.
Mutual arrangement of the
vectors $\boldsymbol{a}$ and $\overrightarrow{T_0 T}$

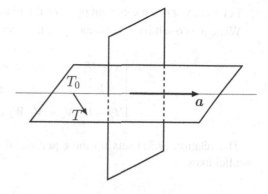

The point $T_0(4, 8, -1)$ belongs to the sought plane since it lies on the line (see Fig. 9.2).

Write the equation of the plane that passes through the point T and is parallel to the two vectors $\boldsymbol{a}$ and $\boldsymbol{n}$ (the vectors $\boldsymbol{a}$, $\boldsymbol{n}$ and $\overrightarrow{T_0 T}$ are coplanar):

$$\begin{vmatrix} x-4 & y-8 & z+1 \\ -8 & -1 & 3 \\ -3 & 1 & 1 \end{vmatrix} = 0, \tag{9.27}$$

hence:

$$-4(x-4) - (y-8) - 11(z+1) = 0. \tag{9.28}$$

Thus, we obtain the sought equation of the plane:

$$4x + y + 11z - 13 = 0. \tag{9.29}$$

$\square$

Example 9.6 Find the equation of the plane that passes through the two parallel lines:

$$\frac{x - U_1}{k} = \frac{y - V_1}{l} = \frac{z - W_1}{m}, \quad \frac{x - U_2}{k} = \frac{y - V_2}{l} = \frac{z - W_2}{m}, \tag{9.30}$$

and at least one of the inequalities is valid: $U_1 \neq U_2$, $V_1 \neq V_2$, $W_1 \neq W_2$.

Solution The sought plane passes through the points $T_1(U_1, V_1, W_1)$ and $T_2(U_2, V_2, W_2)$ and is parallel to the directing vector $\boldsymbol{a} = (k, l, m)$. On the other hand, this plane is parallel to the vector $\overrightarrow{T_1 T_2} = (U_2 - U_1, V_2 - V_2, W_2 - W_1)$.

Let $T(x, y, z)$ be the current point of the plane.

Write the condition of coplanarity of the vectors $\overrightarrow{T_1T}$, a and $\overrightarrow{T_1T_2}$:

$$\begin{vmatrix} x - U_1 & y - V_1 & z - W_1 \\ k & l & m \\ U_2 - U_1 & V_2 - V_1 & W_2 - W_1 \end{vmatrix} = 0. \tag{9.31}$$

The relation (9.31) sets up the equation of the plane that passes through two parallel lines. $\square$

Review Questions

1. Write the equation of a straight line that passes through a specified point parallel to a specified vector.
2. What does the equation of a line in space that passes through two specified points look like?
3. Write the formula for computing an angle between lines in space.
4. How is an angle between a line and a plane computed?
5. Formulate the condition of parallelism of a line and a plane.
6. Write the condition of perpendicularity of a line and a plane.
7. What lines are called skew lines?
8. What is the condition of parallelism of two lines in space?

Problems

9.1. Find the intersection points of the line specified by the equation

$$\begin{cases} 2x + 4y + z + 9 = 0, \\ 4x - 6y - 2z + 1 = 0, \end{cases}$$

with the coordinate planes.

9.2. Find the equation of the plane that passes through the two parallel lines:

$$\frac{x-2}{2} = \frac{y}{3} = \frac{z-1}{-3}, \quad \frac{x}{4} = \frac{y+1}{6} = \frac{z+1}{-6}.$$

9.3. Find the coordinates of the foot of the perpendicular dropped from the point $T(0, 2, -4)$ onto the plane $x + y - z + 3 = 0$.

9.4. Determine whether the plane $8x - 3z + 11 = 0$ and the line $\dfrac{x-2}{2} = \dfrac{y}{3} = \dfrac{z-1}{8}$ are parallel.

9.5. Determine whether the plane $x + 2y - z - 2 = 0$ and the line $\dfrac{x-1}{3} = \dfrac{y+1}{3} = \dfrac{z}{-1}$ are parallel?

9.6. Write the necessary and sufficient condition of perpendicularity of the plane $Ax + By + Cz + D = 0$ and the line $\dfrac{x - x_0}{k} = \dfrac{y - y_0}{l} = \dfrac{z - z_0}{m}$.

9.7. Prove that the condition of belonging of the line $\dfrac{x - x_0}{k} = \dfrac{y - y_0}{l} = \dfrac{z - z_0}{m}$ to the plane $Ax + By + Cz + D = 0$ has the form

$$\begin{cases} Ax_0 + By_0 + Cz_0 + D = 0, \\ Ak + Bl + Cm = 0. \end{cases} \qquad (9.32)$$

∗9.8. Show that the distance from the point $T_0(x_0, y_0, z_0)$ to the line $\dfrac{x - x_1}{k} = \dfrac{y - y_1}{l} = \dfrac{z - z_1}{m}$ can be calculated by the formula

$$d = \sqrt{\dfrac{F_1^2 + F_2^2 + F_3^2}{k^2 + l^2 + m^2}},$$

where the following designations are introduced

$$F_1 = \begin{vmatrix} l & m \\ y_1 - y_0 & z_1 - z_0 \end{vmatrix}, \; F_2 = \begin{vmatrix} m & k \\ z_1 - z_0 & x_1 - x_0 \end{vmatrix}, \; F_3 = \begin{vmatrix} k & l \\ x_1 - x_0 & y_1 - y_0 \end{vmatrix}.$$

9.9. Find the acute angle between the lines specified by the equations $\dfrac{x+1}{2} = \dfrac{y-9}{-5} = \dfrac{z+2}{2}$ and $\dfrac{x}{3} = \dfrac{y+1}{-6} = \dfrac{z+4}{3}$.

9.10. Find the obtuse angle between the lines $\dfrac{x+1}{-1} = \dfrac{y-9}{-5} = \dfrac{z+2}{7}$ and $\dfrac{x-3}{-3} = \dfrac{y-9}{3} = \dfrac{z}{5}$.

9.11. Find the equation of the plane that passes through the line

$$\dfrac{x - 10}{-7} = \dfrac{y}{3} = \dfrac{z + 10}{-1} \qquad (9.33)$$

parallel to the line

$$\frac{x+1}{2} = \frac{y+2}{4} = \frac{z-6}{8}. \tag{9.34}$$

*9.12. Compute the distance between the skew lines $\dfrac{x-x_1}{k_1} = \dfrac{y-y_1}{l_1} = \dfrac{z-z_1}{m_1}$

and $\dfrac{x-x_2}{k_2} = \dfrac{y-y_2}{l_2} = \dfrac{z-z_2}{m_2}$.

Answers and Solutions

9.1 *Answer:*
$$\left(-\frac{29}{14}, -\frac{17}{14}, 0\right), \left(-\frac{19}{8}, 0, -\frac{17}{4},\right), \left(0, -\frac{19}{2}, 29\right).$$

9.2 *Answer:*
$$9x - 10y - 4z - 14 = 0.$$

9.3 *Solution.*
   Write the equation of the line that passes through the point T perpendicular to the plane:

$$\frac{x}{1} = \frac{y-2}{1} = \frac{z+4}{-1}. \tag{9.35}$$

Represent the equation of this line in parametric form:

$$x = u, \ y = u + 2, \ z = -u - 4, \tag{9.36}$$

where $u \in \mathbb{R}$.
   Substitute these coordinates into the equation of the plane:

$$u + (u + 2) - (-u - 4) + 3 = 0, \tag{9.37}$$

hence, we obtain $u = -3$.
   Knowing u, find the coordinates of the intersection point of the line and the plane:

$$x = -3, \ y = -1, \ z = -1. \tag{9.38}$$

So, the intersection point of the perpendicular dropped from the point T onto the plane has the coordinates $(-3, -1, -1)$.

9.4 *Solution.*

The condition of parallelism of the line $\dfrac{x - x_0}{k} = \dfrac{y - y_0}{l} = \dfrac{z - z_0}{m}$ and the plane $Ax + By + Cz + D = 0$ reduces to fulfilment of the equality $Ak + Bl + Cm = 0$ that reflects orthogonality of the directing vector $\boldsymbol{\tau} = (k, l, m)$ and the normal to the plane $\boldsymbol{n} = (A, B, C)$.

According to the problem statement, $\boldsymbol{\tau} = (2, 3, 8)$, $\boldsymbol{n} = (8, 0, -3)$.

Since the equality to zero of the scalar product $(\boldsymbol{\tau} \cdot \boldsymbol{n}) = 0$ is valid, then the line and the plane are parallel.

9.5 *Answer:* no, the plane $x + 2y - z - 2 = 0$ and the line $\dfrac{x - 1}{3} = \dfrac{y + 1}{3} = \dfrac{z}{-1}$ are not parallel.

9.6 *Solution.*

The normal vector of the plane $Ax + By + Cz + D = 0$ has the coordinates (A, B, C). At the same time, the directing vector of the line $\dfrac{x - x_0}{k} = \dfrac{y - y_0}{l} = \dfrac{z - z_0}{m}$ is equal to (k, l, m).

In order for the line to be perpendicular to the plane, it is necessary and sufficient that the normal vector of the plane be collinear with the directing vector of the line:

$$\frac{A}{k} = \frac{B}{l} = \frac{C}{m}.$$

9.7 *Proof.*

Indeed, the first equation (9.32) means that the point (x_0, y_0, z_0), through which the line passes, belongs to the plane. The second equation reflects the fact of parallelism of the line and the plane (see Problem **9.4**).

9.8 *Proof.*

From the equation of a line, we find the coordinates of its directing vector $s = (l, m, n)$. Denote the point (x_1, y_1, z_1) that lies on this line by $T_1(x_1, y_1, z_1)$.

It is known from the properties of the vector product that the modulus of the vector product of vectors is equal to the area of the parallelogram constructed on these vectors:

$$S = |\overrightarrow{T_0 T_1} \times s|.$$

On the other hand, the area of the parallelogram is equal to the product of its side by the height drawn to this side: $S = |s|d$.

In our case, the height of the parallelogram is equal to the distance d from the point to the plane, and its side is equal to the modulus of the directing vector $|s|$.

Having equated the expressions for the area, it is easy to obtain a formula of the distance from the point to the line:

$$d = \frac{|\overrightarrow{T_0 T_1} \times s|}{|s|}.$$

Hence, we find

$$\overrightarrow{T_0 T_1} \times s = \begin{vmatrix} i & j & k \\ x_1 - x_0 & y_1 - y_0 & z_1 - z_0 \\ k & l & m \end{vmatrix}$$

$$= (m(y_1 - y_0) - l(z_1 - z_0))i$$
$$+ (k(z_1 - z_0) - m(x_1 - x_0))j$$
$$+ (l(x_1 - x_0) - k(y_1 - y_0))k.$$

Thus,

$$|\overrightarrow{T_0 T_1} \times s|$$

$$= \sqrt{\begin{vmatrix} l & m \\ y_1 - y_0 & z_1 - z_0 \end{vmatrix}^2 + \begin{vmatrix} m & k \\ z_1 - z_0 & x_1 - x_0 \end{vmatrix}^2 + \begin{vmatrix} k & l \\ x_1 - x_0 & y_1 - y_0 \end{vmatrix}^2},$$

where

$$|s| = \sqrt{k^2 + l^2 + m^2}.$$

Therefore,

$$d = \frac{\sqrt{\begin{vmatrix} l & m \\ y_1 - y_0 & z_1 - z_0 \end{vmatrix}^2 + \begin{vmatrix} m & k \\ z_1 - z_0 & x_1 - x_0 \end{vmatrix}^2 + \begin{vmatrix} k & l \\ x_1 - x_0 & y_1 - y_0 \end{vmatrix}^2}}{\sqrt{k^2 + l^2 + m^2}}.$$

9.9 *Solution.*

The lines are specified in canonical form.

Find the angle α between their directing vectors $a = (2, -5, 2)$ and $b = (3, -6, 3)$:

$$\cos \alpha = \frac{(a \cdot b)}{|a| \cdot |b|};$$

$$\cos\alpha = \frac{2\cdot 3 + (-5)\cdot(-6) + 2\cdot 3}{\sqrt{2^2 + (-5)^2 + 2^2}\cdot\sqrt{3^2 + (-6)^2 + 3^2}} = \frac{14}{3\sqrt{22}}.$$

The sought angle is $\alpha = \arccos\left(\dfrac{14}{3\sqrt{22}}\right)$.

9.10 *Solution.*

Find the angle α between the directing vectors $\boldsymbol{a} = (-1, -5, 7)$ and $\boldsymbol{b} = (-3, 3, 5)$ of the specified lines:

$$\cos\alpha = \frac{\boldsymbol{a}\cdot\boldsymbol{b}}{|\boldsymbol{a}||\boldsymbol{b}|}.$$

We substitute here the numeric values from the problem statement and obtain

$$\cos\alpha = \frac{(-1)\cdot(-3) + (-5)\cdot 3 + 7\cdot 5}{\sqrt{(-1)^2 + (-5)^2 + 7^2}\sqrt{(-3)^2 + 3^2 + 5^2}} = \frac{23}{5\sqrt{129}}.$$

Therefore, the sought angle is $\pi - \alpha = \pi - \arccos\left(\dfrac{23}{5\sqrt{129}}\right)$.

9.11 *Solution.*

The sought plane is parallel to the directing vectors $\boldsymbol{a}_1 = (-7, 3, -1)$ and $\boldsymbol{a}_2 = (2, 4, 8)$ and passes through the point $T_0(10, 0, -10)$ that lies on the first line. Thus, the vectors $\boldsymbol{a}_1, \boldsymbol{a}_2$ and $\overrightarrow{T_0 T}$ are coplanar.

Expanding the determinant that expresses the scalar triple product $(\overrightarrow{T_0 T}, \boldsymbol{a}_1, \boldsymbol{a}_2)$, for example, in the first row, we obtain

$$\begin{vmatrix} x-10 & y & z+10 \\ -7 & 3 & -1 \\ 2 & 4 & 8 \end{vmatrix} = 28(x-10) + 54y - 34(z+10) = 0,$$

and we finally obtain

$$14x + 27y - 17z - 310 = 0.$$

9.12 *Solution.*

Introduce the designations $T_1(x_1, y_1, z_1)$, $T_2(x_2, y_2, z_2)$, $\boldsymbol{a}_1 = (k_1, l_1, m_1)$, $\boldsymbol{a}_2 = (k_2, l_2, m_2)$.

Consider the parallelogram constructed on the vectors $\overrightarrow{T_1 T_2}$, $\boldsymbol{a}_1$ and $\boldsymbol{a}_2$, and compute its volume V.

On the one hand, the volume V is equal to the product of the module of the vector product $|\boldsymbol{a}_1 \times \boldsymbol{a}_2|$ and the height of the parallelogram. On the other hand, $V = \text{abs}(\overrightarrow{T_1 T_2}, \boldsymbol{a}_1, \boldsymbol{a}_2)$.

Since the sought distance d between the skew lines is equal to the height, then

$$d = \frac{\mathrm{abs}(\overrightarrow{T_1 T_2}, \boldsymbol{a}_1, \boldsymbol{a}_2)}{|\boldsymbol{a}_1 \times \boldsymbol{a}_2|}, \quad \text{or}$$

$$d = \frac{\mathrm{abs} \begin{vmatrix} x_2 - x_1 & y_2 - y_1 & z_2 - z_1 \\ k_1 & l_1 & m_1 \\ k_2 & l_2 & m_2 \end{vmatrix}}{\sqrt{\begin{vmatrix} k_1 & l_1 \\ k_2 & l_2 \end{vmatrix}^2 + \begin{vmatrix} l_1 & m_1 \\ l_2 & m_2 \end{vmatrix}^2 + \begin{vmatrix} m_1 & k_1 \\ m_2 & k_2 \end{vmatrix}^2}}.$$

Chapter 10
Bilinear and Quadratic Forms

10.1 Bilinear Forms

Assume that in a n-dimensional vector space L_n a basis $B = (e_1, e_2, \ldots, e_n)$ is specified. Consider two vectors belonging to the space L_n:

$$x = \sum_{i=1}^{n} x_i e_i, \quad y = \sum_{i=1}^{n} y_i e_i, \tag{10.1}$$

where $x_i, y_i \in \mathbb{R}$ for all $i = 1, 2, \ldots, n$.

The linear combination of all possible products of the projection of the vectors x and y on the basic normalized vectors

$$\sum_{i,j=1}^{n} a_{ij} x_i y_j, \tag{10.2}$$

where a_{ij} are arbitrary real numbers, is referred to as the **bilinear form** of $A(x, y)$ defined on the basis B. The matrix $A = (a_{ij})$, where $1 \leqslant i, j \leqslant n$, is called the **matrix of bilinear form**. As is easy to see, an arbitrary bilinear form can be written with the help of the matrix multiplication operation as

$$A(x, y) = x^T A \, y. \tag{10.3}$$

The following **properties of linearity** of the form are fulfilled for each of its arguments:

$\forall x, y, z \in L_n$ and $\forall \alpha \in \mathbb{R}$

1. $A(x + y, z) = A(x, z) + A(y, z)$,
2. $A(\alpha x, z) = \alpha A(x, z)$,

© Springer Nature Switzerland AG 2021
S. Kurgalin, S. Borzunov, *Algebra and Geometry with Python*,
https://doi.org/10.1007/978-3-030-61541-3_10

3. $A(x, y + z) = A(x, y) + A(x, z)$,
4. $A(x, \alpha z) = \alpha A(x, z)$.

A bilinear form is referred to as **symmetric**, if for any $x, y \in L_n$ the condition $A(x, y) = A(y, x)$ is fulfilled.

A bilinear form is referred to as **positively definite**, if $\forall x \in L_n, x \neq 0$, the inequality $A(x, x) > 0$ is fulfilled.

Example 10.1 Consider the vectors $x = (x_1, x_2, x_3)$ and $y = (y_1, y_2, y_3)$ of some three-dimensional vector space, and the bilinear form $A(x, y) = x_1 y_1 + 3 x_2 y_2 + 8 x_3 y_3$, defined on the basis of this space. The matrix A of bilinear form will in this case have the following diagonal form:

$$A = \begin{bmatrix} 1 & 0 & 0 \\ 0 & 3 & 0 \\ 0 & 0 & 8 \end{bmatrix}. \tag{10.4}$$

Represent A in matrix notation:

$$A(x, y) = x^T A y = [x_1, x_2, x_3] \begin{bmatrix} 1 & 0 & 0 \\ 0 & 3 & 0 \\ 0 & 0 & 8 \end{bmatrix} \begin{bmatrix} y_1 \\ y_2 \\ y_3 \end{bmatrix}$$

$$= [x_1, 3 x_2, 8 x_3] \begin{bmatrix} y_1 \\ y_2 \\ y_3 \end{bmatrix} = x_1 y_1 + 3 x_2 y_2 + 8 x_3 y_3. \tag{10.5}$$

It is easy to show that it has the properties of symmetry and positive definiteness:

$$A(x, y) = x_1 y_1 + 3 x_2 y_2 + 8 x_3 y_3 = y_1 x_1 + 3 y_2 x_2 + 8 y_3 x_3$$

$$= A(y, x) \Rightarrow \text{the form is symmetric;} \tag{10.6}$$

$$A(x, x) = x_1^2 + 3 x_2^2 + 8 x_3^2 \geqslant 0 \text{ for all non-zero vectors } x$$

$$\Rightarrow A(x, y) \text{is a positively definite form.} \tag{10.7}$$

□

Example 10.2 $M(x, y) = x_1 y_1 - x_2 y_2 - x_3 y_3 - x_4 y_4$ is a bilinear form on $\mathbb{R}^4$. As is easy to see, it is symmetric, but not positively definite. Note that this form defines the space-time metric in the special theory of relativity [45].   □

10.2 Quadratic Forms

Quadratic form will be referred to as the expression

$$\omega(x) = \mathcal{A}(x, x), \tag{10.8}$$

where $\mathcal{A}(x, x)$ is some bilinear form. The name "quadratic form" is associated with
the fulfilment for this expression of the property of the second degree homogeneity
in the argument of the form: $\forall \alpha \in \mathbb{R}$ the following equation is valid:

$$\omega(\alpha x) = \alpha^2 \omega(x). \tag{10.9}$$

Example 10.3 In the basis of the bilinear form $\mathcal{M}(x, y)$, defined in Example 10.2
on page 336, we can construct the quadratic form

$$\mu(x) = \mathcal{M}(x, x) = x_1^2 - x_2^2 - x_3^2 - x_4^2 \tag{10.10}$$

that depends on four variables: x_1, x_2, x_3 and x_4.   □

If in the n-dimensional vector space L_n the basis is specified and the vector
$x = [x_1, x_2, \ldots, x_n]^T$ is selected, then

$$\omega(x) = x^T A x = \sum_{i,j=1}^{n} a_{ij} x_i x_j, \tag{10.11}$$

which allows interpreting the quadratic form as a function specified on the set of all
possible vectors x.

The matrix $A = (a_{ij})$ is referred to as the **matrix of quadratic form** $\omega(x)$. This
matrix can be deemed symmetric, since the expression of the form $a_{ij} x_i x_j + a_{ji} x_j x_i$,
due to commutativity of multiplication of real numbers, can always be represented
as

$$a_{ij} x_i x_j + a_{ji} x_j x_i \equiv \tilde{a}_{ij} x_i x_j + \tilde{a}_{ji} x_j x_i, \tag{10.12}$$

where the designation $\tilde{a}_{ij} = \tilde{a}_{ji} = (a_{ij} + a_{ji})/2$ is introduced.

So, an arbitrary quadratic form in some basis can be specified in matrix form:

$$x^T A x, \tag{10.13}$$

where $x = (x_1, x_2, \ldots, x_n)$ is the column composed of variables, and the matrix
A of quadratic form always allows a notation in symmetric form: $a_{ji} = a_{ij}$ for all
$i, j = 1, \ldots, n$.

The symmetric bilinear form $\mathcal{A}(x, y)$ is referred to as **polar** to the quadratic form $\omega(x) = \mathcal{A}(x, x)$. For the quadratic form, the formula is valid:

$$A(x, y) = \frac{1}{2}[\omega(x + y) - \omega(x) - \omega(y)]. \tag{10.14}$$

Example 10.4 Find the bilinear form polar to $\omega(x) = x_1x_2 + x_2x_3 + x_1x_3$.

Solution According to the definition of polar form (10.14), we obtain

$$A(x, y) = \frac{1}{2}[(x_1 + y_1)(x_2 + y_2) + (x_2 + y_2)(x_3 + y_3) + (x_1 + y_1)(x_3 + y_3)$$
$$- (x_1x_2 + x_2x_3 + x_1x_3) - (y_1y_2 + y_2y_3 + y_1y_3)]. \tag{10.15}$$

After algebraic transformations, we find the expression for the sought bilinear form:

$$A(x, y) = \frac{1}{2}(x_1y_2 + x_1y_3 + x_2y_1 + x_2y_3 + x_3y_1 + x_3y_2). \tag{10.16}$$

□

When passing to a new basis, i.e. during nondegenerate change of the variables $x_1, x_2, \ldots, x_n$ with the change matrix C, the matrix A' of quadratic form in the new basis will take the form:

$$A' = C^T A C. \tag{10.17}$$

It is known that changing the basis does not result in the change of the rank of the matrix in quadratic form.

Rank of the matrix in quadratic form is referred to as the **rank of quadratic form**. If this matrix has a rank equal to the dimension of the vector space, i.e. the number of variables n, then the quadratic form is called **nondegenerate**, and if the rank is less than n, then it is called **degenerate**.

Example 10.5 Find the matrix of quadratic form of the three variables:

$$\omega(x) = 3x_1^2 + 2x_1x_2 - 5x_1x_3 - x_2^2 + x_3^2. \tag{10.18}$$

Solution The diagonal elements a_{ii} of the matrix of quadratic form $\omega(x)$ are defined as the coefficients of the quadratic summands x_i^2, and the non-diagonal ones a_{ij}, where $i \neq j$, are twice smaller than the respective coefficients of the summands of the form x_ix_j:

$$A = \begin{bmatrix} 3 & 1 & -5/2 \\ 1 & -1 & 0 \\ -5/2 & 0 & 1 \end{bmatrix}. \tag{10.19}$$

Using any of the known methods for computing the rank of the matrix, for example, the method of bringing to echelon form, we find rk $A = 3$. Since the tank of the matrix A is equal to the number of variables, then this quadratic form is nondegenerate.

In matrix notation, the quadratic form can be represented in the form:

$$\omega(x) = [x_1, x_2, x_3] \begin{bmatrix} 3 & 1 & -5/2 \\ 1 & -1 & 0 \\ -5/2 & 0 & 1 \end{bmatrix} \begin{bmatrix} x_1 \\ x_2 \\ x_3 \end{bmatrix}. \tag{10.20}$$

$\square$

10.3 Bringing the Quadratic Form to the Canonical Form

If, in some basis, the matrix of quadratic form takes diagonal form

$$\Lambda = \begin{bmatrix} \lambda_1 & 0 & 0 \dots & 0 & 0 \\ 0 & \lambda_2 & 0 \dots & 0 & 0 \\ & & \dots\dots\dots\dots\dots & & \\ 0 & 0 & 0 \dots & \lambda_{n-1} & 0 \\ 0 & 0 & 0 \dots & 0 & \lambda_n \end{bmatrix}, \tag{10.21}$$

then, as is easy to see, the quadratic form is formed by a linear combination of squares of the variables $x_1, x_2, \dots, x_n$:

$$\omega(x) = \lambda_1 x_1^2 + \lambda_2 x_2^2 + \cdots + \lambda_n x_n^2. \tag{10.22}$$

The crossing summands of the form $x_i x_j$ for $i \neq j$ are in this case not included into the expression for $\omega(x)$.

Among the coefficients λ_i, where $i = 1, 2, \dots, n$, there can be positive numbers, negative numbers and numbers equal to zero.

Theorem 10.1 *For each quadratic form, there exists a basis, in which it has the canonical form:*

$$\omega(x) = \sum_{i=1}^{n} \lambda_i x_i^2. \tag{10.23}$$

Attention should be paid to the fact that the canonical basis is defined non-uniquely.

10.3.1 Lagrange's Method of Separating Perfect Squares

In applications, we often come across a problem of bringing a quadratic form to a diagonal form. Several methods have been suggested for solving this problem.

Lagrange's method of bringing the quadratic form to the canonical form consists in the following.

Assume that the quadratic form $\omega(x) = \sum\limits_{i,j=1}^{n} a_{ij}x_ix_j$ is given, defined in the basis of the space L_n.

Depending on the presence in the sum of the summands of the form $a_{ii}x_i^2$, consider two cases.

1. *At least one of the coefficients a_{tt} of the quadratic summands is not equal to zero.*

The main idea of the method consists in separating the perfect square, uniting all the summands that contain the variable x_t. The obtained perfect square is used as the basis for change of the variables, excluding the terms linear in x_t. If in the form still remain crossing variables of the form x_ix_j, then we return to the beginning of the procedure. Otherwise, the form contains only quadratic summands, and the solution is complete.

As we can see, without loss of generality, we may assume that $a_{11} \neq 0$.

Consider the first step of Lagrange's method in more detail.

Denote by S the sum of all summands containing x_1:

$$S = a_{11}x_1^2 + 2a_{12}x_1x_2 + \cdots + 2a_{1n}x_1x_n, \tag{10.24}$$

and complete the square of S. We obtain

$$S = a_{11}\left(x_1 + \frac{a_{12}}{a_{11}}x_2 + \cdots + \frac{a_{1n}}{a_{11}}x_n\right)^2 - R, \tag{10.25}$$

where the expression R does not contain x_1 in its notation.

Then, change the variables

$$\begin{cases} x_1' = x_1 + \dfrac{a_{12}}{a_{11}}x_2 + \cdots + \dfrac{a_{1n}}{a_{11}}x_n, \\ x_i' = x_i \quad \text{for } i = 2, \ldots, n. \end{cases} \tag{10.26}$$

Then, the quadratic form in the new basis will take the form:

$$\omega(x) = a_{11}(x_1')^2 + \sum_{i,j=2}^{n} a_{ij}'x_i'x_j'. \tag{10.27}$$

From here on, the same method can be applied to the variables x_2, x_3 and so on, to finally exclude all summands of the form $x_i x_j$ for $i \neq j$.

2. *All coefficients are $a_{ii} = 0$.*

In this case, select $a_{ij} \neq 0$ for some $i \neq j$ and change the variables:

$$\begin{cases} x_i = x'_i + x'_j, \\ x_j = x'_i - x'_j, \\ x_k = x'_k \text{ for } k \neq i, j. \end{cases} \tag{10.28}$$

As a result, each product $x_i x_j$ will be presented in the form of a linear combination of the quadratic summands $x_i x_j = (x'_i)^2 - (x'_j)^2$, and we will arrive at the first case.

After step (1) and, when necessary, step (2), bringing of the quadratic form to the diagonal form will be completed.

Example 10.6 Using Lagrange's method, bring the following form to the canonical one

$$\omega(x) = 2x_1 x_2 - 6x_1 x_3 - x_2^2 + 5x_3^2. \tag{10.29}$$

Solution Collect the summands that contain x_2:

$$\underbrace{-x_2^2 + 2x_1 x_2}_{S} - 6x_1 x_3 + 5x_3^2 = -\underbrace{(x_2^2 - 2x_1 x_2 + x_1^2)}_{\text{perfect square}} + \underbrace{x_1^2}_{-R} - 6x_1 x_3 + 5x_3^2$$

$$= -(x_1 - x_2)^2 + x_1^2 - 6x_1 x_3 + 5x_3^2. \tag{10.30}$$

Change the variables:

$$\begin{cases} y_1 = x_1, \\ y_2 = x_1 - x_2, \\ y_3 = x_3. \end{cases} \tag{10.31}$$

As a result, we obtain the form

$$\omega(y) = -y_2^2 + y_1^2 - 6y_1 y_3 + 5y_3^2. \tag{10.32}$$

Now collect the summands that contain y_1, and complete the square of them

$$\omega(y) = -y_2^2 + (y_1^2 - 6y_1 y_3 + 9y_3^2) - 9y_3^2 + 5y_3^2$$

$$= -y_2^2 + (y_1 - 3y_3)^2 - 4y_3^2. \tag{10.33}$$

Perform the second change of the variables:

$$\begin{cases} z_1 = y_1 - 3y_3, \\ z_2 = y_2, \\ z_3 = y_3. \end{cases} \tag{10.34}$$

We finally obtain $\omega(z) = z_1^2 - z_2^2 - 4z_3^2$. $\square$

Example 10.7 Using Lagrange's method, bring the following form to the canonical one

$$\omega(x) = x_1 x_2 + x_1 x_3 + x_2 x_3. \tag{10.35}$$

Solution Since there are no summands of the form x_i^2 in this expression, we change the variables

$$\begin{cases} x_1 = x_1' + x_2', \\ x_2 = x_1' - x_2', \\ x_3 = x_3', \end{cases} \tag{10.36}$$

as a result of which we obtain

$$\omega(x') = \left[(x_1')^2 + 2x_1' x_3' + (x_3')^2 \right] - (x_2')^2 - (x_3')^2 \tag{10.37}$$

$$= (x_1' + x_3')^2 - (x_2')^2 - (x_3')^2. \tag{10.38}$$

The new change

$$\begin{cases} y_1 = x_1' + x_3', \\ y_2 = x_2', \\ y_3 = x_3' \end{cases} \tag{10.39}$$

results in the form

$$\omega(y) = y_1^2 - y_2^2 - y_3^2, \tag{10.40}$$

where $y_1 = \dfrac{1}{2}(x_1 + x_2) + x_3$, $y_2 = \dfrac{1}{2}(x_1 - x_2)$, $y_3 = x_3$. $\square$

10.3.2 Jacobi Method

Let the following determinants composed of the elements of the matrix $A = (a_{ij})$ of the quadratic form $\omega(x)$ be other than zero:

$$\Delta_1 = a_{11}, \quad \Delta_2 = \begin{vmatrix} a_{11} & a_{12} \\ a_{21} & a_{22} \end{vmatrix}, \quad \dots, \tag{10.41}$$

$$\Delta_n = \begin{vmatrix} a_{11} & a_{12} & \dots & a_{1n} \\ a_{21} & a_{22} & \dots & a_{2n} \\ \vdots & \vdots & \ddots & \vdots \\ a_{n1} & a_{n2} & \dots & a_{nn} \end{vmatrix}. \tag{10.42}$$

Then, there exists the basis $B = (e_1, e_2, \dots, e_n)$, in which $\omega(x)$ is presented in the form:

$$\omega(z) = \frac{\Delta_1}{\Delta_0} z_1^2 + \frac{\Delta_2}{\Delta_1} z_2^2 + \dots + \frac{\Delta_n}{\Delta_{n-1}} z_n^2, \tag{10.43}$$

where z_i, $i = 1, 2, \dots, n$, denote the coordinates of the vector x in the new basis B, and for uniformity, we assume that $\Delta_0 = 1$. This is the essence of the **Jacobi method** of bringing the quadratic form to the canonical form.

In comparison with Lagrange's method, the Jacobi method has an advantage that the transition to the basis B is direct, without any intermediate steps.

The transition from the basis $(e_1, e_2, \dots, e_n)$ to the canonical basis $(c_1, c_2, \dots, c_n)$ is performed by the formulae

$$c_i = \sum_{j=1}^{i} \eta_{ij} e_j, \quad i = 1, 2, \dots, n, \tag{10.44}$$

$$\eta_{ij} = (-1)^{i+j} \frac{\Delta_{i-1,j}}{\Delta_{i-1}}, \tag{10.45}$$

where $\Delta_{i-1,j}$ is the minor of the matrix formed by the elements from (a_{ij}) that are situated at the intersection of the rows numbered $k = 1, 2, \dots, i-1$ with the columns numbered $k = 1, 2, \dots, j-1, j+1, \dots, i$.

Note that the introduced determinants $\Delta_1, \Delta_2, \dots, \Delta_n$ are called the **corner minors** of the matrix of quadratic form.

Rank of quadratic form is the number of non-zero coefficients $\lambda_i \neq 0$.

Let us introduce the following designations:

- n_+—number of positive coefficients of $\lambda_i > 0$,
- n_-—number of negative coefficients of $\lambda_i < 0$,
- n_0—number of coefficients equal to zero of $\lambda_i = 0$.

The ordered set of integral non-negative numbers (n_+, n_-, n_0) is called the **signature** of quadratic form.

The quadratic form $\omega(x)$ is referred to as **positively definite**, if for any non-zero x the inequality

$$\omega(x) > 0 \tag{10.46}$$

is fulfilled, and **negatively definite**, if $\omega(x) < 0$ is fulfilled.

The quadratic form $\omega(x)$ is referred to as **alternating**, if there exist such x_1 and x_2, that the following inequalities occur

$$\omega(x_1) > 0, \quad \omega(x_2) < 0. \tag{10.47}$$

Sylvester's[1] Criterion *For the positive definiteness of the quadratic form $\omega(x)$, it is necessary and sufficient that all the primary minors of its matrix should be positive:* $\Delta_1 > 0, \Delta_2 > 0, \ldots, \Delta_n > 0$.

For the negative definiteness of the quadratic form $\omega(x)$, it is necessary and sufficient that the signs of the corner minors of its matrix should alternate, and $\Delta_1 < 0$.

The Law of Inertia *The number of summands with positive (negative) coefficients of a quadratic form brought to the canonical form does not depend on the method used to obtain such a representation.*

Review Questions

1. Define bilinear form.
2. Enumerate the property of linearity of bilinear forms.
3. What bilinear forms are called symmetric and positively defined?
4. Define quadratic form.
5. Explain why a matrix of an arbitrary quadratic form can always be deemed to be symmetric.
6. What bilinear form is called polar to the quadratic form?
7. How is the rank of a quadratic form determined?
8. What is the difference between degenerate and nondegenerate quadratic forms?

---

[1] James Joseph Sylvester (1814–1897), English mathematician.

9. Explain the methods of bringing quadratic forms to diagonal form: Lagrange's method and the Jacobi method.
10. How are the corner minors of a matrix of quadratic form determined?
11. Formulate Sylvester's criterion.
12. What is the law of inertia of quadratic forms?

Problems

10.1. Which of the following functions $F(x, y)$ of the vectors $x = (x_1, x_2) \in \mathbb{R}^2$ and $y = (y_1, y_2) \in \mathbb{R}^2$ are bilinear forms?

(1) $F(x, y) = x_1 y_1 + x_2 y_2$;
(2) $F(x, y) = \sqrt{2}$;
(3) $F(x, y) = (x_1 - y_1)^2 - (x_2 - y_2)^2$;
(4) $F(x, y) = 4x_2 y_2$.

10.2. Find the bilinear form that is polar to the form:

(1) $x_1^2 - 2x_1 x_2 + 3x_1 x_3 + 7x_3^2$;
(2) $-2x_1^2 + 3x_1 x_3 + x_2 x_3$;
(3) $3x_1 x_3 + x_2^2$;
(4) $x_1^2 + 4x_1 x_2 + 4x_1 x_3 - 4x_2^2 - 2x_2 x_3 - x_3^2$.

10.3. Bring to the canonical form the quadratic form of three variables:

(1) $x_1^2 + 2x_1 x_2 + 4x_1 x_3 - 4x_3^2$;
(2) $x_1 x_2 + x_1 x_3 - 6x_2 x_3$;
(3) $-2x_1^2 - 7x_1 x_3$;
(4) $x_1^2 - 4x_1 x_2 + 12x_1 x_3 - x_2^2 + 4x_2 x_3 + 3x_3^2$.

10.4. Bring to the canonical form the quadratic form of four variables:

(1) $x_1^2 + 2x_1 x_2 + 2x_1 x_3 + 2x_1 x_4 + 3x_2^2 + 6x_2 x_3 + 8x_2 x_4 + x_3^2 + 2x_3 x_4 + x_4^2$;
(2) $x_1 x_2 + x_1 x_4 + x_2 x_3 + x_3 x_4$;
(3) $-5x_1^2 + 2x_1 x_4 + 3x_2^2 + x_2 x_3 - 2x_2 x_4 - x_3^2 + 2x_3 x_4 + x_4^2$;
(4) $x_1^2 + x_1 x_2 + x_1 x_3 + x_1 x_4 + x_2^2 + x_2 x_3 + x_2 x_4 + x_3^2 + x_3 x_4 + x_4^2$.

10.5. Which of the following quadratic forms are positively definite, negatively definite and alternating?

(1) $x_1^2 + x_1 x_2 + x_2^2$;
(2) $x_1^2 - 9x_1 x_2 + x_2^2$;
(3) $x_1 x_2 + 2x_1 x_4 + 3x_2 x_3 + 4x_3 x_4$;
(4) $x_1 x_2 + x_3 x_4$.

∗10.6. At what values of the real parameter a is the quadratic form:

$$\omega(x) = ax_1^2 + 2x_1x_2 + (10 - a)x_2^2 \qquad (10.48)$$

positively definite and negatively definite?

∗10.7. Show that the form $\displaystyle\sum_{i,j=1}^{n} \frac{x_ix_j}{i+j-1}$ is positively definite.

Answers and Solutions

10.1 *Answer:*

The functions from items (1) and (4) are bilinear forms.

10.2 *Solution.*

Use the formula (10.14) for the polar form:

(1)

$$A(x, y) = \frac{1}{2}\left[\omega(x + y) - \omega(x) - \omega(y)\right]$$

$$= \frac{1}{2}\left[(x_1 + y_1)^2 - 2(x_1 + y_1)(x_2 + y_2)\right.$$

$$+ 3(x_1 + y_1)(x_3 + y_3) + 7(x_3 + y_3)^2$$

$$\left. - (x_1^2 - 2x_1x_2 + 3x_1x_3 + 7x_3^2) - (y_1^2 - 2y_1y_2 + 3y_1y_3 + 7y_3^2)\right]$$

$$= x_1y_1 - x_1y_2 + \frac{3}{2}x_1y_3 - x_2y_1 + \frac{3}{2}x_3y_1 + 7x_3y_3;$$

(2)

$$A(x, y) = \frac{1}{2}\left[-2(x_1 + y_1)^2 + 3(x_1 + y_1)(x_3 + y_3) + (x_2 + y_2)(x_3 + y_3)\right.$$

$$\left. - (-2x_1^2 + 3x_1x_3 + x_2x_3) - (-2y_1^2 + 3y_1y_3 + y_2y_3)\right]$$

$$= -2x_1y_1 + \frac{3}{2}x_1y_3 + \frac{1}{2}x_2y_3 + \frac{3}{2}x_3y_1 + \frac{1}{2}x_3y_2;$$

(3)

$$A(x, y) = \frac{1}{2}\left[3(x_1 + y_1)(x_3 + y_3) + (x_2 + y_2)^2\right.$$

$$\left. - (3x_1x_3 + x_2^2) - (3y_1y_3 + y_2^2)\right]$$

$$= \frac{3}{2}x_1y_3 + x_2y_2 + \frac{3}{2}x_3y_1;$$

(4)

$$
\begin{aligned}
\mathcal{A}(x, y) = \frac{1}{2}\big[&(x_1 + y_1)(x_1 + y_1) + 4(x_1 + y_1)(x_2 + y_2) \\
&+ 4(x_1 + y_1)(x_3 + y_3) \\
&- 4(x_2 + y_2)^2 - 2(x_2 + y_2)(x_3 + y_3) - (x_3 + y_3)(x_3 + y_3) \\
&- (x_1^2 + 4x_1x_2 + 4x_1x_3 - 4x_2^2 - 2x_2x_3 - x_3^2) \\
&- (y_1^2 + 4y_1y_2 + 4y_1y_3 - 4y_2^2 - 2y_2y_3 - y_3^2)\big] \\
= x_1y_1 &+ 2x_1y_2 + 2x_1y_3 + 2x_2y_1 - x_2y_3 - 4x_2y_2 \\
&+ 2x_3y_1 - x_3y_2 - x_3y_3.
\end{aligned}
$$

10.3 *Solution.*

(1) Transform the expression:

$$
\begin{aligned}
x_1^2 + 2x_1x_2 &+ 4x_1x_3 - 4x_3^2 \\
&= x_1^2 + 2x_1(x_2 + 2x_3) + (x_2 + 2x_3)^2 - (x_2 + 2x_3)^2 - 4x_3^2 \\
&= (x_1 + x_2 + 2x_3)^2 - (x_2 + 2x_3)^2 - (2x_3)^2.
\end{aligned}
$$

Change the variables:

$$
\begin{cases}
y_1 = x_1 + x_2 + 2x_3, \\
y_2 = x_2 + 2x_3, \\
y_3 = 2x_3.
\end{cases}
$$

We obtain the quadratic form in the canonical form:

$$
y_1^2 - y_2^2 - y_3^2.
$$

(2) Change the variables:

$$
\begin{cases}
x_1 = x_1' + x_2', \\
x_2 = x_1' - x_2', \\
x_3 = x_3'.
\end{cases}
$$

Transform the expression:

$$x_1x_2 + x_1x_3 - 6x_2x_3$$
$$= (x_1' + x_2')(x_1' - x_2') + (x_1' + x_2')x_3' - 6(x_1' - x_2')x_3'$$
$$= (x_1')^2 - (x_2')^2 + x_1'x_3' + x_2'x_3' - 6x_1'x_3' + 6x_2'x_3'$$
$$= (x_1')^2 - (x_2')^2 - 5x_1'x_3' + 7x_2'x_3'$$
$$= (x_1')^2 - 5x_1'x_3' + \left(\frac{5}{2}x_3'\right)^2 - \left(\frac{5}{2}x_3'\right)^2 + 7x_2'x_3' - (x_2')^2$$
$$= \left(x_1' - \frac{5}{2}x_3'\right)^2 + 7x_2'x_3' - (x_2')^2 - \left(\frac{5}{2}x_3'\right)^2$$
$$= \left(x_1' - \frac{5}{2}x_3'\right)^2 - \left(x_2' - \frac{7}{2}x_3'\right)^2 - \left(\frac{5}{2}x_3'\right)^2 - \left(\frac{7}{2}x_3'\right)^2$$
$$= \left(x_1' - \frac{5}{2}x_3'\right)^2 - \left(x_2' - \frac{7}{2}x_3'\right)^2 + 6(x_3')^2.$$

Change the variables:

$$\begin{cases} y_1 = x_1' - \dfrac{5}{2}x_3' = \dfrac{1}{2}(x_1 + x_2 - 5x_3), \\ y_2 = x_2' - \dfrac{7}{2}x_3' = \dfrac{1}{2}(x_1 - x_2 - 7x_3), \\ y_3 = \sqrt{6}x_3'. \end{cases}$$

We obtain the quadratic form in the canonical form:

$$y_1^2 - y_2^2 + y_3^2.$$

(3) Transform the expression:

$$-2x_1^2 - 7x_1x_3$$
$$= -\left(\sqrt{2}x_1\right)^2 - 7x_1x_3 - \left(\frac{7\sqrt{2}}{4}x_3\right)^2 + \left(\frac{7\sqrt{2}}{4}x_3\right)^2$$
$$= \left(\sqrt{2}x_1 + \frac{7\sqrt{2}}{4}x_3\right)^2 + \left(\frac{7\sqrt{2}}{4}x_3\right)^2.$$

Change the variables:

$$\begin{cases} y_1 = \sqrt{2}x_1 + \dfrac{7\sqrt{2}}{4}x_3, \\ y_2 = \dfrac{7\sqrt{2}}{4}x_3. \end{cases}$$

We obtain the quadratic form in the canonical form:

$$-y_1^2 + y_2^2.$$

(4) Transform the expression:

$$x_1^2 - 4x_1x_2 + 12x_1x_3 - x_2^2 + 4x_2x_3 + 3x_3^2$$

$$= x_1^2 - 4x_1(x_2 - 3x_3) + (2x_2 - 6x_3)^2$$

$$- (2x_2 - 6x_3)^2 - x_2^2 + 4x_2x_3 + 3x_3^2$$

$$= (x_1 - 2x_2 + 6x_3)^2 - 4x_2^2 + 24x_2x_3 - 36x_3^2 - x_2^2 + 4x_2x_3 + 3x_3^2$$

$$= (x_1 - 2x_2 + 6x_3)^2 - 5x_2^2 + 28x_2x_3 - 33x_3^2$$

$$= (x_1 - 2x_2 + 6x_3)^2 - \left(\sqrt{5}x_2\right)^2 + 28x_2x_3$$

$$- \left(\frac{14\sqrt{5}}{5}x_3\right)^2 + \left(\frac{14\sqrt{5}}{5}x_3\right)^2 - 33x_3^2$$

$$= (x_1 - 2x_2 + 6x_3)^2 - \left(\sqrt{5}x_2 - \frac{14\sqrt{5}}{5}x_3\right)^2 + \frac{31}{5}x_3^2.$$

Change the variables:

$$\begin{cases} y_1 = x_1 - 2x_2 + 6x_3, \\ y_2 = \sqrt{5}\left(x_2 - \dfrac{14}{5}x_3\right), \\ y_3 = \sqrt{\dfrac{31}{5}}x_3. \end{cases}$$

We obtain the quadratic form in the canonical form:

$$y_1^2 - y_2^2 + y_3^2.$$

10.4 *Solution.*

(1) Transform the expression:

$$x_1^2 + 2x_1x_2 + 2x_1x_3 + 2x_1x_4 + 3x_2^2 + 6x_2x_3 + 8x_2x_4 + x_3^2 + 2x_3x_4 + x_4^2$$

$$= x_1^2 + 2x_1(x_2 + x_3 + x_4) + (x_2 + x_3 + x_4)^2 - (x_2 + x_3 + x_4)^2$$

$$+ 3x_2^2 + 6x_2x_3 + 8x_2x_4 + x_3^2 + 2x_3x_4 + x_4^2$$

$$= (x_1 + x_2 + x_3 + x_4)^2 - x_2^2 - 2x_2x_3 - 2x_2x_4 - x_3^2 - 2x_3x_4 - x_4^2$$

$$+ 3x_2^2 + 6x_2x_3 + 8x_2x_4 + x_3^2 + 2x_3x_4 + x_4^2$$

$$= (x_1 + x_2 + x_3 + x_4)^2 + 2x_2^2 + 4x_2x_3 + 6x_2x_4$$

$$= (x_1 + x_2 + x_3 + x_4)^2 + \left(\sqrt{2}x_2\right)^2 + 2x_2(2x_3 + 3x_4)$$

$$+ \frac{1}{2}(2x_3 + 3x_4)^2 - \frac{1}{2}(2x_3 + 3x_4)^2$$

$$= (x_1 + x_2 + x_3 + x_4)^2 + \left(\sqrt{2}x_2 + \sqrt{2}x_3 + \frac{3\sqrt{2}}{2}x_4\right)^2$$

$$- \left(\sqrt{2}x_3 + \frac{3\sqrt{2}}{2}x_4\right)^2.$$

Change the variables:

$$\begin{cases} y_1 = x_1 + x_2 + x_3 + x_4, \\ y_2 = \sqrt{2}x_2 + \sqrt{2}x_3 + \dfrac{3\sqrt{2}}{2}x_4, \\ y_3 = \sqrt{2}x_3 + \dfrac{3\sqrt{2}}{2}x_4. \end{cases}$$

We obtain the quadratic form in the canonical form:

$$y_1^2 + y_2^2 - y_3^2.$$

(2) Change the variables:

$$\begin{cases} x_1 = x_1' + x_2', \\ x_2 = x_1' - x_2', \\ x_3 = x_3' + x_4', \\ x_4 = x_3' - x_4'. \end{cases}$$

Transform the expression:

$$x_1x_2 + x_1x_4 + x_2x_3 + x_3x_4$$

$$= (x_1' + x_2')(x_1' - x_2') + (x_1' + x_2')(x_3' - x_4')$$

$$+ (x_1' - x_2')(x_3' + x_4') + (x_3' + x_4')(x_3' - x_4')$$

$$= (x_1')^2 - (x_2')^2 + 2x_1'x_3' - 2x_2'x_4' + (x_3')^2 - (x_4')^2$$

$$= (x_1')^2 + 2x_1'x_3' + (x_3')^2 - (x_2')^2 - 2x_2'x_4' - (x_4')^2$$

$$= (x_1' + x_3')^2 - (x_2' + x_4')^2.$$

Change the variables:

$$\begin{cases} y_1 = x_1' + x_3' = \dfrac{1}{2}(x_1 + x_2 + x_3 + x_4), \\[2mm] y_2 = x_2' + x_4' = \dfrac{1}{2}(x_1 - x_2 + x_3 - x_4). \end{cases}$$

We obtain the quadratic form in the canonical form:

$$y_1^2 - y_2^2.$$

(3) Transform the expression:

$$- 5x_1^2 + 2x_1x_4 + 3x_2^2 + x_2x_3 - 2x_2x_4 - x_3^2 + 2x_3x_4 + x_4^2$$

$$= x_4^2 + 2x_4(x_1 - x_2 + x_3) + (x_1 - x_2 + x_3)^2$$

$$- (x_1 - x_2 + x_3)^2 - 5x_1^2 + 3x_2^2 + x_2x_3 - x_3^2$$

$$= (x_1 - x_2 + x_3 + x_4)^2 - 6x_1^2 + 2x_1x_2 + 2x_2^2 - 2x_1x_3 + 3x_2x_3 - 2x_3^2$$

$$= (x_1 - x_2 + x_3 + x_4)^2 - \left(\sqrt{6}x_1\right)^2 + 2x_1(x_2 - x_3) - \left(\frac{\sqrt{6}}{6}(x_2 - x_3)\right)^2$$

$$+ \left(\frac{\sqrt{6}}{6}(x_2 - x_3)\right)^2 + 2x_2^2 + 3x_2x_3 - 2x_3^2$$

$$= (x_1 - x_2 + x_3 + x_4)^2 - \left(\sqrt{6}x_1 - \frac{\sqrt{6}}{6}(x_2 - x_3)\right)^2$$

$$+ \frac{13}{6}x_2^2 + \frac{8}{3}x_2x_3 - \frac{11}{6}x_3^2$$

$$= (x_1 - x_2 + x_3 + x_4)^2 - \left(\sqrt{6}x_1 - \frac{\sqrt{6}}{6}(x_2 - x_3) \right)^2$$

$$+ \left(\sqrt{\frac{13}{6}}x_2 \right)^2 + \frac{8}{3}x_2x_3 + \left(\frac{8\sqrt{78}}{78}x_3 \right)^2 - \left(\frac{8\sqrt{78}}{78}x_3 \right)^2 - \frac{11}{6}x_3^2$$

$$= (x_1 - x_2 + x_3 + x_4)^2 - \left(\sqrt{6}x_1 - \frac{\sqrt{6}}{6}(x_2 - x_3) \right)^2$$

$$+ \left(\sqrt{\frac{13}{6}}x_2 + \frac{8\sqrt{78}}{78}x_3 \right)^2 - \left(\sqrt{\frac{69}{26}}x_3 \right)^2 .$$

Change the variables:

$$\begin{cases} y_1 = x_1 - x_2 + x_3 + x_4, \\ y_2 = \sqrt{6}x_1 - \dfrac{\sqrt{6}}{6}(x_2 - x_3), \\ y_3 = \sqrt{\dfrac{13}{6}}x_2 + \dfrac{8\sqrt{78}}{78}x_3, \\ y_4 = \sqrt{\dfrac{69}{26}}x_3. \end{cases}$$

We obtain the quadratic form in the canonical form:

$$y_1^2 - y_2^2 + y_3^2 - y_4^2.$$

(4) Transform the expression:

$$x_1^2 + x_1x_2 + x_1x_3 + x_1x_4 + x_2^2 + x_2x_3 + x_2x_4 + x_3^2 + x_3x_4 + x_4^2$$

$$= x_1^2 + x_1(x_2 + x_3 + x_4) + \frac{1}{2}(x_2 + x_3 + x_4)^2 - \frac{1}{2}(x_2 + x_3 + x_4)^2$$

$$+ x_2^2 + x_2x_3 + x_2x_4 + x_3^2 + x_3x_4 + x_4^2$$

$$= \left(x_1 + \frac{1}{2}(x_2 + x_3 + x_4) \right)^2 + \frac{3}{4}x_2^2 + \frac{1}{2}x_2x_3 + \frac{1}{2}x_2x_4$$

$$+ \frac{3}{4}x_3^2 + \frac{1}{2}x_3x_4 + \frac{3}{4}x_4^2$$

$$= \left(x_1 + \frac{1}{2}(x_2 + x_3 + x_4) \right)^2 + \left(\frac{\sqrt{3}}{2}x_2 \right)^2$$

$$+ \frac{1}{2}x_2(x_3 + x_4) + \left(\frac{\sqrt{3}}{6}(x_3 + x_4) \right)^2$$

$$= \left(x_1 + \frac{1}{2}(x_2 + x_3 + x_4) \right)^2 + \left(\frac{\sqrt{3}}{2}x_2 + \frac{\sqrt{3}}{6}(x_3 + x_4) \right)^2$$

$$+ \frac{2}{3}x_3^2 + \frac{1}{3}x_3x_4 + \frac{2}{3}x_4^2$$

$$= \left(x_1 + \frac{1}{2}(x_2 + x_3 + x_4) \right)^2 + \left(\frac{\sqrt{3}}{2}x_2 + \frac{\sqrt{3}}{6}(x_3 + x_4) \right)^2$$

$$+ \left(\sqrt{\frac{2}{3}}x_3 \right)^2 + \frac{1}{3}x_3x_4 + \left(\frac{\sqrt{6}}{12}x_4 \right)^2 - \left(\frac{\sqrt{6}}{12}x_4 \right)^2 + \frac{2}{3}x_4^2$$

$$= \left(x_1 + \frac{1}{2}(x_2 + x_3 + x_4) \right)^2 + \left(\frac{\sqrt{3}}{2}x_2 + \frac{\sqrt{3}}{6}(x_3 + x_4) \right)^2$$

$$+ \left(\sqrt{\frac{2}{3}}x_3 + \frac{\sqrt{6}}{12}x_4 \right)^2 + \left(\frac{\sqrt{10}}{4}x_4 \right)^2 .$$

Change the variables:

$$\begin{cases} y_1 = x_1 + \dfrac{1}{2}(x_2 + x_3 + x_4), \\[2mm] y_2 = \dfrac{\sqrt{3}}{2}x_2 + \dfrac{\sqrt{3}}{6}(x_3 + x_4), \\[2mm] y_3 = \sqrt{\dfrac{2}{3}}x_3 + \dfrac{\sqrt{6}}{12}x_4, \\[2mm] y_4 = \dfrac{\sqrt{10}}{4}x_4. \end{cases}$$

We obtain the quadratic form in the canonical form:

$$y_1^2 + y_2^2 + y_3^2 + y_4^2.$$

10.5 *Solution.*

(1) Write the matrix of quadratic form:

$$\begin{bmatrix} 1 & \dfrac{1}{2} \\[2mm] \dfrac{1}{2} & 1 \end{bmatrix}.$$

Use Sylvester's criterion. For this, compute the corner minors:

$$\left| 1 \right| = 1 > 0, \qquad \begin{vmatrix} 1 & \dfrac{1}{2} \\[2mm] \dfrac{1}{2} & 1 \end{vmatrix} = 1 - \left(\dfrac{1}{2}\right)^2 > 0.$$

Therefore, the form is positively definite.

(2) Write the matrix of quadratic form:

$$\begin{bmatrix} 1 & -\dfrac{9}{2} \\[2mm] -\dfrac{9}{2} & 1 \end{bmatrix}.$$

Use Sylvester's criterion. For this, compute the corner minors:

$$\left| 1 \right| = 1 > 0, \qquad \begin{vmatrix} 1 & -\dfrac{9}{2} \\[2mm] -\dfrac{9}{2} & 1 \end{vmatrix} = 1 - \left(\dfrac{9}{2}\right)^2 < 0.$$

Therefore, the form is alternating.

(3) Assume that $x_3 = x_4 = 0$. In this case, only one non-zero summand $x_1 x_2$ remains in the quadratic form. It is obvious that this summand can take both positive and negative values. Therefore, the form is alternating.

(4) Assume that $x_3 = 0$. In this case, only one non-zero summand $x_1 x_2$ remains in the quadratic form. It is obvious that this summand can be both positive and negative. Therefore, the form is alternating.

10.6 *Solution.*

Write the matrix of quadratic form:

$$\begin{bmatrix} a & 1 \\ 1 & 10 - a \end{bmatrix}.$$

Use Sylvester's criterion. For this, compute the corner minors:

$$\left| a \right| = a; \qquad \begin{vmatrix} a & 1 \\ 1 & 10 - a \end{vmatrix} = a(10 - a) - 1.$$

In order for the form to be positively definite, it is necessary and sufficient that both determinants should be positive:

$$\begin{cases} a > 0, \\ a(10 - a) - 1 > 0; \end{cases} \Rightarrow \begin{cases} a > 0, \\ a^2 - 10a + 1 < 0; \end{cases}$$

$$\Rightarrow \begin{cases} a > 0, \\ a \in \left(5 - 2\sqrt{6}, 5 + 2\sqrt{6}\right). \end{cases}$$

In order for the form to be negatively definite, it is necessary and sufficient that the corner minor of the first order should be negative, and that of the second order should be positive:

$$\begin{cases} a < 0, \\ a \in \left(5 - 2\sqrt{6}, 5 + 2\sqrt{6}\right). \end{cases}$$

This system is inconsistent; therefore, the form cannot be negatively definite at any values of a.

Consider the cases when the corner minors are equal to zero:

$$a = 0, \quad \omega(x) = 2x_1 x_2 + 10x_2^2;$$

$\omega(x)$ can take a negative value, for example, at $x_1 = -10$, $x_2 = 1$;

$$a = 5 - 2\sqrt{6},$$

$$\omega(x) = \left(5 - 2\sqrt{6}\right) x_1^2 + 2x_1 x_2 + \left(5 + 2\sqrt{6}\right) x_2^2$$

$$= \left(\sqrt{5 - 2\sqrt{6}} x_1\right)^2 + 2x_1 x_2 + \left(\sqrt{5 + 2\sqrt{6}} x_2\right)^2$$

$$= \left(\sqrt{5 - 2\sqrt{6}} x_1 + \sqrt{5 + 2\sqrt{6}} x_2\right)^2.$$

It is clear that $\forall x \neq 0$ the inequality $\omega(x) > 0$ is fulfilled.

At $a = 5 + 2\sqrt{6}$, the reasoning is similar.

We obtain the final answer: the form is positively definite if and only if $a \in \left[5 - 2\sqrt{6}, 5 + 2\sqrt{6}\right]$.

10.7 *Solution.*

Transform the expression, having introduced integration by the auxiliary variable t:

$$\sum_{i,j=1}^{n} \frac{x_i x_j}{i+j-1} = \sum_{i,j=1}^{n} \int_0^1 x_i x_j t^{i+j-2} dt = \int_0^1 \left(\sum_{i=1}^{n} x_i t^{i-1} \right)^2 dt.$$

As is easy to see, the expression $\left(\sum_{i=1}^{n} x_i t^{i-1} \right)^2$ is always greater than zero at the non-zero values of $x_1, x_2, \ldots, x_n$ and $t > 0$.

The initial quadratic form is positively definite, since it is the integral of the positively definite form.

Chapter 11
Curves of the Second Order

The general **equation of curve of the second order** has the form

$$Ax^2 + 2Bxy + Cy^2 + 2Dx + 2Ey + F = 0, \qquad (11.1)$$

where the real coefficients A, B, C are not equal to zero simultaneously.

The expression $Ax^2 + 2Bxy + Cy^2$ is referred to as the **quadratic term** of the equation of this curve, $2Dx + 2Ey$ is referred to as the **linear term**, and F is referred to as the **constant term**.

In the so-called **canonical system of coordinates**, the equation of curve of the second order takes the simplest **canonical** form [8].

Let us consider the classification of curves of the second order based on their canonical form.

11.1 Ellipse

Ellipse is a curve of the second order that in some Cartesian rectangular system of coordinates is defined by the equation

$$\frac{x^2}{a^2} + \frac{y^2}{b^2} = 1, \qquad (11.2)$$

where $a \geqslant b > 0$. The numbers a and b are referred to as the **major** and **minor** **semiaxes** of the ellipse, respectively.

The points $(\pm a, 0)$ and $(0, \pm b)$ are called the **vertices** of the ellipse, and $(\pm c, 0)$, where $c = \sqrt{a^2 - b^2}$, are its **foci**. We will denote the foci by F_1 and F_2. Figure 11.1 schematically shows an ellipse in the canonical system of coordinates.

In case of equality of the constants $a = b$, the ellipse degenerates into a **circle**.

© Springer Nature Switzerland AG 2021
S. Kurgalin, S. Borzunov, *Algebra and Geometry with Python*,
https://doi.org/10.1007/978-3-030-61541-3_11

Fig. 11.1 Ellipse
$\frac{x^2}{a^2} + \frac{y^2}{b^2} = 1$. The directrices
$x = \pm a/\varepsilon$ are shown by the
dotted lines

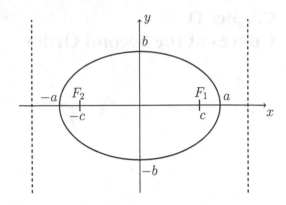

From the geometrical point of view, the ellipse is a set of points of a plane, for each of which the sum of distances to two specified points F_1 and F_2 (foci) is constant and equal to $2a > F_1 F_2$. The **focal distance**, i.e. the distance between F_1 and F_2, is equal to $2c$.

The non-negative number $\varepsilon = c/a < 1$ defines the degree of "compression" of the ellipse on the abscissa axis and is referred to as **eccentricity**. The greater is the value ε, the more pronounced is the "compression" (see Fig. 11.2). As is easy to see, the eccentricity of the circle is equal to $\varepsilon = \sqrt{a^2 - b^2}/a = 0$. In this connection, we can say that the ellipse is obtained from the circle through its compression on the Ox-axis, when the ordinates of all its points decrease in the same proportion b/a.

Theorem 11.1 *The distance from an arbitrary point $P(x, y)$, which belongs to the ellipse, to each focus, is equal to*

$$r_1 = PF_1 = a - \varepsilon x, \quad r_2 = PF_2 = a + \varepsilon x. \tag{11.3}$$

The ellipse has two **directrices** — the straight lines of the form $x = \pm a/\varepsilon$. They are shown in Fig. 11.1 as dotted lines. Directrices are not defined for the circle. The

Fig. 11.2 Ellipses with
different eccentricities: $\varepsilon = 0$
(solid line), $\varepsilon = 0.7$ (dotted
line), $\varepsilon = 0.9$ (dash-and-dot
line)

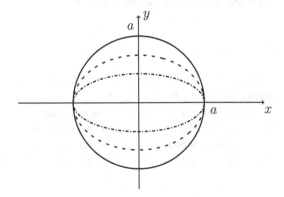

directrix and the focus that lie on the same side of the axis Oy will be considered to be corresponding to each other.

Theorem 11.2 *For any point that belongs to the ellipse, the ratio of its distance to the focus to its distance to the corresponding directrix is equal to the eccentricity of the ellipse.*

Let us consider the mutual arrangement of the ellipse and the straight line. An arbitrary straight line either does not intersect the ellipse or intersects it at one or two points. If there is the only common point, such a straight line is a **tangent**. Exactly one tangent passes through any point on the ellipse.

The equation of tangent to the ellipse at the point $P_0(x_0, y_0)$ has the form:

$$\frac{xx_0}{a^2} + \frac{yy_0}{b^2} = 1. \tag{11.4}$$

The ellipse has the following **optical property**: the light beams originating from one focus, after mirror reflection from the ellipse, pass through another focus, i.e. focus on it. This explains the origin of the term "focus" borrowed from optics.

Example 11.1 Let us find the intersection points of the coordinate axes and the tangent drawn to the ellipse $\frac{x^2}{9} + \frac{y^2}{4} = 1$ at the point with the coordinates $(3/2, \sqrt{3})$.

Solution Denote the tangency point by P_0 and verify that it belongs to the ellipse:

$$\frac{(3/2)^2}{9} + \frac{(\sqrt{3})^2}{4} = 1, \text{ or } 1 = 1 - \text{true.} \tag{11.5}$$

Taking into account that the semiaxes of the ellipse are equal to $a = 3$ and $b = 2$, substitute the coordinates $P_0(x_0, y_0)$ into the equation of tangent (11.4):

$$\frac{(3/2)x}{9} + \frac{\sqrt{3}y}{4} = 1. \tag{11.6}$$

Therefore, the general equation of straight line has the form $2x + 3\sqrt{3}y - 12 = 0$ (see Fig. 11.3). The intersection points of this line with the coordinate axes are $(6, 0)$ and $(0, 4/\sqrt{3})$. □

The idea of the ellipse as the compressed circle results in an alternative method of specifying the ellipse **in parametric form.**

Transform the coordinates in accordance with the formulae

$$x' = x, \quad y' = \frac{b}{a}y. \tag{11.7}$$

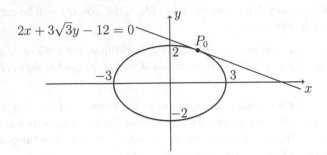

$2x + 3\sqrt{3}y - 12 = 0$

Fig. 11.3 To the Example 11.1 The ellipse $\dfrac{x^2}{9} + \dfrac{y^2}{4} = 1$ and the tangent to it at the point $P_0(3/2, \sqrt{3})$

After such a transformation, the circle $x^2 + y^2 = a^2$, in the new system of coordinates, turns into the ellipse $\dfrac{(x')^2}{a^2} + \dfrac{(y')^2}{b^2} = 1$. As is known, for example, from the course of mathematical analysis [76], any circle can be specified in parametric form as

$$\begin{cases} x = a\cos t, \\ y = a\sin t, \end{cases} \tag{11.8}$$

where the real number t takes the values that belong to the interval $[0, 2\pi)$.

From (11.7) follows that the parametric representation of the ellipse with the semiaxes a and b will have the form:

$$\begin{cases} x = a\cos t, \\ y = b\sin t, \end{cases} \quad \text{where } 0 \leqslant t < 2\pi. \tag{11.9}$$

The parameter t is referred to as the **anomaly of eccentricity**.

11.2   Hyperbola

Hyperbola is a curve of the second order, which in some Cartesian rectangular system of coordinates is defined by the equation

$$\frac{x^2}{a^2} - \frac{y^2}{b^2} = 1, \tag{11.10}$$

where $a, b > 0$ (see Fig. 11.4). The numbers a and b are called **real** and **imaginary** **semiaxes** of the hyperbola, respectively.

Unlike the ellipse, which is a connected curve, the hyperbola consists of two connected components: left and right branches (sheets of the hyperbola).

Vertices of hyperbola are the points $(\pm a, 0)$. **Foci** of hyperbola are the points $F_1(-c, 0)$ and $F_2(c, 0)$, where $c = \sqrt{a^2 + b^2}$.

A hyperbola has **asymptotes** $y = \pm \dfrac{b}{a} x$ that define the run of the curve for infinitely great values of coordinates.

A hyperbola contains only those points of Cartesian plane, the modulus of the difference of distances of each of which to the two given points F_1 and F_2 (foci) is constant and equal to $2a < F_1 F_2$. The focal distance is equal to $2c$.

For the hyperbola, a concept of **eccentricity** $\varepsilon = c/a > 1$ is also introduced.

Theorem 11.3 *The distances r_1, r_2 from an arbitrary point $P(x, y)$ of the hyperbola to each focus are equal*

$$r_1 = PF_1 = abs(a - \varepsilon x), \quad r_2 = PF_2 = abs(a + \varepsilon x). \tag{11.11}$$

The **directrices** of the hyperbola are specified in the canonical system of coordinates by the equations $x = \dfrac{a}{\varepsilon}$ and $x = -\dfrac{a}{\varepsilon}$ (see Fig. 11.4, where the directrices are shown by dotted lines). The directrix and the focus that lie one the same side of the axis Oy will be considered to be corresponding to each other.

Theorem 11.4 *For an arbitrary point that lies on the hyperbola, the ratio of its distance to the focus to the distance to the corresponding directrix is equal to the eccentricity of the hyperbola.*

The equation of tangent to hyperbola at the point $P_0(x_0, y_0)$ has the form:

$$\frac{x x_0}{a^2} - \frac{y y_0}{b^2} = 1. \tag{11.12}$$

Optical property of hyperbola: the light from the source situated at one of the hyperbola's foci is reflected by the second branch of the hyperbola so that the continuations of reflections of the beams intersect at the second focus.

Fig. 11.4 Hyperbola $\dfrac{x^2}{a^2} - \dfrac{y^2}{b^2} = 1$. The asymptotes $y = \pm \dfrac{b}{a} x$ are shown by thin solid lines, the directrices $x = \pm \dfrac{a}{\varepsilon}$—by dotted lines

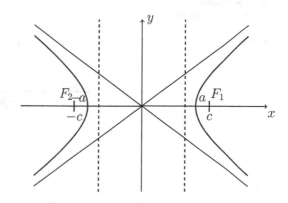

Both for the ellipse and the hyperbola, in the canonical system of coordinates, the origin of coordinates $O(0, 0)$ is the centre of symmetry of the curve, this is why the ellipse and the hyperbola belong to the class of **central** curves.

In conclusion of the section devoted to the hyperbola, we will provide its representation in parametric form:

$$\begin{cases} x = \pm a \cosh t, \\ y = \quad b \sinh t, \end{cases} \tag{11.13}$$

where $t \in \mathbb{R}$, $\cosh x = (e^x + e^{-x})/2$ is the hyperbolic cosine, and $\sinh x = (e^x - e^{-x})/2$ is the hyperbolic sine. In the first equation of this system, the sign "+" corresponds to the right branch of the hyperbola, and the sign "−" corresponds to the left branch.

11.3 Parabola

Parabola is the noncentral curve of the second order, defined by the canonical equation in Cartesian rectangular system of coordinates

$$y^2 = 2px, \tag{11.14}$$

where $p > 0$ is the **focal parameter** of the parabola, or simply the **parameter**.

The **vertex** of the parabola is the origin of coordinates $(0, 0)$, the **focus** is the point $F(p/2, 0)$. The **directrix** of the parabola is the straight line specified by the equation $x = -p/2$ (see Fig. 11.5).

From the geometrical point of view, the parabola is the set of the points of the plane, for each of which the distance to the focus F is equal to the distance to the directrix. The distance from the focus to the directrix is equal to the parameter.

Fig. 11.5 Parabola
$y^2 = 2px$. The directrix
$x = -p/2$ is shown by a
dotted line

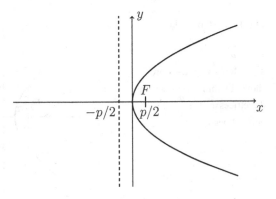

Theorem 11.5 *The distance from the point $P(x, y)$ lying on the parabola to its focus is equal to*

$$r = x + p/2. \tag{11.15}$$

The parabola is assigned the eccentricity equal to one.

For the parabola, the analogue of theorems 11.2 and 11.4 repeats its geometrical property, which allows formulating the following generalized statement.

Theorem 11.6 *For an arbitrary point that lies on the ellipse, hyperbola or parabola, the ratio of the distance from this point to the focus to the distance to the corresponding directrix (to the only directrix in case of parabola) is equal to the eccentricity of the curve.*

The equation of tangent to parabola at the point $P_0(x_0, y_0)$ has the form:

$$yy_0 = p(x + x_0). \tag{11.16}$$

Optical property of parabola: the light beams originating from the focus, after mirror reflection from the parabola, will be directed parallel to its axis of symmetry. Note that this property of parabola underlies the arrangement of parabolic mirrors and parabolic antennas.

11.4 Degenerate Curves

Among the **degenerate** curves of the second order are the curves whose canonical form is different from the equation of ellipse, parabola or hyperbola. There exist the following types of such curves: imaginary ellipse, pair of intersecting lines, pair of imaginary intersecting lines, pair of parallel lines, pair of imaginary parallel lines and pair of coincident lines. Thus, **nondegenerate** curves of the second order are ellipse, hyperbola and parabola (and they alone).

11.4.1 Imaginary Ellipse

Imaginary ellipse is described in the canonical system of coordinates by the equation

$$\frac{x^2}{a^2} + \frac{y^2}{b^2} = -1, \tag{11.17}$$

where $a \geqslant b > 0$.

Since the sum of squares of real numbers cannot be equal to a negative number, this curve contains no points. The term "imaginary ellipse" is associated with the fact that when changing the variables $x' = ix$, $y' = iy$, where $i = \sqrt{-1}$, in the new coordinates we obtain the equation of ellipse (11.2).

11.4.2 Pair of Intersecting Lines

The equation of the form

$$\frac{x^2}{a^2} - \frac{y^2}{b^2} = 0 \tag{11.18}$$

corresponds to the pair of intersecting lines. Their intersection point is the origin of coordinates.

Having written the Eq. (11.18) in the form

$$\left(\frac{x}{a} - \frac{y}{b}\right)\left(\frac{x}{a} + \frac{y}{b}\right) = 0, \tag{11.19}$$

we conclude that it is satisfied by all points of two lines $y = bx/a$ and $y = -bx/a$ intersecting at the origin of coordinates.

11.4.3 Pair of Imaginary Intersecting Lines

Pair of imaginary intersecting lines is described by the equation

$$\frac{x^2}{a^2} + \frac{y^2}{b^2} = 0. \tag{11.20}$$

The sum of squares is equal to zero if and only if each of the summands is equal to zero: $\frac{x}{a} = 0$ and $\frac{y}{b} = 0$, i.e. $x = y = 0$. This condition specifies the only point that coincides with the origin of coordinates.

11.4.4 Pair of Parallel Lines

The equation

$$x^2 - a^2 = 0, \text{ where } a \neq 0, \tag{11.21}$$

defines two parallel vertical lines $x = a$ and $x = -a$.

11.4.5   Pair of Imaginary Parallel Lines

The curve of the second order, specified in the canonical system of coordinates by the equation

$$x^2 + a^2 = 0 \qquad (11.22)$$

under condition $a \neq 0$, contains no real points.

11.4.6   Pair of Coincident Lines

And the final type of the general equation of curve of the second order is

$$x^2 = 0. \qquad (11.23)$$

This equation is satisfied by the coordinates of all points lying on the ordinate axis.

As a result, based on the canonical form of curves of the second order, they are subdivided into nine above classes, the most practically important being ellipse, hyperbola and parabola.

11.5   Algorithms for Computing the Coordinates of the Tangent Points of Second Order Curve and the Straight Line

Consider a nondegenerate curve of the second order, for example, the ellipse $\dfrac{x^2}{a^2} + \dfrac{y^2}{b^2} = 1$, and an arbitrary point of Cartesian plane $P_1(x_1, y_1)$. Let us provide the algorithm that computes the coordinates of the tangency points of the said curve and the lines passing through P_1 (see Fig. 11.6).

We will begin with deducing the analytical relations for the coordinates of the sought points. Denote the tangency points by $T_1(x_{t1}, y_{t1})$ and $T_2(x_{t2}, y_{t2})$. In Chap. 7 "Equation of a Straight Line on a Plane" it is shown that the equation of the family of lines passing through the set point $P(x_1, y_1)$ can be written in the form $y - y_1 = k(x - x_1)$, where $k < \infty$ is the slope. The case of a vertical tangent will be discussed separately.

Fig. 11.6 Tangents to the ellipse $\dfrac{x^2}{a^2} + \dfrac{y^2}{b^2} = 1$, passing through a point P_1

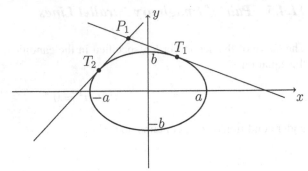

It is known that the tangent to the ellipse intersects it exactly at one point (see page 359). Therefore, the unknown slope k takes such a value that the nonlinear system of equations

$$\begin{cases} y - y_1 = k(x - x_1), \\ \dfrac{x^2}{a^2} + \dfrac{y^2}{b^2} = 1 \end{cases} \tag{11.24}$$

has a unique solution. Having substituted the first equation of this system into the second one, we obtain the quadratic equation

$$(a^2k^2 + b^2)x^2 - 2a^2k(kx_1 - y_1)x + a^2(kx_1 - y_1)^2 - a^2b^2 = 0. \tag{11.25}$$

The necessary and sufficient condition for the quadratic equation to have the only root is the discriminant's equality to zero

$$D = a^4k^2(kx_1 - y_1)^2 + a^2\left(a^2k^2 + b^2\right)\left(b^2 - (kx_1 - y_1)^2\right)$$

$$= a^2b^2\left(a^2k^2 + b^2 - (kx_1 - y_1)^2\right) = 0. \tag{11.26}$$

Solution of the equation $D = 0$ relative to the variable k allows writing the values of the slope:

$$k = \frac{x_1y_1 \pm \sqrt{b^2x_1^2 + a^2y_1^2 - a^2b^2}}{x_1^2 - a^2}. \tag{11.27}$$

Further, let us consider three cases depending on the mutual arrangement of the point P_1 and the ellipse.

1. The point P_1 lies inside the ellipse, i.e. $\dfrac{x_1^2}{a^2} + \dfrac{y_1^2}{b^2} < 1$. The radical expression $x_1^2 b^2 + y_1^2 a^2 - a^2 b^2$ in (11.27) is negative, and there are no real k in this case. Therefore, it is impossible to draw a tangent through the point P_1, which is situated inside the ellipse.

2. The point P_1 lies on the ellipse, i.e. $\dfrac{x_1^2}{a^2} + \dfrac{y_1^2}{b^2} = 1$. Then the Eq. (11.26) has the only root $k = \dfrac{x_1 y_1}{x_1^2 - a^2}$, and P_1 is a tangency point.

3. The point P_1 lies outside the ellipse, $\dfrac{x_1^2}{a^2} + \dfrac{y_1^2}{b^2} > 1$. In this case, there are two possible values of k that satisfy two tangents. We will find the coordinates of the tangency points by substituting (11.27) into the Eq. (11.25):

$$x_{t1,2} = \frac{a^2 k(kx_1 - y_1)}{b^2 + a^2 k^2}, \qquad y_{t1,2} = -\frac{b^2(kx_1 - y_1)}{b^2 + a^2 k^2}. \tag{11.28}$$

Now, let us consider a special case when $x_1 = \pm a$, and one of the tangents is vertical (see Fig. 11.7).

In this case, one of the tangency points has the coordinates $(a, 0)$ or $(-a, 0)$, and the second one is computed based on the equation $D = 0$, which leads to the condition $k = \pm \dfrac{y_1^2 - b^2}{2ay_1}$ and, consequently, to the coordinates $\left(\pm a \dfrac{b^2 - y_1^2}{b^2 + y_1^2}, \dfrac{2b^2 y_1}{b^2 + y_1^2} \right)$, if $y_1 \neq 0$, and to the equation $x = \pm a$, if $y_1 = 0$.

As a result, we formulate the algorithm for finding the coordinates of the tangency points of the straight line and the ellipse.

1. If $\dfrac{x_1^2}{a^2} + \dfrac{y_1^2}{b^2} < 1$, then there are no tangency points.

2. If $\dfrac{x_1^2}{a^2} + \dfrac{y_1^2}{b^2} = 1$, then there is the only tangency point (x_1, y_1).

Fig. 11.7 Tangents to the ellipse $\dfrac{x^2}{a^2} + \dfrac{y^2}{b^2} = 1$, when one of the them is vertical

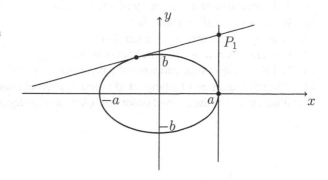

3. If $x_1 = \pm a$, then there are two tangency points

$(x_1, 0)$ and $\left(\pm a \dfrac{b^2 - y_1^2}{b^2 + y_1^2}, \dfrac{2b^2 y_1}{b^2 + y_1^2} \right)$.

In all other cases, k_1 and k_2 are calculated by the formulae:

$$k_1 = \frac{x_1 y_1 + \sqrt{b^2 x_1^2 + a^2 y_1^2 - a^2 b^2}}{x_1^2 - a^2}, k_2 = \frac{x_1 y_1 - \sqrt{b^2 x_1^2 + a^2 y_1^2 - a^2 b^2}}{x_1^2 - a^2},$$

(11.29)

and the sought tangency points will have the coordinates:

$$\left(\frac{a^2 k_1 (k_1 x_1 - y_1)}{b^2 + a^2 k_1^2}, -\frac{b^2 (k_1 x_1 - y_1)}{b^2 + a^2 k_1^2} \right), \left(\frac{a^2 k_2 (k_2 x_1 - y_1)}{b^2 + a^2 k_2^2}, -\frac{b^2 (k_2 x_1 - y_1)}{b^2 + a^2 k_2^2} \right).$$

(11.30)

Example 11.2 Let us apply the above algorithm to the values of the parameters $a = 8, b = 5$ and $P_1(6, 5)$. We obtain the coordinates of the two points of tangency: $T_1 \left(\dfrac{192}{25}, \dfrac{7}{5} \right)$ and $T_2(0, 5)$.   □

Review Questions

1. Write the general equation of a curve of the second order.
2. What curve is called an ellipse?
3. What is the eccentricity of an ellipse?
4. Define directrices of an ellipse.
5. Tell about the optical property of an ellipse.
6. What curve is called hyperbola?
7. What is the eccentricity of a hyperbola?
8. Define directrices of a hyperbola.
9. Tell about the optical property of a hyperbola.
10. What curve is called parabola?
11. What is the eccentricity of a parabola?
12. How are directrices of a parabola defined.
13. Tell about the optical property of a parabola.
14. Write the equation of a tangent to ellipse, hyperbola and parabola.
15. What curves of the second order are referred to as degenerate ones?

Problems

11.1. For the ellipse $\dfrac{x^2}{18} + \dfrac{y^2}{9} = 1$, compute the eccentricity and write the equation of directrices.

11.2. Find the shortest distance from the ellipse $\dfrac{x^2}{4} + \dfrac{y^2}{3} = 1$ to the line:

(1) $x + 2y - 5 = 0$;
(2) $2x + y - 5 = 0$.

11.3. The ellipse $x^2 + 4y^2 = 5$ is given. Find the equation of the line that is tangent to this ellipse at the point with the coordinates $x = 1$, $y = -1$.

11.4. Find the angle between the tangents drawn from the point $(-8, 2)$ to the ellipse $\dfrac{x^2}{24} + \dfrac{y^2}{12} = 1$.

∗11.5. From what points of Cartesian coordinate plane is the ellipse $\dfrac{x^2}{a^2} + \dfrac{y^2}{b^2} = 1$ seen at a right angle?

11.6. The semiaxes of an **equilateral hyperbola** are equal: $a = b$. Find the eccentricity of the equilateral hyperbola.

11.7. Find the angle between the asymptotes to the hyperbola $\dfrac{x^2}{a^2} - \dfrac{y^2}{b^2} = 1$.

11.8. Find the distance from the point $(4, 0)$ to the curve $y^2 - 2x = 0$.

11.9. Write the equation of tangent to the parabola $y^2 = 5x$ at the point nearest to the point $M_0(2, 1/2)$.

11.10. The representation of hyperbola in parametric form (11.13) has a disadvantage consisting in that its left and right branches are described by different expressions with different signs. Find the parametric specification of hyperbola that does not have the said disadvantage.

∗11.11. Assume that it is known that the line $Ax + By + C = 0$ is tangent to the ellipse whose focal distance is equal to $2c$. Set up the equation of this ellipse.

11.12. Find at what points the line $-3x + 3y - 2 = 0$ and the ellipse $\dfrac{x^2}{2} + \dfrac{y^2}{4} = 1$ intersect.

11.13. At what points do the line $x - 3y - 2 + 3\sqrt{3} = 0$ and the hyperbola $x^2 - y^2 = 1$ intersect?

11.14. At what points do the line $x + y - 5\sqrt{5} = 0$ and the parabola $y^2 = 6x$ intersect?

11.15. The eccentricity of Mercury's orbit is equal to 0.2; the major semiaxis is equal to 0.39 astronomical units (a. u.). Compute the greatest and the least distances of the planet from the Sun.

11.16. The eccentricity of the Earth's orbit is equal to 0.017; the major semiaxis is equal to one astronomical unit. What are the greatest and the least distances from the Earth to the Sun?

Fig. 11.8 The region
bounded by the ellipse and
the right branch of the
hyperbola

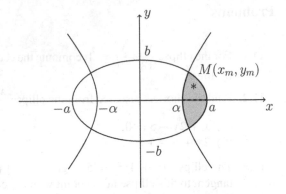

11.17. Assume that it is known that the eccentricity of the ellipse is $\varepsilon = 1/2$. In
the canonical system of coordinates, one of its directrices is specified by the
equation $x = 12$. Compute the distance in this coordinate system from the
point M_1 of the ellipse with the abscissa equal to -1 to the focus unilateral
with this directrix.

11.18. Assume that it is known that the eccentricity of the hyperbola is $\varepsilon = 3/2$. In
the canonical system of coordinates, one of its directrices is specified by the
equation $x = -4$. Compute the distance from the point N_1 of the hyperbola
with the ordinate equal to 9 to the focus unilateral with this directrix.

∗11.19. Find the area of the regions bounded by the ellipse $\dfrac{x^2}{a^2} + \dfrac{y^2}{b^2} = 1$ and the
right branch of the hyperbola $\dfrac{x^2}{\alpha^2} - \dfrac{y^2}{\beta^2} = 1$ (see Fig. 11.8, where the said
region is colour-highlighted).

Answers and Solutions

11.1 *Solution.*
   The semiaxes of the ellipse are equal to $a = \sqrt{18}$ and $b = 3$. Compute the
eccentricity of the ellipse:

$$\varepsilon = \frac{c}{a} = \frac{\sqrt{a^2 - b^2}}{a} = \frac{\sqrt{18 - 9}}{\sqrt{18}} = \frac{1}{\sqrt{2}}.$$

   Write the equations of directrices:

$$x = \pm\frac{a}{\varepsilon} \quad \text{or} \quad x = \pm 6.$$

11.2 *Solution.*

(1) Find the equations of two lines tangent to the ellipse and parallel to the line $x + 2y - 5 = 0$. For this, write the equation of tangent to the ellipse:

$$\frac{x_0 x}{a^2} + \frac{y_0 y}{b^2} = 1,$$

where (x_0, y_0) is the point of tangency, $a^2 = 4$, $b^2 = 3$. Thus, the equation of tangent takes the form:

$$y = -\frac{3x_0 x}{4 y_0} + \frac{3}{y_0}.$$

The slope of the line $x + 2y - 5 = 0$ is equal to $k = -\frac{1}{2}$. As is known, the slopes of parallel lines coincide. Therefore, the following equality is valid: $-\frac{3x_0 x}{4 y_0} = -\frac{1}{2}$. Moreover, the point (x_0, y_0) belongs to the ellipse. As a result, we obtain the system of equations:

$$\begin{cases} -\dfrac{3x_0}{4 y_0} = -\dfrac{1}{2}, \\ \dfrac{x_0^2}{4} + \dfrac{y_0^2}{3} = 1. \end{cases}$$

The set of its solutions has the form $\{(1, 3/2), (-1, -3/2)\}$.

We obtain the equations of tangents:

$$x + 2y - 2 = 0, \quad x + 2y + 2 = 0.$$

Note that the first of them is located closer to the initial line, since the following inequality is valid: $\mathrm{abs}((-2) - (-5)) < \mathrm{abs}(2 - (-5))$ (see Problem **7.13**).

It only remains to find the distance from the point $(1, 3/2)$ to the line $x + 2y - 5 = 0$.

Let us use the formula (7.37):

$$d = \mathrm{abs}(\delta) = \frac{\mathrm{abs}(A x_0 + B y_0 + C)}{\sqrt{A^2 + B^2}} = \frac{\mathrm{abs}(1 \cdot 1 + 2 \cdot \frac{3}{2} - 5)}{\sqrt{1^2 + 2^2}} = \frac{\sqrt{5}}{5}.$$

So, the shortest distance from the ellipse $\dfrac{x^2}{4} + \dfrac{y^2}{3} = 1$ to the line $x + 2y - 5 = 0$ is equal to $\dfrac{\sqrt{5}}{5}$.

(2) Find the equations of two lines, tangent to the given ellipse and parallel to the line $2x + y - 5 = 0$. Similarly to the previous item of this problem, we obtain the system of equations relative to the unknown coordinates of the tangency point x_0 and y_0:

$$\begin{cases} -\dfrac{3x_0}{4y_0} = -2, \\ \dfrac{x_0^2}{4} + \dfrac{y_0^2}{3} = 1. \end{cases}$$

We obtain the set of solutions $\left\{ \left(\dfrac{8}{\sqrt{19}}, \dfrac{3}{\sqrt{19}} \right), \left(-\dfrac{8}{\sqrt{19}}, -\dfrac{3}{\sqrt{19}} \right) \right\}$. Find the equation of the tangents:

$$y = -2x + \sqrt{19}, \quad y = -2x - \sqrt{19}.$$

Note that the first of them is closer to the initial line, since the following inequality is valid: $\mathrm{abs}(-\sqrt{19} - (-5)) < \mathrm{abs}(\sqrt{19} - (-5))$ (see Problem **7.13**). The distance from the point $\left(\dfrac{8}{\sqrt{19}}, \dfrac{3}{\sqrt{19}} \right)$ to the line $2x + y - 5 = 0$ will be found by the formula (7.37):

$$d = \frac{\mathrm{abs}(Ax_0 + By_0 + C)}{\sqrt{A^2 + B^2}} = \frac{\mathrm{abs}\left(2 \cdot \dfrac{8}{\sqrt{19}} + 1 \cdot \dfrac{3}{\sqrt{19}} - 5\right)}{\sqrt{2^2 + 1^2}} = \frac{5\sqrt{5} - \sqrt{95}}{5}.$$

As a result, the shortest distance from the ellipse $\dfrac{x^2}{4} + \dfrac{y^2}{3} = 1$ to the line $2x + y - 5 = 0$ is equal to $\sqrt{5} - \dfrac{1}{5}\sqrt{95}$.

11.3 *Solution.*

Use the equation of tangent to ellipse (11.4):

$$\frac{xx_0}{a^2} + \frac{yy_0}{b^2} = 1.$$

In our case, $a^2 = 5$, $b^2 = \dfrac{5}{4}$, $x_0 = 1$, $y_0 = -1$, and the equation of tangent takes the form $\dfrac{x \cdot 1}{5} + \dfrac{y \cdot (-1)}{5/4} = 1$, or $x - 4y - 5 = 0$.

11.4 *Solution.*

According to the problem statement, through the point $(-8, 2)$ pass both tangents to the ellipse.

As is shown in Sect. 11.5 (formula (11.27)), the slopes of the tangents k_1 and k_2 are computed by the formula

$$k_{1,2} = \frac{x_1 y_1 \pm \sqrt{b^2 x_1^2 + a^2 y_1^2 - a^2 b^2}}{x_1^2 - a^2}.$$

We obtain $k_1 = 1/5$, $k_2 = -1$.
Based on the (7.17), we find the angle between the lines.

$$\varphi = \arctan \frac{k_1 - k_2}{1 + k_1 k_2} = \arctan \frac{1/5 - (-1)}{1 + (1/5)(-1)} = \frac{3}{2}.$$

Therefore, the angle between the tangents to the ellipse is equal to $\varphi = \arctan \dfrac{3}{2}$.

11.5 Solution.

Since the ellipse seen at a right angle from some point $M(x, y)$, then the angle between the tangents drawn from M to the ellipse is equal to $\pi/2$. Based on the formula (11.27) on page 366 the slopes of the tangents k_1 and k_2 are equal to

$$k_{1,2} = \frac{x_1 y_1 \pm \sqrt{b^2 x_1^2 + a^2 y_1^2 - a^2 b^2}}{x_1^2 - a^2}.$$

Compute the angle φ between the tangents using the equation (7.17):

$$\tan \varphi = \frac{k_1 - k_2}{1 + k_1 k_2},$$

therefore,

$$\varphi = \arctan \frac{k_1 - k_2}{1 + k_1 k_2} = \arctan \frac{2\sqrt{b^2 x_1^2 + a^2 y_1^2 - a^2 b^2}}{x_1^2 + y_1^2 - a^2 - b^2}.$$

It follows from the obtained equation that the angle $\varphi = \pi/2$, if the denominator of fraction under the arctangent is equal to zero: $x_1^2 + y_1^2 - a^2 - b^2 = 0$.

As a result, the ellipse $\dfrac{x^2}{a^2} + \dfrac{y^2}{b^2} = 1$ is seen at a right angle from all points of the plane that satisfy the equation

$$x^2 + y^2 = a^2 + b^2.$$

Such points, as is easy to see, form a circle of radius $\sqrt{a^2 + b^2}$.

11.6 *Solution.*
    The eccentricity of the hyperbola is equal to $\varepsilon = \dfrac{c}{a}$, where $c = \sqrt{b^2 + a^2}$ is half of the interfocal distance, a, b are semiaxes of the hyperbola.
    If $a = b$, then

$$\varepsilon = \frac{\sqrt{b^2 + a^2}}{a} = \frac{\sqrt{a^2 + a^2}}{a} = \sqrt{2}.$$

Thus, we obtain that the eccentricity of the equilateral hyperbola is equal to $\sqrt{2}$.

11.7 *Solution.*
    The asymptotes to the hyperbola $\dfrac{x^2}{a^2} - \dfrac{y^2}{b^2} = 1$, as is known (see page 361), are defined by the equations $y = \pm\dfrac{b}{a}x$. The slopes of these lines are equal to $\pm\dfrac{b}{a}$. Therefore, the tangent of half of the angle α between the asymptotes is equal to $\tan(\alpha/2) = \dfrac{b}{a}$. Let us express, from the obtained relation, the sought angle:

$$\alpha = 2\arctan\frac{b}{a}.$$

Note that for the equilateral hyperbola (see Problem **11.6**), the angle between the asymptotes is equal to $\pi/2$.

11.8 *Solution.*
    In order to find the distance from the point (x_0, y_0) to the curve $y^2 - 2x = 0$, find the least value of the functions

$$d_1(x) = \sqrt{(x - x_0)^2 + (f_1(x) - y_0)^2} \text{ and}$$

$$d_2(x) = \sqrt{(x - x_0)^2 + (f_2(x) - y_0)^2},$$

where $x_0 = 4$, $y_0 = 0$, and by $f_{1,2}(x) = \pm\sqrt{2x}$ are denoted two branches of the parabola $y^2 - 2x = 0$.
    First, consider the function $d_1(x)$: $d_1(x) = \sqrt{(x - 4)^2 + 2x}$. Its derivative is equal to $d_1'(x) = \dfrac{x - 3}{\sqrt{(x - 4)^2 + 2x}}$. The point suspected of being the extremum is the solution of the equation $d_1'(x) = 0$, or $x - 3 = 0$. This is the minimum point. Therefore, $\min d_1(x) = \sqrt{(3 - 4)^2 + 2 \cdot 3} = \sqrt{7}$.
    Similarly, we find the least value of the function $d_2(x)$: $\min d_2(x) = \sqrt{7}$.
    As a result, we obtain that the distance from the point $(4, 0)$ to the parabola is equal to $\sqrt{7}$.

11.9 *Answer:*
    $5x - 2\sqrt{10}y + 10 = 0$.

11.10 *Answer:*

$$\begin{cases} x = \dfrac{a}{\cos t}, \\ y = b \tan t, \end{cases}$$

where $t \in (0, 2\pi)$.

11.11 *Solution.*

Denote the tangency point by $P(x_0, y_0)$. The equation of tangent to ellipse at this point has the form:

$$\frac{xx_0}{a^2} + \frac{yy_0}{b^2} = 1,$$

where a and b are the major and minor semiaxes of the ellipse, respectively. On the other hand, according to the problem statement, the general equation of line has the form $Ax + By + C = 0$, and it can be rewritten as

$$\left(-\frac{A}{C}\right)x + \left(-\frac{B}{C}\right)y = 1.$$

Comparing the obtained equations, we obtain $\dfrac{x_0}{a^2} = -\dfrac{A}{C}$, $\dfrac{y_0}{b^2} = -\dfrac{B}{C}$, therefore, the coordinates of the tangency point P are equal to $\left(-Aa^2/C, -Bb^2/C\right)$. It is clear that the coordinates of the point P satisfy the equation of the ellipse $\dfrac{x_0^2}{a^2} + \dfrac{y_0^2}{b^2} = 1$. Moreover, from the definition of focus follows the equality $a^2 - b^2 = c^2$, where c is the focal parameter. Hence, we obtain the system of equations relative to the parameters a, b:

$$\begin{cases} \dfrac{1}{a^2}\left(-\dfrac{Aa^2}{C}\right)^2 + \dfrac{1}{b^2}\left(-\dfrac{Bb^2}{C}\right)^2 = 1, \\ a^2 - b^2 = c^2, \end{cases}$$

or:

$$\begin{cases} \dfrac{a^2 A^2}{C^2} + \dfrac{b^2 B^2}{C^2} = 1, \\ a^2 - b^2 = c^2. \end{cases}$$

The solution of this system is: $a = \left(\dfrac{C^2 - c^2 B^2}{A^2 + B^2}\right)^{1/2}$, $b = \left(\dfrac{C^2 - c^2 A^2}{A^2 + B^2}\right)^{1/2}$.

Thus, the sought equation of ellipse has the form:

$$\frac{A^2 + B^2}{C^2 - c^2 B^2} x^2 + \frac{A^2 + B^2}{C^2 - c^2 A^2} y^2 = 1.$$

11.12 *Solution.*

Express y from the equation of line: $y = x + \frac{2}{3}$.

Substitute y into the equation of ellipse:

$$\frac{x^2}{2} + \frac{1}{4}\left(\frac{4}{9} + \frac{4}{3} \cdot x + x^2\right) = 1,$$

or $27x^2 + 12x - 32 = 0$, hence, we obtain $x_1 = \frac{8}{9}$, $y_1 = \frac{14}{9}$; $x_2 = -\frac{4}{3}$, $y_2 = -\frac{2}{3}$.

So, the intersection points of the line and the ellipse are $(8/9, 14/9)$, $(-4/3, -2/3)$.

11.13 *Solution.*

Express the variable x: $x = 3y + 2 - 3\sqrt{3}$.

In order to find the points at which the line and the hyperbola intersect, substitute the obtained expression for x into the equation of hyperbola:

$$9y^2 + 6y(2 - 3\sqrt{3}) + 31 - 12\sqrt{3} - y^2 - 1 = 0,$$

or

$$8y^2 + y\left(12 - 18\sqrt{3}\right) + 30 - 12\sqrt{3} = 0,$$

hence: $y_1 = \dfrac{5\sqrt{3} - 6}{4}$, $x_1 = \dfrac{3\sqrt{3} - 10}{4}$; $y_2 = \sqrt{3}$, $x_2 = 2$.

Thus, the sought intersection points are $\left(\dfrac{3\sqrt{3} - 10}{4}, \dfrac{5\sqrt{3} - 6}{4}\right)$, $(2, \sqrt{3})$.

11.14 *Solution.*

In order to find the intersection points of the line and the parabola, solve the system of equations:

$$\begin{cases} x + y - 5\sqrt{5} = 0, \\ y^2 = 6x \end{cases} \Rightarrow \begin{cases} x = -y + 5\sqrt{5}, \\ y^2 = -6y + 30\sqrt{5} \end{cases} \Rightarrow y^2 + 6y - 30\sqrt{5} = 0,$$

The obtained quadratic equation has the roots

$$y_{1,2} = -3 \pm \sqrt{9 + 30\sqrt{5}}.$$

Substitute the values $y_{1,2}$ into the equation of a straight line $x + y - 5\sqrt{5} = 0$. We obtain

$$x_{1,2} = 3 \mp \sqrt{9 + 30\sqrt{5}} + 5\sqrt{5}.$$

Thus, the coordinates of the intersection points are:
$$M_1(3 - \sqrt{9 + 30\sqrt{5}} + 5\sqrt{5}, -3 + \sqrt{9 + 30\sqrt{5}}),$$
$$M_2(3 + \sqrt{9 + 30\sqrt{5}} + 5\sqrt{5}, -3 - \sqrt{9 + 30\sqrt{5}}).$$

11.15 *Solution.*

According to the first law of Kepler,[1] all planets of the Solar system move along an ellipse, in one of the foci of which the Sun is situated [10]. The closest to the Sun point of the orbit P is referred to as the perihelion, and the farthest from the Sum point of the orbit A is referred to as the aphelion.

In order to compute these values for the planet Mercury, introduce the notations: $F_1(c, 0)$ are the coordinates of the Sun, $P(a, 0)$ are the coordinates of the perihelion, $A(-a, 0)$ are the coordinates of the aphelion and $O(0, 0)$ is the centre of symmetry of the ellipse. Here, a is the major semiaxis, and c is the focal parameter.

Then:

$$r_a = AF_1 \text{ --- the greatest distance from Mercury to the Sun,}$$

$$r_p = PF_1 \text{ --- the least distance from Mercury to the Sun,}$$

$$c = OF_1.$$

Geometrically, we obtain

$$\begin{array}{ll} a = r_p + c, \\ a = r_a - c; \end{array} \Rightarrow \begin{array}{ll} r_p = a - c, \\ r_a = a + c; \end{array}$$

The eccentricity of the ellipse ε will be computed by the formula:

$$\varepsilon = \frac{\sqrt{a^2 - b^2}}{a},$$

where $\sqrt{a^2 - b^2} = c$,

$$\varepsilon = \frac{c}{a} \Rightarrow c = a\varepsilon.$$

---

[1] Johannes Kepler (1571–1630), German mathematician and astronomer.

Hence,

$$r_p = a - a\varepsilon,$$
$$r_a = a + a\varepsilon;$$

$$r_p = a(1 - \varepsilon),$$
$$r_a = a(1 + \varepsilon).$$

Substitute the numeric values from the problem statement:

$$r_p = 0.39 \cdot (1 - 0.2) \Rightarrow r_p = 0.312 \text{ (a. u.)},$$
$$r_a = 0.39 \cdot (1 + 0.2) \Rightarrow r_a = 0.468 \text{ (a. u.)}.$$

So, the greatest distance from Mercury to the Sun is equal to 0.468 a.u., the least distance is equal to 0.312 a.u.

11.16 *Solution.*

Denote the points $F_1(c, 0)$ as the coordinates of the Sun, $P(a, 0)$ as the coordinates of the perihelion, $A(-a, 0)$ as the coordinates of the aphelion, $O(0, 0)$ as the centre of symmetry of the ellipse. Here, a is the major semiaxis, and c is the focal parameter of the Earth's orbit. Then (see previous problem):

$$r_a = AF_1,$$
$$r_p = PF_1, \Rightarrow$$
$$c = OF_1;$$

$$r_p = a(1 - \varepsilon),$$
$$r_a = a(1 + \varepsilon);$$

where $\varepsilon = 0.017$ is the eccentricity of the Earth's orbit, $a = 1$ a.u.

Perform the necessary computations:

$$r_p = 1 \cdot (1 - 0.017) \Rightarrow r_p = 0.983 \text{ (a. u.)},$$
$$r_a = 1 \cdot (1 + 0.017) \Rightarrow r_a = 1.017 \text{ (a. u.)}.$$

As a result we obtain that the greatest distance from the Earth to the Sun is equal to 1.017 a.u., and the least distance is equal to 0.983 a.u.

11.17 *Solution.*

The point M_1 has the coordinates $(-1, y_1)$ and is situated on the left of Oy-axis, and the focus is on the right. The sought distance is equal to the length of the hypotenuse r of the triangle with the vertices $F_1(c, 0)$, $M_1(-1, y_1)$ and $L(-1, 0)$.

According to the Pythagorean theorem:

$$r = \sqrt{(c + 1)^2 + y^2}.$$

Using the definition of directrix of ellipse $x = \pm\dfrac{a}{\varepsilon}$, we find

$$x_d = \frac{a}{\varepsilon}, \quad \Rightarrow \quad a = \varepsilon x_d,$$

where, according to the problem statement, $x_d = 12$.

Express the focal parameter c of the ellipse:

$$c = a\varepsilon = \varepsilon^2 x_d.$$

The coordinate y_1 will be determined using the equation of the ellipse $\dfrac{(-1)^2}{a^2} + \dfrac{y_1^2}{b^2} = 1$:

$$y_1 = \pm\frac{b}{a}\sqrt{a^2 - 1}.$$

Since for the ellipse the following relation is valid: $b^2 = a^2 - c^2 = a^2(1 - \varepsilon^2)$, then

$$y_1 = \pm\sqrt{(1 - \varepsilon^2)(a^2 - 1)}.$$

Therefore, the length of the hypotenuse $F_1 M_1$ of the triangle $F_1 L M_1$ is equal to

$$r = \sqrt{(c + 1)^2 + y_1^2} = \sqrt{(a\varepsilon + 1)^2 + (1 - \varepsilon^2)(a^2 - 1)} = a + \varepsilon = \varepsilon(x_d + 1).$$

Substitute here the numeric values from the problem statement, and we obtain the final answer: $r = \dfrac{13}{2}$.

11.18 *Solution.*

The sought distance is equal to the length of the hypotenuse of the triangle with the vertices $F_2(-c, 0)$, $N_1(x_1, y_1)$ and $L(x_1, 0)$, where, according to the problem statement, $y_1 = 9$. Depending on whether the point $N_1(x_1, y_1)$ is situated on the right or on the left branch of the hyperbola, we obtain two possible values of r:

$$r = \sqrt{(x_1 \pm c)^2 + y_1^2}.$$

According to the definition of directrix of the hyperbola $x = \pm\dfrac{a}{\varepsilon}$, we have

$$a = \varepsilon \, \mathrm{abs}(x_d), \quad \text{where } x_d = -3.$$

The parameter c of the hyperbola is equal to

$$c = a\varepsilon = \varepsilon^2 \mathrm{abs}(x_d).$$

Since $\dfrac{(-1)^2}{a^2} - \dfrac{y_1^2}{b^2} = 1$ and $c^2 = a^2 - b^2$, then

$$r^2 = (x_1 \pm c)^2 + b^2\Big(\frac{x_1^2}{a^2} - 1\Big) = (a \pm \varepsilon x_1)^2 = \varepsilon^2(\mathrm{abs}(x_d) \pm x_1)^2,$$

where $x_1 = \dfrac{a}{b}\sqrt{y_1^2 + b^2} = \sqrt{\dfrac{y_1^2}{\varepsilon^2 - 1} + \varepsilon^2 x_d^2}$. Having substituted the values from

the problem statement, we obtain $r = 9\sqrt{\dfrac{14}{5}} + 6$ or $r = 9\sqrt{\dfrac{14}{5}} - 6$.

11.19 *Solution.*

Denote the region bounded by the ellipse and the hyperbola by D. Due to the properties of symmetry of ellipse and hyperbola, the sought area is expressed as the doubled area of the region D_0:

$$S(D) = 2S(D_0).$$

In Fig. 11.8 the region D_0 is marked with the symbol "$*$".

The equation of the part of the ellipse lying in the first quadrant has the form $y_{\mathrm{ell}} = \dfrac{b}{a}\sqrt{a^2 - x^2}$, where $0 < x < a$.

The equation of the part of the right branch of the hyperbola, which is situated above the abscissa axis, is: $y_{\mathrm{hyp}} = \dfrac{\beta}{\alpha}\sqrt{\alpha^2 - x^2}$, where $x > \alpha$.

Assume that x_M is the abscissa of the point M of intersection of the curves y_{ell} and y_{hyp}, which point lies in the first quadrant.

Note that when the condition $a \geqslant \alpha$ is satisfied, such a point always exists. Otherwise, when $a < \alpha$, the region D is empty, and $S(D) = 0$.

So, under the condition $a \geqslant \alpha$, the following equality is valid:

$$S(D_0) = \int_{\alpha}^{x_M} y_{hyp}(x)\,dx + \int_{x_M}^{a} y_{ell}(x)\,dx. \tag{11.31}$$

Compute the coordinate x_M:

$$y_{ell}(x_M) = y_{hyp}(x_M) \;\Rightarrow\; \frac{b}{a}\sqrt{a^2 - x_M^2} = \frac{\beta}{\alpha}\sqrt{x_M^2 - a^2}.$$

Solution of this equation that satisfies the conditions $x > 0$, $y > 0$, has the form:

$$x_M = a\alpha\sqrt{\frac{b^2 + \beta^2}{a^2\beta^2 + b^2\alpha^2}}.$$

Substitution of x_M into the formula (11.31) for $S(D_0)$ results in:

$$S(D_0) = \frac{1}{2}\left[ab\arctan\frac{\beta}{\alpha}\sqrt{\frac{a^2 + \alpha^2}{b^2 + \beta^2}} \right.$$

$$\left. -\alpha\beta\ln\left(b\sqrt{\frac{a^2 - \alpha^2}{a^2\beta^2 + \alpha^2 b^2}} + a\sqrt{\frac{b^2 + \beta^2}{a^2\beta^2 + \alpha^2 b^2}} \right) \right].$$

Finally, we obtain the area of the region bounded by the ellipse and the right branch of the hyperbola:

$$S(D) = \begin{cases} 0, \text{ if } a \leqslant \alpha, \\[2mm] ab\arctan\dfrac{\beta}{\alpha}\sqrt{\dfrac{a^2 + \alpha^2}{b^2 + \beta^2}} - \\[4mm] \quad -\alpha\beta\ln\left(b\sqrt{\dfrac{a^2 - \alpha^2}{a^2\beta^2 + \alpha^2 b^2}} + a\sqrt{\dfrac{b^2 + \beta^2}{a^2\beta^2 + \alpha^2 b^2}} \right), \text{ if } a > \alpha. \end{cases}$$

Chapter 12
Elliptic Curves

Elliptic curve is a plane curve that has no singular points and is defined by an equation of the form

$$y^2 = x^3 + ax + b, \qquad (12.1)$$

where a and b are real numbers. The requirement of absence of singular points means that the curve must not have any self-intersection and cusps.[1] This condition will be satisfied if and only if the **discriminant** of the equation

$$\Delta = -16(4a^3 + 27b^2) \neq 0 \qquad (12.2)$$

is other than zero. Of course, the constant factor 16 does not influence the sign of the discriminant; this factor is introduced for convenience of investigation of further and deeper properties of the curve.

The name "elliptic curve" goes back to the problem of computing the length of the ellipse arc, leading to computation of the definite integral of the form

$$\int_{x_1}^{x_2} \frac{R(x)}{\sqrt{x^3 + ax + b}}\, dx \qquad (12.3)$$

for some rational function $R(x)$ [7, 32]. Morphological similarity of the terms "ellipse" and "elliptic curve" is due to historic reasons. Let us emphasize that these terms belong to different mathematical concepts.

---

[1] An example of self-intersection see on Fig. 12.6 at page 403. An example of a cusp is shown on Fig. 12.1e at page 385.

© Springer Nature Switzerland AG 2021
S. Kurgalin, S. Borzunov, *Algebra and Geometry with Python*,
https://doi.org/10.1007/978-3-030-61541-3_12

As a sphere of application of elliptic curves, we should mention the proof of Fermat's[2] Last Theorem and the discovery of a new and rapidly developing scientific field about the confidential data transfer methods—elliptic-curve cryptography. There are other known applications of such curves in mathematics and adjacent fields, in particular, in number theory [13, 14].

Depending on the sign of the discriminant, the elliptic curve is formed by one or two connected components:

- if $\Delta > 0$, then the curve graph consists of two connected components;
- if $\Delta < 0$, then the curve graph consists of one connected component.

Note If the conditions $\Delta = 0$ and $a \neq 0$ are satisfied, a point of self-intersection appears on the curve, and if the equalities $a = b = 0$ occur, then a cusp appears.

Example 12.1 In Fig. 12.1 are shown several elliptic curves for the values of the parameters $a \in \{-4, 0, 4\}$, $b \in \{-2, 0, 2\}$. Note that at $a = b = 0$ we obtain the equation $y^2 = x^3$, for which $\Delta = 0$, therefore, such a curve does not belong to the class of elliptic curves. In Fig. 12.1 it is located on the panel *e*) and is shown by the dotted line.   □

12.1   Operation of Multiplication of the Elliptic Curve Points

Having fixed two points located on an arbitrary elliptic curve $\Gamma = \{(x, y) \in \mathbb{R}^2 \colon y^2 = x^3 + ax + b\}$, we can deduce a rule for constructing the third point. Such an operation will be referred to as "addition" of points on an elliptic curve.

In order to perform the operation of addition of points, the Cartesian plane $\mathbb{R}^2$ should be expanded by introducing a **point at infinity** ∞. As will be clear from the further discussion, the point ∞ has the properties of a zero element in the set $\mathbb{R}^2 \cup \{\infty\}$ with the operation of multiplication of points defined on it. Due to this, the designation $\mathcal{O} \equiv \infty$ is also used for the point at infinity.

Assume that any elliptic curve passes through the point $\mathcal{O}$. We can say that two vertical lines intersect at this point.

So, let the following elliptic curve be specified

$$\Gamma = \{(x, y) \in \mathbb{R}^2 \colon y^2 = x^3 + ax + b\} \cup \{\mathcal{O}\}. \tag{12.4}$$

The main idea of determining the sum $A \oplus B$ of the two points $A = (x_A, y_A)$ and $B = (x_B, y_B)$ on the elliptic curve consists in computing the coordinates of the intersection point of the line AB and Γ:

$$C' = A \oplus B = \{(x_C, y_C) \colon C \in AB \text{ and } C \in \Gamma\} \tag{12.5}$$

---

[2]Pierre de Fermat (1601–1665), French mathematician and lawyer.

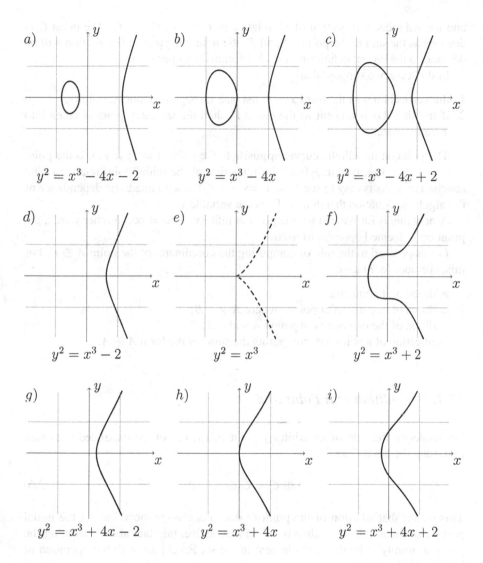

Fig. 12.1 Elliptic curves for the values of the parameters $a \in \{-4, 0, 4\}$, $b \in \{-2, 0, 2\}$. Rectangular mesh on the panels (**a**)–(**i**) has a step equal to 2. The point $(0, 0)$ on panel (**e**) is the cusp for the curve $y^2 = x^3$

and the subsequent reflection of C' relative to Ox-axis: $C' \to C$. The point C is deemed as the sum of the points A and B; the remaining part of this section will be devoted to the formal definition of such a summation operation.

In this case, we suppose that

1. the vertical lines of the form $x = \text{const}$ pass through the point at infinity;
2. if the line AB is tangent to the curve Γ, then the tangency point is taken into account twice.

The point on the elliptic curve, **opposite** to the point $A = (x_A, y_A)$, is the point $-A = (x_A, -y_A)$, resulting from the reflection of the initial point relative to the abscissa axis. As is easy to see, $A \in \Gamma \Rightarrow -A \in \Gamma$ due to quadratic dependence of the algebraic equation that defines Γ on the variable y.

A new rule is introduced for the point at infinity: $-\infty = \infty$, in other words, the point ∞ is deemed opposite to itself.

Let us proceed to the rule of computing the coordinates of the point $A \oplus B$. For this, consider four cases:

1. addition of a point and $\mathcal{O}$;
2. addition of two different points, where $A \neq -B$;
3. addition of the two opposite points A and $-A$;
4. duplication of a point, i.e. computing the sums of the form $A \oplus A$.

12.1.1 Addition of a Point and $\mathcal{O}$

Let us define the sum of an arbitrary point $A(x_A, y_A)$ of the expanded Cartesian plane and the point $\mathcal{O}$ as

$$A \oplus \mathcal{O} = \mathcal{O} \oplus A = A. \tag{12.6}$$

This means that addition of the point $\mathcal{O}$ does not change the values of the initial point coordinates. As was already mentioned above, this fact allows deeming the point at infinity to be the zero element in the set $\mathbb{R}^2 \cup \{\infty\}$ with the operation of addition defined on it.

12.1.2 Addition of Two Different Points

In order to compute the sum $A \oplus B$ on the condition $A \neq -B$ we should find the intersection point of the line AB with the specified elliptic curve Γ. Denote the respective intersection point by C'. Then we find the point $C = -C'$ opposite to C', reflecting the point C' relative to the Ox-axis. This, the coordinates of the point C and C' are connected by the relations $x_C = x_{C'}$, $y_C = -y_{C'}$.

Fig. 12.2 Addition of two different points A and B on the elliptic curve

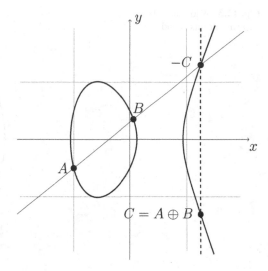

The sum of the points A and B is the point C, constructed in the above manner:

$$C = A \oplus B. \tag{12.7}$$

In Fig. 12.2 is shown a geometric method of constructing the point $A \oplus B$. Solution of the system

$$\begin{cases} (-y_C)^2 = x_C^3 + ax_C + b, \\ (-y_C) - y_A = \dfrac{y_B - y_A}{x_B - x_A}(x_C - x_A) \end{cases} \tag{12.8}$$

reduces to solving the equation of the third degree with real coefficients. Of course, the roots of such an arbitrary equation can be found using known Cardano formulae [41]; however, in this case, we can perform easier computations.

Using Viète's formulae (see Problem **4.31**), we may say that the sought coordinates of the point C are determined by the formulae:

$$\begin{cases} x_C = \varkappa^2 - x_A - x_B, \\ y_C = -y_A + \varkappa(x_A - x_C), \end{cases} \tag{12.9}$$

where the designation $\varkappa = \dfrac{y_B - y_A}{x_B - x_A}$ is introduced for the slope of the line AB.

Fig. 12.3 Addition of two
opposite points A and $-A$

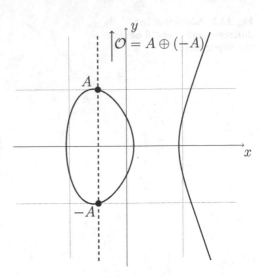

12.1.3 Addition of Two Opposite Points

Let us proceed to the case when the summands in the sum of the points are opposite
points, for example, A and $-A$.

Draw a line that passes through two opposite points. As can be seen from the
Fig. 12.3, such a line will be positioned vertically, and the third point, which is
common for the elliptic curve and the drawn line, can only be positioned at infinity.
Based on this fact, the sum of the opposite points is defined as $A \oplus (-A) = \infty$. The
Fig. 12.3 illustrates the formulated rule.

12.1.4 Duplication of a Point

The computation of the sum of the points of the form $A \oplus A$, i.e. **duplication of a
point**, will require the operation of passage to the limit:

$$A \oplus A = \lim_{B \to A} A \oplus B. \qquad (12.10)$$

Geometrically it means that we are drawing a line through two different points
$A(x_A, y_A)$ and $B(x_B, y_B)$ of the elliptic curve, and the second point lies in the small
neighbourhood of the first one:

$$\lim_{B \to A} A \oplus B = \lim_{\substack{x_B \to x_A, \\ y_B \to y_A}} A \oplus B. \qquad (12.11)$$

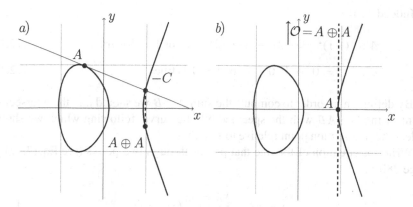

Fig. 12.4 Duplication of a point A. On the panel (**a**), the case $y_A \neq 0$ is schematically shown; on the panel (**b**) the case $y_A = 0$

As is seen from the Fig. 12.4, this case corresponds to drawing a tangent at the point A to the elliptic curve.

Suppose that the tangent intersects the curve at the point $-C$. Note that if the ordinate of the point A is equal to zero, then the said tangent will be positioned vertically. Based on the previous case described in Sect. 12.1.3, we obtain $-C = \infty$, $A \oplus A = A \oplus (-A) = \mathcal{O}$. For all other values of the ordinate $y_A \neq 0$, as a result of reflection of the point $-C$ relative to the horizontal axis, a point $C = A \oplus A$ will be constructed.

Let us write the respective analytical relations:

$$\begin{cases} x_C = \varkappa^2 - 2x_A, \\ y_C = -y_A + \varkappa(x_A - x_C), \end{cases} \tag{12.12}$$

where the value $\varkappa = \dfrac{3x_A^2 + a}{2y_A}$ is equal to the tangent of the slope of the tangent line relative to the horizontal axis.

Two other cases of point duplication discussed above are illustrated in Fig. 12.4a and b.

Analysis of the formulated addition procedure leads to the criterion of equality $\mathcal{O}$ of the sum of three points.

Theorem 12.1 *A sum of three points is equal to $\mathcal{O}$ if and only if they lie on the same straight line.*

Example 12.2 Compute the sum $(-2, -1) \oplus (0, 1)$ of two points on the elliptic curve defined by the equation $y^2 = x^3 - 4x + 1$.

Solution Denote $A = (-2, -1)$, $B = (0, 1)$ and $C = A \oplus B$. Verify that the points A and B lie on the curve $y^2 = x^3 - 4x + 1$.

Indeed,

$$A: \quad (-1)^2 = (-2)^3 - 4 \cdot (-2) + 1, \text{ or } 1 = 1 \text{ is true,} \tag{12.13}$$

$$B: \quad 1^2 = 0^3 - 4 \cdot 0 + 1, \text{ or } 1 = 1 \text{ is true.} \tag{12.14}$$

By definition, in order to compute the sum $A \oplus B$ we should find the intersection point of the line AB with the specified elliptic curve, following which we should reflect the intersection point relative to Ox-axis.

Write the equation of the line that passes through two points (see Eq. (7.12) on page 280):

$$y - y_A = \frac{y_B - y_A}{x_B - x_A}(x - x_A), \tag{12.15}$$

where (x_A, y_A) and (x_B, y_B) are the coordinates of the initial points.

Having substituted the numeric values of the coordinates into Eq. (12.15), we arrive at the equation of the line $y = x + 1$. The line specified by this equation intersects the elliptic curve $y^2 = x^3 - 4x + 1$ at a point with the abscissa x_C, for which the following equality is valid

$$(x_C + 1)^2 = x_C^3 - 4x_C + 1, \tag{12.16}$$

or

$$x_C^3 - x_C^2 - 6x_C = 0. \tag{12.17}$$

As is easy to see, the obtained cubic equation has the roots

$$(x_C)_1 = -2, \quad (x_C)_2 = 0, \quad (x_C)_3 = 3. \tag{12.18}$$

The first two values satisfy the abscissas of the initial points A and B; the value $x_C = 3$ is the abscissa of the intersection point of the line AB and the elliptic curve. Denote this point as $C': x_{C'} = 3$.

Having substituted $x_{C'}$ into the equation $y = x + 1$, we obtain $y_{C'} = 4$—the ordinate of the point C'.

The coordinates of the sought point $C = A \oplus B$ are equal to

$$\begin{cases} x_C = x_{C'}, \\ y_C = -y_{C'}. \end{cases} \tag{12.19}$$

As a result, $x_C = 3$, $y_C = -4$, therefore, on the curve $y^2 = x^3 - 4x + 1$ the following equality is valid

$$(-2, -1) \oplus (0, 1) = (3, -4). \tag{12.20}$$

Note that the use of the relations (12.9) also leads to the desired result. □

Example 12.3 Compute the sum $(-2, 3) \oplus (3, -3)$ of two points on the elliptic curve specified by the equation $y^2 = x^3 - 7x + 3$.

Solution Verify that the points $(-2, 3)$ and $(3, -3)$ belong to the curve.

The first point: $3^2 = (-2)^3 - 7(-2) + 3$, or $9 = 9$ is true;

the second point: $(-3)^2 = 3^3 - 7 \cdot 3 + 3$, or $9 = 9$ is true.

In order to compute the sum, use the formulae (12.9), into which the values $x_A = -2, y_A = 3, x_B = 3, y_B = -3$ should be substituted:

$$\varkappa = \frac{-3 - 3}{3 - (-2)} = -\frac{6}{5}, \tag{12.21}$$

$$x_C = \left(-\frac{6}{5}\right)^2 - (-2) - 3 = \frac{11}{25}, \tag{12.22}$$

$$y_C = -3 + \left(-\frac{6}{5}\right)\left(-2 - \frac{11}{25}\right) = -\frac{9}{125}. \tag{12.23}$$

So, by algebraic method, we have obtained the equality $(-2, 3) \oplus (3, -3) = \left(\frac{11}{25}, -\frac{9}{125}\right)$. □

For the sum of the points of the form $C = \underbrace{A \oplus A \oplus A \oplus \cdots \oplus A}_{n \text{ times}}$ the designation $C = n A$ is used.

For $n < 0$ the point $n A$ is defined as the element opposite to $(-n) A$.

Multiplication by zero results in a point at infinity $\mathcal{O}$.

Thus, the point $n A$, which is the sum n of the points A, is defined for all integral n in accordance with the following rule:

$$n A = \begin{cases} \underbrace{A \oplus A \oplus A \oplus \cdots \oplus A}_{n \text{ times}}, & \text{if } n > 0, \\ \mathcal{O}, & \text{if } n = 0, \\ -|n|A, & \text{if } n < 0. \end{cases} \tag{12.24}$$

Example 12.4 Compute the sum of four points $4(5, 11)$ on the elliptic curve, defined by the equation $y^2 = x^3 - 2x + 6$.

Solution Verify that the point $(5, 11)$ belongs to the elliptic curve:

$$11^2 = 5^3 - 2 \cdot 5 + 6, \text{ or } 121 = 121 \text{ is true.} \tag{12.25}$$

Use the equality

$$4(5, 11) = (5, 11) + (5, 11) + (5, 11) + (5, 11) = 2(5, 11) + 2(5, 11). \tag{12.26}$$

By the duplication formulae (12.12) we obtain

$$x = \frac{73}{22}, \quad 2(5, 11) = (489/484, 23835/10648). \tag{12.27}$$

Applying the duplication formula again and using a program in Python for computing, we arrive at the final answer:

$$4(5, 11) = (- 2\,160\,508\,643\,999/1\,099\,855\,587\,600,$$

$$- 1\,767\,794\,172\,358\,992\,751/1\,153\,462\,548\,939\,624\,000). \tag{12.28}$$

As is seen from the considered example, the arithmetic operations with the elliptic curve points are in many cases very labour-intensive and it is almost impossible to obtain the result without application of computing systems. □

Theorem 12.2 *The operation of addition of the points defined on the elliptic curve* Γ *on the set* $\mathbb{R}^2 \cup \{\mathcal{O}\}$ *has the following properties:*

1. $A \oplus \mathcal{O} = \mathcal{O} \oplus A = A \quad \forall A \in \Gamma;$
2. $A \oplus (-A) = \mathcal{O} \quad \forall A \in \Gamma;$
3. $A \oplus B = B \oplus A \quad \forall A, B \in \Gamma;$
4. $(A \oplus B) \oplus C = A \oplus (B \oplus C) \quad \forall A, B, C \in \Gamma.$

The result of the operation "$\oplus$" belongs to the same set $\mathbb{R}^2 \cup \{\mathcal{O}\}$.

In other words, the operation "$\oplus$" is commutative and associative. The role of the opposite to A is played by $-A$; the role of zero (neutral element) is played by the point $\mathcal{O} = \infty$.

12.2 Elliptic Curves with Rational Points

In the previous sections of this chapter, as the universal set was considered a Cartesian plane completed with a point at infinity ∞:

$$U = \mathbb{R}^2 \cup \{\mathcal{O}\}. \tag{12.29}$$

Recall that the universal set, by definition, contains the entire set of values that the variables can take in a certain problem [41]. For cryptography and number theory, the most significant are the properties of the elliptic curves on the set of points with the rational coordinates $\mathbb{Q}^2 \cup \{\infty\}$, where $\mathbb{Q} = \{p/q: p, q \in \mathbb{Z}, q \neq 0\}$ is a set of rational numbers. As in the case of a standard Cartesian plane, an additional point ∞ is introduced, which is considered to be rational.

So, let us fix as the universal set

$$U = \mathbb{Q}^2 \cup \{\infty\}. \tag{12.30}$$

The operation of multiplication of points is defined by the same rule as in the previous section. Formulate this rule in the form of an algorithm.

Algorithm of addition of points on the elliptic curve *The sum of the two points* $A(x_A, y_A)$ *and* $B(x_B, y_B)$ *on the elliptic curve*

$$\Xi = \{(x, y) \in \mathbb{Q}^2 \colon y^2 = x^3 + ax + b\} \cup \{\mathcal{O}\} \tag{12.31}$$

is the point $C(x_C, y_C)$, *whose coordinates are found by the following rule.*

1. *If* $A = \mathcal{O}$, *then* $C = A \oplus B = B$.
2. *If* $B = \mathcal{O}$, *then* $C = A \oplus B = A$.
3. *If* $x_A = x_B$ *and* $y_A = -y_B$, *then* $C = A \oplus B = \mathcal{O}$.
4. *In other cases, the parameter* $\varkappa$ *is computed:*

$$\varkappa = \begin{cases} \dfrac{y_B - y_A}{x_B - x_A}, & \text{if } A \neq B, \\[2mm] \dfrac{3x_A^2 + a}{2y_A}, & \text{if } A = B, \end{cases} \tag{12.32}$$

and the coordinates of the point C *will be equal to*

$$x_C = \varkappa^2 - x_A - x_B, \tag{12.33}$$

$$y_C = \varkappa(x_A - x_C) - y_A. \tag{12.34}$$

Here, the procedure of addition of the points is defined specifically, since it does not take the result beyond the extended set of rational numbers. Indeed, all usual arithmetic operation—addition, subtraction, multiplication and division—are performed in an algorithm on rational operands, which eventually results in rational numbers. Division by zero in an algorithm cannot take place due to preliminary processing of the exceptional case $B = -A$. Therefore, a theorem similar to theorem 12.2 occurs.

Theorem 12.3 *The operation of addition of the points defined on the elliptic curve* Ξ *on the set* $\mathbb{Q}^2 \cup \{\mathcal{O}\}$ *has the following properties:*

1. $A \oplus \mathcal{O} = \mathcal{O} \oplus A = A \quad \forall A \in \Xi;$
2. $A \oplus (-A) = \mathcal{O} \quad \forall A \in \Xi;$
3. $A \oplus B = B \oplus A \quad \forall A, B \in \Xi;$
4. $(A \oplus B) \oplus C = A \oplus (B \oplus C) \quad \forall A, B, C \in \Xi.$

The result of the operation "$\oplus$" belongs to the same extended set of points with rational coordinates.

As we can see, addition of rational point on an arbitrary elliptic curve has all the properties of a regular operation of addition of rational numbers.

12.3 Implementation of the Addition Algorithm

Let us consider a software implementation of an algorithm of addition of rational numbers on an elliptic curve. The program's input data are the coordinates of the points A and B; the coordinates of the sum $A \oplus B$ on the set $\mathbb{Q}^2 \cup \{\mathcal{O}\}$ are entered into the resulting file.

Listing 12.1

```
 1  class RationalFraction(object):
 2      def __init__(self, n, d):
 3          self.n = n
 4          self.d = d
 5
 6      def __eq__(self, other):
 7          return self.n == other.n and \
 8                 self.d == other.d
 9
10      def __neg__(self):
11          return RationalFraction(-self.n, self.d)
12
13
14  class RationalPoint(object):
15      def __init__(self, x, y):
16          self.x = x
17          self.y = y
18
19      def __eq__(self, other):
20          return self.x == other.x and \
21                 self.y == other.y
22
23
24  # Denominators of coordinates of the point at
       infinity
25  # are equated to zero
26  O = RationalPoint(RationalFraction(1, 0), \
27                    RationalFraction(1, 0))
28
29  # Parameter a of an elliptic curve
30  a = RationalFraction(-4, 1)
31
32
33  def add(p1, p2):
34      if p1.x.d == 0 or p1.y.d == 0:
35          return p2
```

```
36      elif p2.x.d == 0 or p2.y.d == 0:
37          return p1
38      elif p1.x == p2.x and p1.y == (-p2.y):
39          return O
40      else:
41          k = RationalFraction(0, 0)
42          c = RationalPoint(RationalFraction(0, 0), \
43                           RationalFraction(0, 0))
44
45          if p1 == p2:
46              k.n = p1.y.d * \
47                    (3 * p1.x.n * p1.x.n * a.d + \
48                     a.n * p1.x.d * p1.x.d)
49              k.d = 2 * p1.y.n * p1.x.d * \
50                    p1.x.d * a.d
51
52              if k.d < 0:
53                  k.n = -k.n
54                  k.d = -k.d
55
56              c.x.n = \
57                  (k.n * k.n * p1.x.d * p2.x.d - \
58                   p1.x.n * k.d * k.d * p2.x.d - \
59                   p2.x.n * k.d * k.d * p1.x.d)
60              c.x.d = k.d * k.d * p1.x.d * p2.x.d
61
62              c.y.n = \
63                  k.n * p1.y.d * \
64                  (p1.x.n * c.x.d - c.x.n * p1.x.d) -
                     \
65                  k.d * p1.x.d * c.x.d * p1.y.n
66              c.y.d = k.d * p1.x.d * c.x.d * p1.y.d
67
68              fraction_reduce(c.x)
69              fraction_reduce(c.y)
70
71              if c.x.d < 0:
72                  c.x.n = -c.x.n
73                  c.x.d = -c.x.d
74
75              if c.y.d < 0:
76                  c.y.n = -c.y.n
77                  c.y.d = -c.y.d
78
```

```
79          else:
80              # Points are different
81              k.n = p1.x.d * p2.x.d * \
82                  (p2.y.n * p1.y.d - p1.y.n * p2.y.d)
83              k.d = p1.y.d * p2.y.d * \
84                  (p2.x.n * p1.x.d - p1.x.n * p2.x.d)
85
86              if k.d < 0:
87                  k.n = -k.n
88                  k.d = -k.d
89
90              c.x.n = \
91                  (k.n * k.n * p1.x.d * p2.x.d - \
92                  p1.x.n * k.d * k.d * p2.x.d - \
93                  p2.x.n * k.d * k.d * p1.x.d)
94              c.x.d = k.d * k.d * p1.x.d * p2.x.d
95
96              c.y.n = k.n * p1.y.d * \
97                  (p1.x.n * c.x.d - c.x.n * p1.x.d) -
                      \
98                  k.d * p1.x.d * c.x.d * p1.y.n
99              c.y.d = k.d * p1.x.d * c.x.d * p1.y.d
100
101             fraction_reduce(c.x)
102             fraction_reduce(c.y)
103
104             if c.x.d < 0:
105                 c.x.n = -c.x.n
106                 c.x.d = -c.x.d
107
108             if c.y.d < 0:
109                 c.y.n = -c.y.n
110                 c.y.d = -c.y.d
111
112         return c
113
114
115 def gcd(a, b):
116     r = 0
117
118     # Euclid's algorithm
119     while b != 0:
120         r = a % b
121         a = b
```

```
122             b = r
123
124       return a
125
126
127 def fraction_reduce(fraction):
128       temp = gcd(fraction.n, fraction.d)
129
130       fraction.n //= temp
131       fraction.d //= temp
132
133
134 p1 = RationalPoint(RationalFraction(0, 0), \
135                    RationalFraction(0, 0))
136 p2 = RationalPoint(RationalFraction(0, 0), \
137                    RationalFraction(0, 0))
138
139 with open('input.txt') as file:
140     p1.x.n, p1.x.d, p1.y.n, p1.y.d = \
141         [int(num) for num in next(file).split()]
142
143     p2.x.n, p2.x.d, p2.y.n, p2.y.d = \
144         [int(num) for num in next(file).split()]
145
146 result = add(p1, p2)
147
148 with open('output.txt', 'w+') as file:
149     output = '(%d / %d, %d / %d )' % \
150         (result.x.n, result.x.d, \
151          result.y.n, result.y.d)
152
153     file.write(output)
```

For operations with rational fractions, a class `RationalFraction` is implemented in the program, which consists of two fields—the numerator and the denominator of the fraction.

An arbitrary rational point of the plane is characterized by its abscissa and ordinate. Each of these coordinates is a rational fraction; this is why the description of the class `RationalPoint` includes two fields that store the objects `RationalFraction`.

Thus, a rational point of the elliptical curve is represented in the program by a class containing the objects of other classes.

For example, the point $P_1(3, -2)$ of the Cartesian plane can be written in the form $\left(\frac{3}{1}, -\frac{2}{1}\right)$ = RationalPoint(RationalFraction(3, 1), and be represented in the memory of computer as

p1 = {{3, 1}, {−2, 1}}.

The fields are referred to as follows:

- p1.x.n—numerator of the abscissa of the point P_1;
- p1.x.d—denominator of the abscissa of the point P_1;
- p1.y.n—numerator of the ordinate of the point P_1;
- p1.y.d—denominator of the ordinate of the point P_1.

The fraction's sign will be the sign of its numerator, while in the intermediary computations the sign of the denominator will be preserved as positive.

We should separately discuss the issue of representing the point at infinity $\mathcal{O}$. In this implementation, it is introduced as an object

O = RationalPoint(RationalFraction(1, 0), RationalFraction(1, 0))

i.e. the denominators of its coordinates are equated to zero.

The main procedure that executes the addition of the points is called add() and it operates as follows. In its code, an algorithm presented on page 393 is directly reflected, and the arithmetic operations on rational fractions are executed separately for the numerator and for the denominator.

In particular, addition of two fractions is presented in the form:

$$\frac{\text{p1.x.n}}{\text{p1.x.d}} + \frac{\text{p2.x.n}}{\text{p2.x.d}} = \frac{\text{p1.x.n*p2.x.d} + \text{p2.x.n*p1.x.d}}{\text{p1.x.d*p2.x.d}}. \qquad (12.35)$$

Similarly, subtraction, multiplication and division are implemented in the program. Such arithmetic operations may result in a reducible fraction. For example, the summation $\frac{2}{3} + \frac{1}{3}$ is performed as follows:

$$\frac{2}{3} + \frac{1}{3} = \frac{2 \cdot 3 + 1 \cdot 3}{3 \cdot 3} = \frac{9}{9}. \qquad (12.36)$$

In this connection, prior to returning the result, the function add() divides the numerator and the denominator of the fraction by their greatest common divisor, gcd. For this purpose, an auxiliary function fraction_reduce() is called.

Computing of the greatest common divisor of two integers is based on the widely known Euclid's[3] algorithm [32].

---

[3] Euclid (Εὐχλείδης) (about 325 BC–before 265 BC), Ancient Greek mathematician.

At the final stage, the program outputs a sum of two rational points in the form

```
( c.x.n / c.x.d, c.y.n / c.y.d )
```

where `c.x.n` is the numerator of the abscissa of the result, `c.x.d` is the denominator of the abscissa of the result, etc.

Review Questions

1. Define elliptic curve.
2. How is the discriminant of the equation of an elliptic curve computed?
3. Explain how one can, by the sign of the discriminant of the elliptic curve, establish the number of connected components of its graph.
4. Enumerate the properties of the point at infinity $\mathcal{O}$.
5. Explain how the sum of the points $A \oplus B$ on the elliptic curve is computed.
6. How can one, knowing the coordinates of some point on the elliptic curve, find the coordinates of the opposite point?
7. Formulate the necessary and sufficient condition of equality $\mathcal{O}$ of the sum of three points on the elliptic curve.
8. Define the sum n of the points on the elliptic curve, where n is an integer.
9. Enumerate the properties of the operation of addition of points on the elliptic curve.

Problems

12.1. Which of the equations listed below specify the elliptic curves on the expanded Cartesian plane $\mathbb{R}^2 \cup \{\mathcal{O}\}$?

(1) $y^2 = x^2 - x + 1$;
(2) $y^2 = x^3 + x + 1$;
(3) $y^2 = x + 1$;
(4) $y^2 = x^3$;
(5) $y^2 = x^3 + 3x + 2$;
(6) $y^2 = x^3 - 3x + 2$;
(7) $y^2 = x^3 - 15x/27$;
(8) $y^2 = x^4 + x^3$.

If the curve belongs to the class of elliptical ones, then compute its discriminant and construct a graph.

12.2. Draw a graph of the curve $S = \{(x, y): y^2 = x^3 - 12x + 16\}$ and explain why it does not belong to the class of elliptic curves.

12.3. Compute the discriminant for each of the following elliptic curves:

(1) $y^2 = x^3 - 2x + 3$;

(2) $y^2 = x^3 - x - 1$;
(3) $y^2 = x^3 + 4x$;
(4) $y^2 = x^3 - 10x + 8$.

How many connected components does the graph of them contain?

*12.4. Prove that the **duplication formula** for computing of the abscissa of the
point $2A$ on the curve $y^2 = x^3 + ax + b$ can be presented in the following
form:

$$x_{2A} = \left(\frac{3x_A^2 + a}{2y_A}\right)^2 - 2x_A = \frac{x_A^4 - 2ax_A^2 - 8bx_A + a^2}{4x_A^3 + 4ax_A + 4b}. \tag{12.37}$$

12.5. Verify that the points $(-2, 5)$, $(-1, -5)$ and $(103, 1045)$ belong to the
elliptic curve $y^2 = x^3 - 7x + 19$.

12.6. Verify that on the curve $y^2 = x^3 + 15$ the following equality

$$2(1, 4) = (-119/64, -1499/512)$$

is valid.

12.7. Compute the following sums of the points on the elliptic curves on the set
$\mathbb{Q}^2$:

(1) $(-3, 3) \oplus (1, 3)$ on the curve $y^2 = x^3 - 7x + 15$;
(2) $(1, 4) \oplus (1, 4)$ on the curve $y^2 = x^3 + x + 14$.

12.8. Verify the validity of the equalities on the elliptic curve $y^2 = x^3 - 7x + 10$:

(1) $(5, 10) \oplus (9, 26) = (2, 2)$;
(2) $(5, 10) \oplus (2, -2) = (9, -26)$;
(3) $(5, 10) \oplus (2, 2) = (1/9, 82/27)$;
(4) $(5, -10) \oplus (1, -2) = (-2, -4)$.

12.9. Compute the sum of the points

$$(-2, -4) \oplus (1, 2) \oplus (2, 2) \oplus (-3, 2)$$

on the curve $y^2 = x^3 - 7x + 10$.

12.10. Perform duplication of the point $(5, 12)$ on the curve $y^2 = x^3 + x + 14$.

12.11. Compute the sum of the points

(1) $2(7, 19)$ on the curve $y^2 = x^3 + x + 11$;
(2) $3(1, 4)$ on the curve $y^2 = x^3 + x + 14$.

*12.12. Compute the sum of the points

$$2(2, 4) \oplus 2(33, 190) \oplus (1, 2)$$

on the curve $y^2 = x^3 + 5x - 2$.

*12.13. Compute the sum of the points

$$(6, 16) \oplus (-1, -2) \oplus (-2)(9, -28)$$

on the curve $y^2 = x^3 + 5x + 10$.

12.14. Write a function that checks whether the rational point $P(x, y)$ belongs to the elliptic curve $y^2 = x^3 + ax + b$.

Answers and Solutions

12.1 *Solution.*

Elliptic curve, as is known, is defined by the relation

$$y^2 = x^3 + ax + b$$

for some $a, b \in \mathbb{R}$ on the condition $4a^3 + 27b^2 \neq 0$ (see page 383). Leaning upon this definition, we obtain

(1) $y^2 = x^2 - x + 1$ is not an elliptic curve, since the right side lacks the summand x^3;

(2) $y^2 = x^3 + x + 1$ is an elliptic curve; the respective values of the parameters are $a = 1, b = 1$, the discriminant is $\Delta = -16(4a^3 + 27b^2) = -496$;

(3) $y^2 = x + 1$ is not an elliptic curve, since it lacks the cubic summand x^3;

(4) $y^2 = x^3$ is not an elliptic curve, since $a = 0$, $b = 0$ and the equality $4a^3 + 27b^2 = 0$ is valid;

(5) $y^2 = x^3 + 3x + 2$ satisfies the definition of elliptic curve with the parameters $a = 3$ and $b = 2$, the discriminant is $\Delta = -3456$;

(6) $y^2 = x^3 - 3x + 2$ is not an elliptic curve, since $a = -3, b = 2$ and $4a^3 + 27b^2 = 0$;

(7) $y^2 = x^3 - 5x/27$ satisfies the definition of elliptic curve, $a = -5/27$ and $b = 0$, $\Delta = 8000/19683$;

(8) the equation $y^2 = x^4 + x^3$ includes a summand of the fourth degree x^4, therefore, the respective curve does not belong to the class of elliptic curves (Fig. 12.5).

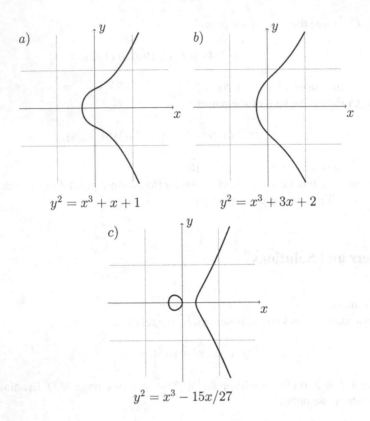

a)

$y^2 = x^3 + x + 1$

b)

$y^2 = x^3 + 3x + 2$

c)

$y^2 = x^3 - 15x/27$

Fig. 12.5 Elliptic curves to Problem **12.1**. The rectangular mesh on the panels (**a**)–(**c**) has a step equal to 2

12.2 *Solution.*

The graph of the curve S is presented in Fig. 12.6. The curve S is not elliptic, since the condition of its discriminant's being other than zero is not valid: $\Delta = -16(4a^3 + 27b^2) = -16(4(-12)^3 + 27 \cdot 16^2) = 0$. The geometric expression of this fact is the presence on the graph of a self-intersection point with the coordinates $(-2, 0)$.

12.3 *Solution.*

Using the formula $\Delta = -16(4a^3 + 27b^2)$, we obtain

(1) $\Delta = -3376$, the graph has one connected component;
(2) $\Delta = -368$, one connected component;
(3) $\Delta = -4096$, one connected component;
(4) $\Delta = 36,352$, the graph consists of two connected components.

Fig. 12.6 The curve S to Problem **12.2**. The rectangular mesh has a step equal to 2. The point $(2, 0)$ is an intersection point for the curve $y^2 = x^3 - 12x + 16$

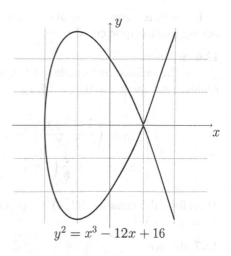

$$y^2 = x^3 - 12x + 16$$

12.4 *Solution.*

Expanding the expression $\left(\dfrac{3x_A^2 + a}{2y_A}\right)^2 - 2x_A$ and collecting similar summands, we arrive at the formula for x_{2A}:

$$
\begin{aligned}
x_{2A} &= \left(\frac{9x_A^4 + 6ax_A^2 + a^2}{4y_A^2}\right)^2 - 2x_A \\[2mm]
&= \frac{9x_A^4 + 6ax_A^2 + a^2 - 8x_A y_A^2}{4y_A^2} \\[2mm]
&= \frac{9x_A^4 + 6ax_A^2 + a^2 - 8x_A(x_A^3 + ax_A + b)}{4(x_A^3 + ax_A + b)^2} \\[2mm]
&= \frac{x_A^4 - 2ax_A^2 - 8bx_A + a^2}{4x_A^3 + 4ax_A + 4b}.
\end{aligned}
$$

In algebraic transformations, the equality $y_A^2 = x_A^3 + ax_A + b$ was used, valid for the coordinates of all points of the elliptic curve.

Thus, the algebraic variation of the duplication formula is proved.

12.5 *Solution.*

Substitute the coordinates of each point into the equation of the curve $y^2 = x^3 - 7x + 19$:

$$5^2 = (-2)^3 - 7(-2) + 19, \text{ or } 25 = 25;$$

$$(-5)^2 = (-1)^3 - 7(-1) + 19, \text{ or } 25 = 25;$$

$$1045^2 = 103^3 - 7(103) + 19, \text{ or } 1\,092\,025 = 1\,092\,025.$$

In all three cases we obtain true equalities, this is why the considered points belong to the elliptic curve $y^2 = x^3 - 7x + 19$.

12.6 *Solution.*

Use the duplication formula (12.12). Substitute in it the numeric values from the problem statement $a = 0$, $b = 15$, $x_A = 1$ and $y_A = 4$. Then we obtain

$$\begin{cases} x_{2A} = \left(\dfrac{3 \cdot 1^2 + 0}{2 \cdot 4}\right)^2 - 2 \cdot 1 = \left(\dfrac{3}{8}\right)^2 - 2 = -\dfrac{119}{64}, \\ y_{2A} = -4 + \dfrac{3}{8}\left(1 - \left(-\dfrac{119}{64}\right)\right) = -\dfrac{1499}{512}. \end{cases}$$

Therefore, the equality $2(1, 4) = (-119/64, -1499/512)$ is valid on the curve $y^2 = x^3 + 15$.

12.7 *Answer:*

(1) $(2, -3)$;
(2) $(-7/4, -21/8)$.

12.9 *Answer:* $(-2, -4)$.

12.10 *Answer:* $(1/36, 809/216)$.

12.11 *Answer:*

(1) $(422/361, 25449/6859)$;
(2) $(793/121, -23132/1331)$.

12.12 *Answer:* $(33, 190)$.

12.13 *Answer:* $\mathcal{O}$.

12.14 *Solution.*

Let us assume that the rational point $P(x, y)$ is represented in the computing system memory as the object of a class `RationalPoint` described in Listing 12.1 (lines 14–21 of the program code).

Let us show the implementation of the function `is_point()` that returns the value `True` or `False` depending on the belonging of the point $P(x, y)$ to the elliptic curve $y^2 = x^3 + ax + b$.

```
def is_point(p):
    # Point at infinity
    if p.x.d == 0 and p.y.d == 0:
        return True
    elif p.x.d == 0 or p.y.d == 0:
        raise ValueError("Zero denomitator of a coordinate")

    # Checking the condition y * y = x * x * x + a * x + b
    temp1 = a.d * b.d * p.x.n * p.x.n * p.x.n + \
            a.n * b.d * p.x.n * p.x.d * p.x.d + \
```

```
              a.d * b.n * p.x.d * p.x.d * p.x.d
temp2 = a.d * b.d * p.x.d * p.x.d * p.x.d
temp3 = temp2 * p.y.n * p.y.n - temp1 * p.y.d * p.y.d

return temp3 == 0
```

Appendix A
Basic Operators in Python and C

This book uses the Python language to write algorithms [68, 71]. Of course, when necessary, all the algorithms presented in the text can be rewritten using any other programming language. In the present Appendix we provide a table of correspondences between the basic constructs of Python 3 and their analogues in the C language (see Table A.1).

Both these languages are high-level programming languages, although the level of abstraction in Python is considered to be higher compared to the C language [50]. As a rule, this results in slower operation of programs in Python.

One of the important differences between the syntaxes of these two languages consists in that the commands in the C end with a comma, while in Python it is not necessary to put a semicolon at the end of the command. Another significant difference is associated with marking out a block of operators: C uses braces for this purpose, while Python uses an indent consisting of exactly four spaces.

A new class in Python can be created as follows:

```
class Point:
    def __init__(self, x, y):
        self.x = x
        self.y = y
```

Such a class was used in Chap. 7 for describing a point of the Cartesian plane. This class contains fields x and y, representing the coordinates of the point. Also, this class includes a class constructor that is called when creating an object and is used for initialization of its fields. In order to refer to the fields, the keyword `self` can be used, which represents the current class instance automatically transferred as an argument into each method of this class.

The C language, unlike Python, is not object-oriented, and C uses structures instead of classes [21, 39].

Let us enumerate some more features of Python reflected in the listings of the programs.

© Springer Nature Switzerland AG 2021
S. Kurgalin, S. Borzunov, *Algebra and Geometry with Python*,
https://doi.org/10.1007/978-3-030-61541-3

Table A.1 Correspondences between the basic operators in Python and C

Command or operation	Python	C
Assignment	`x = y`	`x = y;`
Integer variables	`m, n = 10, -17`	`int m = 10, n = -17;`
Real variables	`v = 0.005` `w = -1.4`	`float v = 0.005;` `double w = -1.4;`
Logic variables	`u = True` `v = False`	`u = 1;` `v = 0;`
String variables	`s = "String text"`	`const char* s = "String text";`
Arrays	`arr = [1, 2, 3]` `arr[2] = 7`	`int arr[3] = {1, 2, 3};` `arr[2] = 7;`
Comparison of variables	`x == y` `x != y`	`x == y;` `x != y;`
Logic operations	`(not A) and (B or C)`	`(!A) && (B || C);`
Arithmetic division	`m/n`	`(double)m/n`
Integer division	`m//n`	`m/n`
Comments	`# comment` `""" Text of multiline comment """`	`// comment` `/* Text of multiline comment */`

Conditional operator	`if a == b:`
	`    # Code1`
	`elif a == c:`
	`    # Code2`
	`else:`
	`    # Code3`
		`if (a == b) {`
		`    // Code1`
		`}`
		`else if (a == c) {`
		`    // Code2`
		`}`
		`else {`
		`    // Code3`
		`}`

Ternary operator	`maxv = a if a>=b else b`	`maxv = (a>=b) ? a : b;`

for loop	`for i in range(n):`
	`    # Code`
		`for(int i=0; i<n; i++) {`
		`    // Code`
		`}`

while loop	`while a == b:`
	`    # Code`
		`while (a == b) {`
		`    // Code`
		`}`

Functions	`def sm(a, b):`
	`    s = a + b`
	`    return s`
		`int sm(int a, int b) {`
		`    int s = a + b;`
		`    return s;`
		`}`

Exchange of values of two variables	`a, b = b, a`	`int c = a;`
		`a = b;`
		`b = c;`

In Python, there is a method for generation of lists (including multidimensional ones):

- creating a list of n numbers filled with the values from 0 to n - 1:
  ```
  V = [ i for i in range(n) ]
  ```
- creating a two-dimensional list (matrix) filled with zeroes:
  ```
  A = [[ 0 for j in range(n) ] \
      for i in range(n) ]
  ```

Similarly to many other languages, Python provides the means for dealing with exceptions, which are useful for processing error situations. So, in order to generate an exception, the keyword raise is used:

```
raise Exception("Exception message")
```

In order to process the exception, the construct try-except is used:

```
try:
    a = 5
    b = 0
    c = a / b
except ZeroDivisionError as e:
    print(e)
```

After executing this code area, the following message will be outputted to the console:

```
division by zero
```

Appendix B
Trigonometric Formulae

In the formulae of this Appendix, unless otherwise specified, $a, b \in \mathbb{R}$ and $k, k' \in \mathbb{Z}$.

$$\sin^2 a + \cos^2 a = 1; \tag{B.1}$$

$$\tan a = \frac{\sin a}{\cos a}, \quad a \neq \frac{\pi}{2} + \pi k; \tag{B.2}$$

$$\cot a = \frac{\cos a}{\sin a}, \quad a \neq \pi k; \tag{B.3}$$

$$1 + \tan^2 a = \frac{1}{\cos^2 a}, \quad a \neq \frac{\pi}{2} + \pi k; \tag{B.4}$$

$$1 + \cot^2 a = \frac{1}{\sin^2 a}, \quad a \neq \pi k; \tag{B.5}$$

$$\sin 2a = 2 \sin a \cos a, \qquad \cos 2a = \cos^2 a - \sin^2 a; \tag{B.6}$$

$$\tan 2a = \frac{2 \tan a}{1 - \tan^2 a}, \quad a \neq \frac{\pi}{4} + \frac{\pi k}{2}, \ a \neq \frac{\pi}{2} + \pi k'; \tag{B.7}$$

$$\sin^2 \frac{a}{2} = \frac{1 - \cos a}{2}, \qquad \cos^2 \frac{a}{2} = \frac{1 + \cos a}{2}; \tag{B.8}$$

$$\sin(a + b) = \sin a \cos b + \cos a \sin b; \tag{B.9}$$

$$\sin(a - b) = \sin a \cos b - \cos a \sin b; \tag{B.10}$$

© Springer Nature Switzerland AG 2021
S. Kurgalin, S. Borzunov, *Algebra and Geometry with Python*,
https://doi.org/10.1007/978-3-030-61541-3

$$\cos(a+b) = \cos a \cos b - \sin a \sin b; \qquad \text{(B.11)}$$

$$\cos(a-b) = \cos a \cos b + \sin a \sin b; \qquad \text{(B.12)}$$

$$\tan(a+b) = \frac{\tan a + \tan b}{1 - \tan a \tan b}, \quad a, b, a+b \neq \frac{\pi}{2} + \pi k; \qquad \text{(B.13)}$$

$$\tan(a-b) = \frac{\tan a - \tan b}{1 + \tan a \tan b}, \quad a, b, a-b \neq \frac{\pi}{2} + \pi k; \qquad \text{(B.14)}$$

$$\sin a + \sin b = 2 \sin\left(\frac{a+b}{2}\right) \cos\left(\frac{a-b}{2}\right); \qquad \text{(B.15)}$$

$$\sin a - \sin b = 2 \cos\left(\frac{a+b}{2}\right) \sin\left(\frac{a-b}{2}\right); \qquad \text{(B.16)}$$

$$\cos a + \cos b = 2 \cos\left(\frac{a+b}{2}\right) \cos\left(\frac{a-b}{2}\right); \qquad \text{(B.17)}$$

$$\cos a - \cos b = -2 \sin\left(\frac{a+b}{2}\right) \sin\left(\frac{a-b}{2}\right); \qquad \text{(B.18)}$$

$$\tan a \pm \tan b = \frac{\sin(a \pm b)}{\cos a \cos b}, \quad a, b \neq \frac{\pi}{2} + \pi k; \qquad \text{(B.19)}$$

$$\cot a \pm \cot b = \frac{\sin(b \pm a)}{\sin a \sin b}, \quad a, b \neq \pi k; \qquad \text{(B.20)}$$

$$\sin a \, \sin b = \frac{1}{2} \left(\cos(a-b) - \cos(a+b)\right); \qquad \text{(B.21)}$$

$$\cos a \, \cos b = \frac{1}{2} \left(\cos(a-b) + \cos(a+b)\right); \qquad \text{(B.22)}$$

$$\sin a \, \cos b = \frac{1}{2} \left(\sin(a-b) + \sin(a+b)\right). \qquad \text{(B.23)}$$

Appendix C
The Greek Alphabet

A, α	alpha		N, ν	nu	
B, β	beta		Ξ, ξ	xi	
Γ, γ	gamma		O, o	omicron	
Δ, δ	delta		Π, π	pi	
E, ε	epsilon		P, ρ	rho	
Z, ζ	zeta		Σ, σ	sigma	
H, η	eta		T, τ	tau	
Θ, θ	theta		Υ, υ	upsilon	
I, ι	iota		Φ, φ	phi	
K, $\varkappa$	kappa		X, χ	chi	
Λ, λ	lambda		Ψ, ψ	psi	
M, μ	mu		Ω, ω	omega	

S. Kurgalin, S. Borzunov, *Algebra and Geometry with Python*,
https://doi.org/10.1007/978-3-030-61541-3

References

1. Anderson, J.A.: Discrete Mathematics with Combinatorics, 2nd edn. Prentice Hall (2003)
2. Arfken, G.B., Weber, H.J.: Mathematical Methods for Physicists, 6th edn. Elsevier Academic Press, Amsterdam (2005)
3. Banchoff, T., Wermer, J.: Linear Algebra Through Geometry. Undergraduate Texts in Mathematics, 2nd edn. Springer, Berlin (1992)
4. Berestetskii, V.B., Lifshitz, E.M., Pitaevskii, L.P.: Quantum Electrodynamics. In: Course of Theoretical Physics, vol. 4, 2nd edn. Pergamon Press, Oxford (1982)
5. Bertsekas, D.P., Tsitsiklis, J.N.: Parallel and Distributed Computation: Numerical Methods. Optimization and Neural Computation; Book 7. Athena Scientific, Belmont (1997)
6. Billig, Y.: Quantum Computing for High School Students. Yuly Billig, Ottawa (2018)
7. Bix, R.: Conics and Cubics: A Concrete Introduction to Algebraic Curves. Undergraduate Texts in Mathematics, 2nd edn. Springer, Berlin (2006)
8. Borsuk, K.: Multidimensional Analytic Geometry. Monografie Matematyczne; T. 50. PWN, Warszawa (1969)
9. Bretscher, O.: Linear Algebra with Applications, 5th edn. Pearson, London (2013)
10. Carroll, B.W., Ostlie, D.A.: An Introduction to Modern Astrophysics. Pearson Addison-Wesley, San Francisco (2007)
11. Cheney, E.W.: Introduction to Approximation Theory, 2nd edn. American Mathematical Society, Providence (2000)
12. Chivers, I., Sleightholme, J.: Introduction to Programming with Fortran, 4th edn. Springer, Berlin (2018)
13. Cohen, H.: Number Theory. Graduate Texts in Mathematics; Vol. 239, Vol. I: Tools and Diophantine Equations. Springer, Berlin (2007)
14. Cohen, H., Frey, G., Avanzi, R., Doche, C., Lange, T., Nguyen, K., Vercauteren, F.: Handbook of Elliptic and Hyperelliptic Curve Cryptography. Discrete Mathematics and Its Applications. Chapman & Hall/CRC, Boca Raton (2006)
15. Cohn, P.M.: Algebra : Volume 1, 2nd edn. Wiley, Chichester (1982)
16. Cormen, T.H., Leiserson, C.E., Rivest, R.L., Stein, C.: Introduction to Algorithms, 3rd edn. The MIT Press, Cambridge (2009)
17. Courant, R., Robbins, H.: What is Mathematics? An Elementary Approach to Ideas and Methods, 2nd edn. Oxford University Press, New York (1996)
18. Cover, T.M., Thomas, J.A.: Elements of Information Theory, 2nd edn. Wiley, New York (2006)
19. Coxeter, H.S.M.: Introduction to Geometry, 2nd edn. Wiley, New York (1969)

© Springer Nature Switzerland AG 2021
S. Kurgalin, S. Borzunov, *Algebra and Geometry with Python*,
https://doi.org/10.1007/978-3-030-61541-3

20. D'Alberto, P., Nicolau, A.: Adaptive Strassen's matrix multiplication. In: Proceedings of the 21st Annual International Conference on Supercomputing, ICS'07, June 18–20, Seattle, Washington, pp. 284–292. ACM, New York (2007)
21. Deitel, P.J., Deitel, H.: C for Programmers with an Introduction to C11. Deitel Developer Series. Pearson Education, London (2013)
22. Diestel, R.: Graph Theory. Graduate Texts in Mathematics, vol. 173, 5th edn. Springer, Berlin (2017)
23. Dubrovin, B.A., Fomenko, A.T., Novikov, S.P.: Modern Geometry — Methods and Applications. Part I. The Geometry of Surfaces, Transformation Groups, and Fields. Graduate Texts in Mathematics, vol. 93, 2nd edn. Springer, Berlin (1992)
24. Gamelin, T.W.: Complex Analysis. Undergraduate Texts in Mathematics. Springer, Berlin (2001)
25. Gantmacher, F.R.: The Theory of Matrices, vol. 1. American Mathematical Society, Providence (2000)
26. Gantmacher, F.R.: The Theory of Matrices, vol. 2. American Mathematical Society, Providence (2000)
27. Golub, G.H., Loan, C.F.V.: Matrix Computations, 4th edn. The Johns Hopkins University Press, Baltimore (2013)
28. Graham, R.L., Knuth, D.E., Patashnik, O.: Concrete Mathematics: A Foundation for Computer Science, 2nd edn. Addison-Wesley, Reading (1994)
29. Haggarty, R.: Discrete Mathematics for Computing. Addison-Wesley, Reading (2002)
30. Halmos, P.R.: Finite-Dimensional Vector Spaces. Undergraduate Texts in Mathematics. Springer, Berlin (1993)
31. Harary, F.: Graph Theory. Addison-Wesley, Reading (1969)
32. Hardy, G.H., Wright, E.M.: An Introduction to the Theory of Numbers, 6th edn. Oxford University Press, Oxford (2008)
33. Hazewinkel, M. (ed.): Encyclopaedia of Mathematics, 10 volume-set. Springer, Berlin (1994)
34. Higham, N.J.: Functions of Matrices: Theory and Computation. SIAM, Philadelphia (2008)
35. Hungerford, T.W.: Algebra. Graduate Texts in Mathematics, vol. 73. Springer, New York (2011)
36. Il'in, V.A., Poznyak, E.G.: Linear Algebra. Collets (1986)
37. Ito, K.: Encyclopedic Dictionary of Mathematics, 2nd edn. The MIT Press; Mathematical Society of Japan, Massachusetts Institute of Technology (1987)
38. Jeffrey, A., Dai, H.H.: Handbook of Mathematical Formulas and Integrals, 4th edn. Academic Press, Cambridge (2008)
39. Kernighan, B.W., Ritchie, D.M.: The C Programming Language, 3rd edn. Prentice Hall, Englewood Cliffs (1988)
40. Knuth, D.E.: The Art of Computer Programming. Fundamental Algorithms, vol. 1, 3rd edn. Addison-Wesley, Reading (1997)
41. Kurgalin, S., Borzunov, S.: The Discrete Math Workbook: A Companion Manual for Practical Study. Texts in Computer Science. Springer, Berlin (2018)
42. Kurgalin, S., Borzunov, S.: A Practical Approach to High-Performance Computing. Springer, Berlin (2019)
43. Kurosh, A.: Higher Algebra, 4th edn. Mir Publishers, Moscow (1984)
44. Landau, L.D., Lifshitz, E.M.: Quantum Mechanics: Non-relativistic Theory. Course of Theoretical Physics, vol. 3, 3rd edn. Pergamon Press, Oxford (1989)
45. Landau, L.D., Lifshitz, E.M.: The Classical Theory of Fields. Course of Theoretical Physics, vol. 2, 4th edn. Butterworth-Heinemann, Oxford (1996)
46. Langtangen, H.P.: A Primer on Scientific Programming with Python. Texts in Computational Science and Engineering, vol. 6, 4th edn. Springer, Berlin (2014)
47. Magnus, J.R., Neudecker, H.: Matrix Differential Calculus with Applications in Statistics and Econometrics. Wiley Series in Probability and Statistics, 3rd edn. Wiley, New York (2019)
48. Markov, A.A.: The theory of algorithms (in Russian). Trudy Mat. Inst. Steklov **42**, 3–375 (1954)

49. Markov, A.A., Nagorny, N.M.: The Theory of Algorithms. Mathematics and its Applications, vol. 23. Springer, Netherlands (1988)
50. Martelli, A., Ravenscroft, A., Holden, S.: Python in a Nutshell, 3rd edn. O'Reilly, Beijing (2017)
51. McConnell, J.J.: Analysis of Algorithms: An Active Learning Approach, 2nd edn. Jones and Bartlett Publishers, Burlington (2008)
52. Miller, R., Boxer, L.: Algorithms Sequential and Parallel: A Unified Approach, 3rd edn. Cengage Learning, Boston (2013)
53. Mirsky, L.: An Introduction to Linear Algebra. Clarendon Press, Oxford (1955)
54. Nielsen, M.A., Chuang, I.L.: Quantum Computation and Quantum Information, 10th anniversary edn. Cambridge University Press (2010)
55. Ore, O.: Theory of Graphs. Colloquium Publications; vol. 38. American Mathematical Society, Providence (1962)
56. Ortega, J.M.: Introduction to Parallel and Vector Solution of Linear Systems. Frontiers in Computer Science. Springer, Berlin (1988)
57. Pérez, F., Granger, B.E.: IPython: a system for interactive scientific computing. Comput. Sci. Eng. **9**(3), 21–29 (2007)
58. Press, W.H., Teukolsky, S.A., Vetterling, W.T., Flannery, B.P.: Numerical Recipes: The Art of Scientific Computing, 3rd edn. Cambridge University Press, Cambridge (2007)
59. Rosen, K.H.: Discrete Mathematics and Its Applications, 7th edn. McGraw-Hill, New York (2012)
60. Rosen, K.H., Michaels, J.G., Gross, J.L. et al. (eds.): Handbook of Discrete and Combinatorial Mathematics. Discrete Mathematics and Its Applications. CRC Press, Boca Raton (2000)
61. Sedgewick, R.: Algorithms in C. Part 5. Graph Algorithms, 3rd edn. Addison-Wesley, Boston (2001)
62. Sedgewick, R., Wayne, K., Dondero, R.: Introduction to Programming in Python: An Interdisciplinary Approach. Addison-Wesley Professional, Princeton University, Princeton (2015)
63. Shafarevich, I.R.: Algebra I: Basic Notions of Algebra. Encyclopaedia of Mathematical Sciences, vol. 11. Springer, Berlin (2005)
64. Shafarevich, I.R., Remizov, A.O.: Linear Algebra and Geometry, 3rd edn. Springer, Berlin (2013)
65. Shilov, G.E.: An Introduction to the Theory of Linear Spaces. Martino Fine Books, Eastford (2013)
66. Sominskii, I.S.: The Method of Mathematical Induction. Popular Lectures in Mathematics, vol. 1. Blaisdell Publishing Company, New York (1961)
67. Steane, A.: Quantum computing. Rep. Progress Phys. **61**, 117–173 (1998)
68. The Python Tutorial (2019). https://docs.python.org/3/tutorial/index.html
69. Trahtenbrot, B.A.: Algorithms and Automatic Computing Machines. Topics in Mathematics. D. C. Heath and Company, Boston (1963)
70. Vorobiev, N.N.: Fibonacci Numbers. Popular Lectures in Mathematics, vol. 2. Springer, Basel AG (2002)
71. Welcome to Python.org (2019). https://www.python.org/
72. Williams, C.P.: Explorations in Quantum Computing. Texts in Computer Science, 2nd edn. Springer, Berlin (2011)
73. Wilson, R.J.: Introduction to Graph Theory, 4th edn. Longman, Harlow (1998)
74. Wirth, N.: Pascal and Its Successors. Springer, Berlin (2002)
75. Zapryagaev, S.A.: Introduction to Quantum Information Systems [in Russian]. Voronezh State University Textbooks. VSU Publishing, Voronezh (2015)
76. Zorich, V.A.: Mathematical Analysis I. Universitext, 2nd edn. Springer, Berlin (2015)

Index

© Springer Nature Switzerland AG 2021   419
S. Kurgalin, S. Borzunov, *Algebra and Geometry with Python*,
https://doi.org/10.1007/978-3-030-61541-3

Printed in the United States
by Baker & Taylor Publisher Services

Printed in the United States
by Baker & Taylor Publisher Services